D1219561

Pleasure with Products

Pleasure with Products

Beyond Usability

Edited by William S. Green and Patrick W. Jordan

London and New York

First published 2002 by Taylor & Francis
11 New Fetter Lane, London EC4P 4EE

Simultaneously published in the USA and Canada
by Taylor & Francis Inc,
29 West 35th Street, New York, NY 10001

Taylor & Francis is an imprint of the Taylor & Francis Group

This book has been prepared from camera-ready copy supplied by the editors

Printed and bound in Great Britain by
TJ International Ltd, Padstow, Cornwall

Every effort has been made to ensure that the advice and information in this book is true and accurate at the time of going to press. However, neither the publisher nor the authors can accept any legal responsibility or liability for any errors or omissions that may be made. In the case of drug administration, any medical procedure or the use of technical equipment mentioned within this book, you are strongly advised to consult the manufacturer's guidelines.

British Library Cataloguing in Publication Data
A catalogue record for this book is available from the British Library

Library of Congress Cataloging in Publication Data
A catalogue record has been requested

ISBN 0-415-23704-1

Contents

Section 2 Design Techniques

INTRODUCTION

Pleasure with Products: Beyond Usability

Human Factors and Design

BILL GREEN

School of Design, Division of Science and Design, University of Canberra,

ACT 2601, Canberra, Australia

INTRODUCTION

In recent years the profile of Ergonomics, or Human Factors, has developed significantly. This is evident in a number of ways, the most important of which is the sharp increase in human factors professionals employed in industry (Green and Jordan 1999). The reasons are several and varied. They range from a response to the so-called Repetitive Strain Injury epidemic in the mid-eighties, which had a dramatic effect on both the public consciousness of ergonomics and on research into office ergonomics, to the increasingly evident technological 'level playing field' which has elevated usability from a kind of optional extra – nice to have but not life and death – to a real cutting-edge differentiator between products.

Industrial Design could be said to have undergone a similar rise in status. It is relatively recently that the professional industrial designer has become recognised as essential for leading manufacturers of consumer, professional and industrial products. Previously, an eclectic mix of artists, architects and mildly reconstructed engineers occupied a poorly defined position somewhere between the sales office and the production manager.

The fact that Human Factors and Industrial Design are growing together is no coincidence, but rather a natural confluence of overlapping knowledge and skills. That it is happening is cause for applause, but integration is a delicate flower and still requires careful tending.

Jordan (op. cit.) has identified three stages of integration, which he has called:

Phase 1 — Being ignored (Self-explanatory)
Phase 2 — 'Bolt on' human factors (Post-facto clean-up of the interface)
Phase 3 — Integrated human factors (H.F. specialists in the design team)

The integration of design and ergonomics/human factors is such an obvious *sine qua non* of design excellence that it seems ridiculous to have to spell it out, but the fact remains that the need was not always so obvious. At the turn of the century neither profession existed anyway, and when ergonomics invented itself as a discipline in the forties and fifties, it assumed the trappings of science and ignored the (apparently scientifically unsupportable) methods of the art-orientated design world. Moving back together has been precipitated by the recognition that design, defined as human/product interaction in its broadest sense, is the most significant product differentiator, moving ahead of technological sophistication and even, in our affluent societies, of price. When every machine has the same cheap chip, what gives a product added value? Black-box technology has shaken the foundations of traditional design approaches. For many current products, the modernist credo that 'form-follows-function' is a very shaky concept; it is difficult to assign, with any degree of certainty, form to a current running through a circuit board. As a consequence we have witnessed unprecedented freedom of formal expression, and along with it the sort of uncertainty which often accompanies such freedoms.

The role of a guiding principle in design has become vacant, various 'ism's' have been tried and found wanting, and there is a clear place for some of the small certainties that ergonomics and human factors provide. Usability, for so long hiding behind the supportable necessities of the lever and the wheel, has been forced out into the open as an issue to be resolved on a clean sheet. The ergonomists and usability professionals have risen to the challenge, and whilst not every interface problem is solved, or even indeed satisfactorily addressed, there is nonetheless an acceptance, by all serious contenders, of usability as a central design requirement. Jordan (1999) has linked the development of ergonomics/human factors to Maslow's heirarchy of need, with the implicit notion that the attainment of one level inevitably and quickly leads to a desire for the next.

Thus is raised the issue of usability and beyond: a holistic view of the human/product interface. A recent publication in the domain is entitled 'Don't think usability: think joy in use' (Overbeeke and Djadjadiningrat 2000). This title tends to raise the hackles of designers who read it, because they say that this is what they have been doing for years. In other words, they see a bunch of human factors specialists or psychologists re-inventing their particular wheel, and there is some basis for such thinking. However, most re-inventions are a function of new vision, new technologies, new social emphases or simply the need for a new generation to establish itself, and the current interest in joy in use, or pleasure with products, has its legitimate roots in all of these. The re-invention is legitimate, and important, for one simple reason: the need for knowledge and understanding of reproducability. *When a design 'works', why does it work?* How can we ensure that the next, different product, will inherit the magic factors? It is clear that the volume of mass-produced goods cannot rely on the relatively few designers of genius with demonstrable ability to integrate aesthetic excellence with accessible and intuitive interfaces. In other words, it would be extremely valuable to be able to provide data for the design profession that assists it to be effective, at the highest possible aesthetic and interactive level, in the absence of that curious attribute we call 'talent'. The generation of such data has not been a priority for designers. In fact, it could be argued that elements of the profession benefit from the mystery, but this is changing. The dilletante artist/designer still exists and is nowhere more evident than in the production of what Beaudrillard called 'models' – defining aesthetic statements which may owe little to functionality or usability. However, the day-to-day business of designing and producing consumer durables is becoming highly structured and, in an engineering sense, reliable. At a purely operational level, there aren't too many bad products produced by the major manufacturers, and their differences become defined by the 'joy in use' factor. The difficulty for all of the disciplines involved is how to generate such a user response to order.

THE CHALLENGE OF A NEW SCIENCE

There is indeed nothing new in the desire to provide joy in use within a context of functional and economic efficiency. This is exactly why the profession of industrial design exists, and it has been rather successful in elevating everyday products into the public consciousness as objects of aesthetic appraisal. So much so that 'poor design' as a descriptor is as likely to mean 'I don't like the design aesthetic' as 'It doesn't function very well'. Jordan (1999) has already described the process whereby usability has moved from being a 'satisfier' when present, to being a 'dissatisfier' when absent, and this is similar to the position of aesthetic considerations in their broadest sense.

What is now happening is the merging of the processes and methods of ergonomics, with its base firmly in the biological sciences, aesthetic design with its base in functional art, and engineering, which has relied on the physical sciences for much of its fundamental data. Such a confluence will never be easy. The very idea of reproduceability is somehow at odds with the freedom associated with the creative process. Art at its best wants to 'show' uniquely: its question is, 'How can this work be elevated beyond analysis?' Science at its best simply wants to know: to grasp knowledge reliably, and its question is, "Is this result capable of being falsified?" It is clear that at a fundamental level there isn't so much difference, and the divergence comes with the sheer difficulty of prediction that human endeavour and behaviour at an individual level brings. The tools that serve physical science well in the laboratory are often not so useful when applied to the creation of a new human artifact, when a major component of its success will be its aesthetic and functional impact on a future consumer or user. The social sciences have more valuable methods and processes, but they are primarily analytical and fall short of having real predictive value. (For a more detailed examination of some of these issues, see Kanis (1999)). Art has, almost by definition, no science at all .... but is that true? We do not think so. The knowledge exists, but undiscovered, uninvented or insufficiently refined to be recognised yet.

PLEASURE WITH PRODUCTS, OR JOY IN USE

What is pleasure? The Concise Oxford says 'a feeling of satisfaction or joy: sensuous enjoyment as an object of life.' The philosophical issues offer tempting possibilities for exploration, but it is better to sidestep them and concentrate on the way in which products may contribute to pleasure in their use. Jordan (1997) has written extensively on the four pleasures – physical, ideological, etc., and when any or all of these pleasures may apply differently to any individual, then the possibilities for generating pleasure with products – or the reverse – become infinite. It is interesting to note that the central defining pleasure state – 'sensuous enjoyment', has received relatively little attention in the design literature. Much has been made of design as a social arbiter, a marketing tool, an aesthetic vehicle, a bridge between art and engineering and so on. It is as though there is a tacit assumption that pleasure in all of these conditions is either a given, not worth discussing, or consigned by reason of complexity to the 'too hard' basket! It is certainly worth discussing and is by no means a given. For example, it is possible to want, and to purchase, and to use, an artefact for the social message which it transmits whilst taking little or no joy from its other attributes.

In spite of the infinite possible variety in the forms of pleasure , there are some conditions under which commonality of response may be expected. Whilst accepting that some humans may well find pleasure in pain, nonetheless a product which causes pain in its use is unlikely to be regarded as contributing to pleasure in the normal course of events. Nor is a product which obstructs the path to goal achievement by having, for example, an incomprehensible interface. Pleasure does not subsume usability, nor does usability alone generate pleasure. The previous examples define pleasure negatively – by its absence. Designers have traditionally attempted positively to invest their products with aesthetic quality, the implicit aim of which is pleasurable experience, and at the same time, have frequently defined an unpleasant experience in use by ignoring basic human factors.

The aesthetics of interaction are even more complex and diverse than use actions. In the first place, there is absolutely no rule that says that every person wants an aesthetic experience when achieving a goal, and there are also no rules about what might be considered a suitable experience. Perhaps we do sometimes want to separate the aesthetic from the functional in our daily lives. Fun may be something that gets in the way for a fair proportion of the time.

In a recent article on the links between product and emotion, Hummels (1999) has described a process whereby an LP record is extracted from its sleeve, cleaned, placed delicately on a turntable, the stylus inspected and placed equally delicately on the record, and so on, in contrast to the modern equivalent of shoving a CD in a drawer and pressing a button. This raises some interesting issues. The first is the specificity of the pleasurable experience, not only inter-, but intra-individually. I easily understand, and have also enjoyed, precisely the same experience that Hummels describes, but it is not one that I seek to replicate very often. Mostly I want the pleasurable experience of the music, and I want it NOW! The CD in the drawer is just fine. In this case, the pleasure focus shifts, but at least a part of it, however small, is still likely to be located in the product use. Does the CD player make a satisfying sound when the drawer opens? Does it say 'high quality device?' Do the controls provide positive and reassuring feedback? The obvious question of "Does it actually play the CD?" is probably taken for granted. A second seductive issue is that of ritual and the way in which products may contribute to one of the most basic of human pleasures. Ritual is deeply embedded in the human condition; a human need which finds expression in simple ways in daily life, where it may be called routine, or may be elevated in esoteric and abstruse ways into forms of art or quasi-religion (tea ceremony, freemasonry, etc.). The product use act can be a number of things: a way of achieving a predetermined goal, a way of exploring goal possibilities, an end in itself which is in some way located by a goal (the tea ceremony is hardly about drinking tea, but that goal serves to locate the ritual), a process by which the goal is experienced in an enhanced way, and so on.

What do we know of such things? At an individual and personal level, quite a lot, although there are many people who find great difficulty in making advance judgements of and for themselves (I only know what I like when I see it). Even those who appear very positive probably experience considerable uncertainty from time to time. At a general level, we know little that provides predictively useful insights to enhance the design process. If this were not the case, then every product would be an unqualified success.

The chapters in this book are contributed by ergonomists, usability specialists, designers and others who share a common belief in the power of research and scientific enquiry. That there is considerable divergence of approach, and different views of what may be considered scientific enquiry, is not coincidence, but a conscious attempt to present a broad spectrum of views. The thread which connects all of them is a desire to know more about the elusive attitudes, skills and techniques for investing future products with the 'beyond usability factor' – pleasure and joy in use.

REFERENCES

Green, W.S. and Jordan, P.W., 1999. Ergonomics, Usability and Product Development: an introduction. In *Human Factors in Product Design: Current practice and Future Trends*. Green, W.S. and Jordan, P.W. (Eds), Taylor & Francis, London.

Hummels, C., 1999. Engaging Contexts to Evoke Experiences. In *Design and Emotion*. Overbeeke, C.J. and Hekkert, P. (Eds), Department of Industrial Design, Delft University of Technology.

Jordan, P.W., 1999. Pleasure with Products: Human Factors for Body, Mind and Soul. In *Human Factors in Product Design: Current Practice and Future Trends*. Green, W.S. and Jordan, P.W. (Eds), Taylor & Francis, London.

Jordan, P.W., 1997. The Four Pleasures – Taking Human Factors beyond Usability. In *From Experience to Innovation*. Proceedings of the 13th Triennial Congress of the IEA Tampere, Finland.

Kanis, H., 1999. Design Centred Research into User Activities. In *Human Factors in Product Design: Current Practice and Future Trends*. Green, W.S. and Jordan, P.W. (Eds), Taylor & Francis, London.

Overbeeke, C.J. and Djadjadiningrat, T., 2000. Don't Think Usability: Think Joy in Use.

Section 1
Beyond Usability

CHAPTER ONE

Beauty in Usability: Forget about Ease of Use!

KEES OVERBEEKE, TOM DJADJADININGRAT,

CAROLINE HUMMELS and STEPHAN WENSVEEN

Delft University of Technology, Department of Industrial Design,

Jaffalaan 9, 2628 BX Delft, The Netherlands

1.1 INTRODUCTION

Despite years of usability research, electronic products do not seem to get any easier to use. The design of electronic products appears to be in a dead-end street. It is time to experiment with fresh approaches. In this paper we offer a new approach based on respect for the user. We all have senses and a body with which we can respond to what our environment affords (Gibson 1979). Why then does interaction design not use these bodily skills more often and make electronic interaction more tangible? And, as humans are emotional beings, why not make interaction a more fun and beautiful experience? This paper focuses on those neglected aspects of human-product interaction: perceptual-motor and emotional skills.

First we give the new background against which the designer operates. Then we give a number of examples from our own and our students' research work. As we go along we make clear why this new approach calls for new methods and what these methods are. The point we wish to make is that to get to new innovative products, the interaction problem should be dealt with on the level of creating a context for experience allowing for rich aesthetics of interaction.

1.2 RESPECT

The shop assistant threw the biscuit at my feet. I bent down and subserviently began to pick up the crumbs. After some fiddling, I managed to get my change out of his clenched fist.

Just imagine you were treated like this in a shop. No doubt you would be most offended. But this is, in fact, the way in which a vending machine treats us when we buy something from it. Somehow we have come to accept a standard of respect in human-machine interaction which is very different from that in human-human interaction.

We believe that respect for man as a whole should be the starting-point for design. For the sake of analysis, man's skills, which are used when interacting with products, may be considered on three levels, the wholly trinity of interaction: cognitive skills, perceptual-motor skills and emotional skills. In other words: knowing, doing and feeling.

Research on human-product interaction, however, has shifted to cognitive skills. This emphasis on the cognitive is easily understood, as there is no electronic counterpart for the mechanical world-view that still dominates Western thinking. We understand the world of moving machines since, to a certain extent, we consider our bodies to be mechanical machines. Take da Vinci as an example. His world of cogwheels and crossbars lends itself to easy experimentation and imaginative play while designing. Even a non-technical person can 'feel' the strength of such designs, even when they don't fully understand the workings. The electronic world is more opaque to us. What happens inside electronic products is intangible: it neither fits the mechanics of our body nor the mechanical view of the world. In contrast with mechanical components, electronic components do not impose specific forms or interactions for a design. Products have become 'intelligent', and intelligence has no form. Design research, quite naturally, turned to the intelligent part of humans. This primacy of rationality is often chosen because it leads to solutions that can be easily implemented into, or simulated on, computers. Interaction design typically starts from what is technically possible and follows the framework of the established sciences. The interaction problem is divided into elements and the relations between these elements, and is often captured in a flow chart. Ease of use is aimed for through rational analysis: **the rational is assumed to lead to ease of use**. Usability is aimed for through logical dialogue using speech recognition, through grouping and colour coding of buttons with related functions, through adding displays with an abundance of text and icons, and through writing logically structured manuals.

This is a very valuable route to follow, but we think it is not the only one that designers should explore.

1.3 THE EXPERIENTIAL

We think it is necessary to include the other two levels of human-product interaction into the picture: perceptual-motor skills and emotional skills.

Perceptual-motor skills, i.e. what people can perceive with their senses and what they can do with their body, require physical interaction, i.e. handling objects instead of icons on a screen. This choice for tangibility nicely fits the newest trends in the human-computer interaction (HCI) community (Cohen *et al.* 1999).

Emphasis on emotional skills is growing too. The Media Lab at MIT is making a study of 'affective computing' (Picard 1997). Damasio's book (Damasio 1994) has shown that pure logic alone, without emotional value, leaves a person, or a machine for that matter, indecisive. Our department organised the First International Conference on Design and Emotion last year (Overbeeke and Hekkert 1999). And emotion has entered the stage not only in academic circles but also in industry. In our faculty, Mitsubishi is funding the 'Designing Emotion' project in which an instrument is developed to measure people's emotional reactions to cars (Desmet, Hekkert and Jacobs 1999).

Where then can the interaction designer seek advice to achieve the integration of the impossible: accelerating technological innovations, human perceptual-motor, cognitive and emotional skills and electronic aesthetics? We think that essentially this problem can be solved by turning to the user's experience, fully respecting all his skills. The designer needs to create a context for experience, rather than just a product. He offers the user a context in which they may enjoy a film, dinner, cleaning, playing, working ... with all their senses. It is his task to make the product's function accessible to the user whilst allowing for interaction with the product in a beautiful way. Aesthetics of interaction is his goal. The user should experience the access to the product's function as aesthetically pleasing. A prerequisite for this is that the user should, at the very least, not be frustrated.

However, we are not promoting 'ease of use' as a design goal. Interfaces should be surprising, seductive, smart, rewarding, tempting, even moody, and thereby exhilarating to use. The interaction with the product should contribute to the overall pleasure found in the function of the product itself. **The experiential is assumed to lead to joy of use.**

The following example should clarify what we mean. Suppose the user wants to watch a movie for his enjoyment. He has to programme his VCR in order to get it working at a later date. VCR manufacturers certainly give the impression of having done everything in their power to make the user as frustrated as possible. Why not make a machine that is a joy to use? We are not saying that 'technical' design with a large number of functions and buttons should be avoided; some people actually like it that way. We call for diversity in product design. Not all VCRs should look the same. Why is there such an experiential diversity in car design and not in VCR design?

To build this stage of emotionally rich interactions; the designer needs new methods to sound out the experiential world of the user. Interaction relabelling and designing for extreme characters are new methods that are illustrated below.

Once the designer gets a feel for the experiential world of the user, he needs to focus on designing the interaction. He needs to stay tuned to the experiential. The following focus supports will also be illustrated in the examples (Djadjadiningrat, Gaver and Frens 2000):

1. Don't think affordances, think temptation.

Ergonomics, HCI and product design have borrowed the term 'affordances' from perception-psychology (Gibson 1979). Affordance is a very useful concept here, because it refers to the inextricability of both perception and action, and a person and his environment. It is about what people can do. Furthermore, it is essentially a non-cognitive and non-representational concept. However, many researchers concentrate on the structural aspects of affordances while neglecting the affective aspects. We lament this clinical interpretation of affordance. People are not invited to act only because a design fits their physical measurements. They can also be tempted to act through the expectation of beauty of interaction.

2. Don't think beauty in appearance, think beauty in interaction.

Usability is generally treated separately from aesthetics. Aesthetics in product design appears to be restricted to making products beautiful in appearance. As the ease of use strategies do not appear to pay off, this has left us in the curious situation that we have products which look good at first sight, but frustrate us as soon as we start interacting with them. Again, we think that the emphasis should shift from a beautiful appearance to beautiful interaction, of which beautiful appearance is a part. Dunne (1999) talks of 'an aesthetics of use': an aesthetics which, through the interactivity made possible by computing, seeks a developing and more nuanced co-operation with the object – a co-operation which, it is hoped, will enhance social contact and everyday experience.

3. Don't think ease of use, think enjoyment of the experience.

Current efforts on improving usability focus on making things easier. However, there is more to usability than ease of use. A user may choose to work with a product despite it being difficult to use, because it is challenging, seductive, playful, surprising, memorable or rewarding, resulting in enjoyment of the experience. No musician learns to play the violin because it is easy. Bringing together 'contexts for experience' and 'aesthetics of interaction' means that we do not strive for making a function as easy to access as possible, but for making the unlocking of the functionality contribute to the overall experience.

1.4 EXAMPLES

The key question for design still remains however: how will these 'experiential' products differ from the 'normal' ones? We will give three examples from research and teaching projects.

Appointment manager

For his masters project Frens designed an appointment manager, a handheld electronic device which aids its user in managing appointments. In ranking the importance of an appointment, Frens' appointment manager not only considers the factual aspects of an appointment, such as time and location, but also the feelings of its user towards an appointment, or towards the person the appointment is with. Frens' approach to the appointment manager acknowledges that emotions are an important consideration in managing our daily lives, often neglected in purely cognitive approaches.

Figure 1.1 (left) Appointment manager, a rotatable ring sits around the top screen

Figure 1.2 (right) Interface example using virtual blades

In his project, Frens used two new methods to explore aesthetics, interaction and role (Djadjadiningrat, Gaver and Frens 2000). In the first method, interaction relabelling, designers choose an existing product that is rich in terms of actions. Then, pretending that this is the object to be designed, they have to explain and act out how it works. For example, Frens offered fellow designers a toy gun, asked them to pretend that it was an appointment manager and to describe and act out how they would interact with it. Interaction relabelling works particularly well with mechanical products as it makes designers aware of how poor electronic products are in terms of actions. Interaction relabelling also sensitises designers to how interaction says something about the user and the relationship between the user and the product. In the second method, designing for extreme characters, designers create products for fictitious characters that are emotional exaggerations. This helps to expose character traits which otherwise remain hidden. For example, one of Frens' extreme characters was a hedonistic, polyandrous, twenty-year old woman. This choice of character required Frens to come up with an appointment manager which allows the woman to maximise the fun in her life and which supports her in juggling appointments with multiple boyfriends who may not know about each other. In his final design, Frens aimed to achieve aesthetics of interaction by treating hardware and on-screen graphics as inseparable. The user navigates through time by means of a ring which sits around the top screen (Figure 1.1). The direct coupling between the rotation of the ring and the flow of characters over the screen makes for a beautiful interaction. Through the positioning of the multiple screens, the woman can rate her boyfriends on various issues such as shopping, dining, sex, etc. in a playful manner. Virtual blades allow her to compare her current view on her boyfriends with the previous one (Figure 1.2). These aspects of the design show respect for the user's perceptual-motor skills, not only from a structural but also from an affective point of view.

To clarify our stance on aesthetics of interaction we may contrast Frens' design with the designs of Emilio Ambasz (Sipek, Poynor and Hudson 1993). Ambasz' products (Figure 1.3) make use of soft materials which are pleasurable to the touch. We value the importance of this approach, how touch is often neglected in electronic products and the influence it has on how enjoyable a product is to work with. However, in our approach the aesthetics of interaction stems from a symbiosis between the physical, the virtual and the resulting interaction rather than the application of pleasurable materials to an otherwise conventional interface.

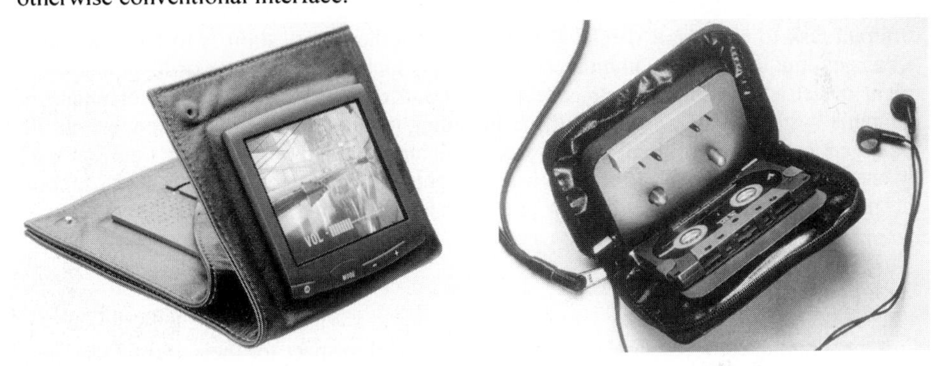

Figure 1.3 Wallet with incorporated television and walkman by Ambasz

Interactive chair

Cheung's masters project is an example of how we try to tackle problems at a behavioural rather than an analytical level. In the context of office chair design, Cheung is researching the relationship between the visible movements of the user in the chair, called macro-movements, and user experience. Macro-movements can be seen as indicators of behaviour. As people's behaviour changes with their emotional state, macro-movements may reflect that state. In an experiment, Cheung compared macro-movements with data obtained through questionnaires and physiological measurements. This information on how macro-movements are related to the user's feelings can be used for two purposes. First, such information could be used to evaluate how users experience existing chair designs. Compared to physiological measurements, detecting macro-movements can be achieved in a less intrusive manner. Compared to questionnaires, macro-movements are more direct and can be taken 'in the act' rather than afterwards.

Figure 1.4 Sketches of interactive chairs by Cheung

Second, this information can lead to the design of interactive chairs. Interactive chairs can be used for a number of purposes. For example, a chair could monitor the user's state of well-being and convince the user to change his posture or to change to a different task to counter RSI-related problems. Another application is to have the chair act as an input device for communication devices, such as a telephone or ICQ. The chair might detect a state of stress or concentration and communicate this to a telephone or computer system, enabling them to filter incoming messages and limit these messages to the urgent or important. Finally, an interactive chair could support the user's current state by changing configuration. When the user is highly concentrated, the chair could support this mode of working by shielding the user from his colleagues (see Figure 1.4).

Personal pagers

Current communication technologies do not show much respect for their users. Telephone calls and internet messages will interrupt you, regardless of how busy or concentrated you are. This can be aggravating for both the sender and the recipient. For an urgent message, a sender may want to get hold of the recipient at any cost, while for a not so important message he only wants to disturb the recipient if time allows. Likewise, the recipient may wish to signal that she is busy, allowing only urgent messages to come through, or that she is more generally available when things are less hectic. For a design exercise, we asked our students to design a pager which would allow the user to communicate the urgency of a message to be sent and his availability for incoming messages. The pager should be able to send the simple message 'I need you' in a non-verbal manner. Furthermore, it should allow the user to contact two specific friends. In its appearance and interaction the pager had to express the users and the functions.

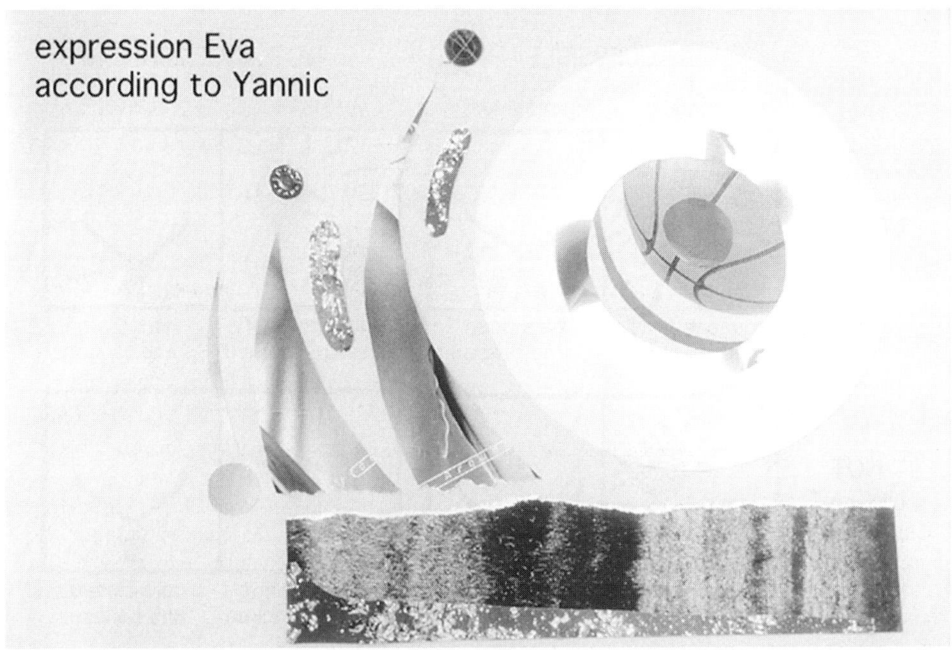

Figure 1.5 The collage expressing Eva according to Yannic

Figure 1.6 The pager

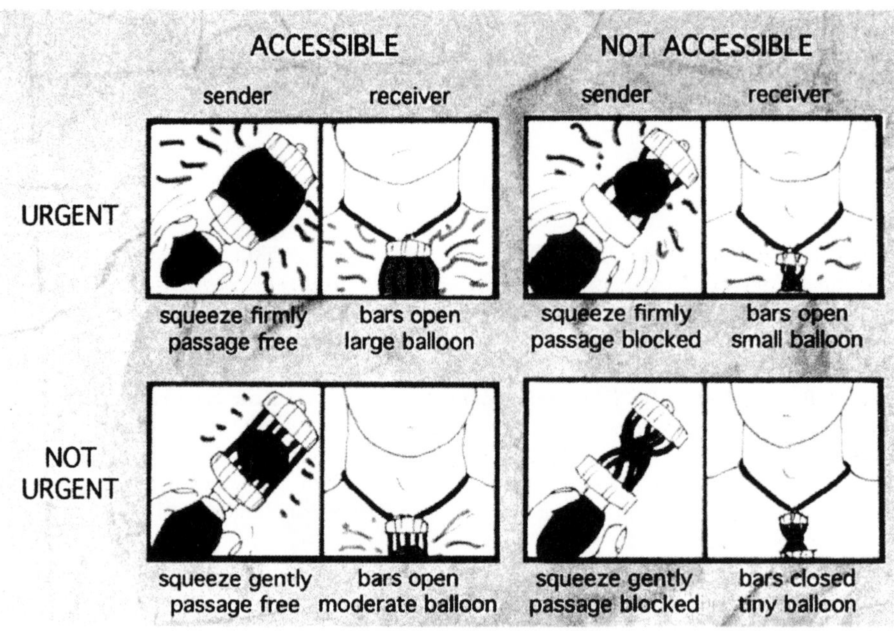

Figure 1.7 How to use the pager

An example of such a pager is the one designed by Yannic which consists of a central unit with bars, a balloon and two personal pumps (Figures 1.5 to 1.7). These pumps are a simplification of the collages Yannic made to express his friends Eva and Cees. The cube is used to contact Cees and the sphere is used to contact Eva. To contact Eva, Yannic places the yellow sphere on the central unit and pumps. The force he uses to squeeze the pump indicates the urgency of his message. A comparable unit owned by Eva is reacting to Yannic's call, by means of a growing balloon that starts to emit a red light. When Eva does not want to be disturbed, she twists the bars from her central unit, thus prohibiting the balloon from growing. Yannic's central unit twists simultaneously, blocking the passage of air and complicating the squeezing activity. These examples show that a context for experience addresses aspects that are ignored in 'traditional' technological products. These contexts for experience have an overall expression in which the appearance and the way of interaction become one. They are adjusted or dynamically adjust to the user and the situation. Finally, the example of the pager shows that intangible electronics do not have to end in intangible products.

1.5 CONCLUSIONS

In this paper we focused on two aspects of human-product interaction design that we think are often neglected: human perceptual-motor and emotional skills. We discussed these issues from the point of view of *respect* for man as a whole. These issues were illustrated with three examples. With the example of the appointment manager we showed how products should have respect for the user's perceptual-motor and emotional skills. The interactive chair example is an illustration of how interaction should be tackled on a behavioural level rather than an analytical one (see also Wensveen, Overbeeke and Djadjadiningrat 2000). The pagers example clarifies how technology can be respectful to the user, and what the importance is of including perceptual-motor skills in products through tangibility.

Last but not least, let us not forget one of the main ingredients in any design: the designer's creativity. Designing itself results from the designer's perceptual-motor, cognitive and emotional skills (q.e.d.).

1.6 REFERENCES

Cohen, J., Withgott, M. and Piernot, P., 1999. Logjam: a tangible multi-person interface for video logging. In *Proceedings of the CHI 99 Conference on Human Factors in Computing Systems*, pp. 128-135.

Damasio, A., 1994. Descartes' error: emotion, reason and the human brain. Gosset / Putnam Press, NY.

Desmet P.M.A, Hekkert P.P.M. and Jacobs J.J., 1999. When a car makes you smile: Development and application of an instrument to measure product emotions. In *Advances in Consumer Research*, **27**. Hoch, S.J. and Meyer, R.J. (Eds), (in press).

Djajadiningrat, J.P., Gaver, W.W. and Frens, J.W., 2000. Interaction relabelling and extreme characters: methods for exploring aesthetic interactions. In *Proceedings DIS2000* (accepted).

Djadjadiningrat, J.P, Overbeeke, C.J. and Wensveen, S.A.G., 2000. Augmenting fun and beauty: a pamphlet. In *Proceedings of DARE'2000*. Mackay, W.E. (Ed.), Elsinore, 12-14 April, Denmark.

Dunne, A., 1999. *Hertzian tales: Electronic products, aesthetic experience and critical design*. RCA CRD Research publications, London.

Gaver, W., Dunne, T. and Pacenti, E., 1999. Cultural probes. In *Interactions*. ACM, Danvers, pp. 21-29.

Gibson, J.J., 1979. *The ecological approach to visual perception*. Lawrence Erlbaum (reprinted in 1986), London.

Overbeeke, C.J. and Hekkert, P.P.M., 1999. Editorial. In *Proceedings of the First International Conference on Design and Emotion*. Overbeeke, C.J. and Hekkert, P.P.M. (Eds), 3-5 November, Delft.

Picard, R.W., 1997. *Affective computing*. MIT Press, Cambridge.

Sipek, B., Poynor, R. and Hudson, J., 1993. *The international design yearbook*. Laurence King.

Wensveen, S.A.G., Overbeeke, C.J. and Djadjadiningrat, J.P., 2000. Touch me, hit me and I know how you feel. A design approach to emotionally rich interaction. In *Proceedings DIS2000* (accepted).

CHAPTER TWO

The Personalities of Products

PATRICK W. JORDAN

Contemporary Trends Institute,

PO Box 31958, London W2 6YD, UK

2.1 INTRODUCTION

This chapter reports a study investigating products as personalities. The concept of products as personalities is founded on the 'New Human Factors' paradigm of products as *living objects* with which *people* have *relationships.* This is as opposed to traditional human factors approaches, which tend to see products as being merely *tools* with which *users* complete *tasks.*

An earlier study (Jordan 1997) had used a technique known as Product Personality Assignment (PPA) in order to investigate the concept of products as personalities, and to try to establish links between the aesthetic qualities of products and their personalities. The study was revealing, establishing a number of links between aesthetics and personality and suggesting that people have a preference for products which they feel reflect their own personality in some way.

However, PPA has been criticised by designers on the grounds that the model of personality used — Briggs-Myers — is not something that is easy for the non-psychologist to understand without explanation. In particular, the terminology does not reflect that which the 'layperson' would use when describing personality. An aim of this study was to identify the terms in which non-psychologists, in this case specifically designers, describe personality and to use this terminology as the basis for designing a questionnaire for rating the personalities of products. The study involved clustering these terms into personality dimensions and then using these dimensions as a means of assigning personalities to a variety of products.

Aside from identifying personality descriptors and designing a questionnaire for measuring product personality, the aims of the study were threefold. To see if product personality was a concept which was meaningful to designers and others in the product design process, to see if designers exhibited a preference for products which they felt reflected their own personalities, and to see whether product personalities were common across different products made by the same manufacturer.

The main outcomes of the study were as follows:

- Seventeen separate dimensions of product personality were identified.
- Product personality appears to be a concept which is meaningful and was perceived in a consistent way by those involved in the study.
- No evidence was found of a preference for products which participants felt matched their own personalities.
- The Braun products were felt by the participants to have a common brand personality.

2.2 THE WORKSHOP

The participants in the workshop were two Industrial Designers, a Human Factors Specialist and an Applications Engineer (someone who tests products in use). [For the sake of simplicity, and notwithstanding debates about what exactly is meant by the term, the participants will be referred to as 'designers' from here on]. Two were men and two were women, all four were in their twenties.

Session 1 — Brainstorm. The aim of this session was to identify the dimensions by which the designers thought of personality. In other words, to identify the terms which they use when describing people's personalities.

Session 2 — Product personality assignment. The aim of the next session was to ask the designers to look at a selection of products and assign personalities to them on the basis of the dimensions that had been identified earlier.

2.3 WORKSHOP PART 1: THE BRAINSTORM SESSION

The aim of the brainstorm was to gain an understanding of the terminology that the designers used in order to describe people. In order to do this, a number of different people were discussed and the words used to describe them were noted down. Each designer gave a verbal description of six characters:

- a personal friend;
- a personal enemy;
- a liked film character;
- a disliked film character;
- a liked public figure, and;
- a disliked public figure.

Clearly, the personal friends and enemies tended to be people who were only known to the designers personally, so there is little point in listing them here.

The film/TV characters and public figures selected for discussion are listed in Table 2.1.

Table 2.1 Film/TV Characters Selected for Discussion

Positive Characters	**Negative Characters**
Film/TV Characters:	
Benny and June	Basil Fawlty (Fawlty Towers)
Rutger Hauer (in 'Bladerunner')	Andy McDowell (her character in 'Four Weddings and a Funeral')
Ferris Bueler	Car salesman in 'Fargo'
Phoebe (from 'Friends')	Uncle Scrooge (from 'Donald Duck')
Public Figures:	
Nelson Mandela	Adolf Hitler
John F. Kennedy	Serial killers (in general)
Howard Stem ('shock' discjockey)	Margaret Thatcher
(only three characters selected)	La Toya Jackson

In discussing these characters a total of 209 different personality descriptor words were used. Each descriptor word was noted on a flipchart.

2.4 PERSONALITY DIMENSION ANALYSIS

Having collated 209 separate personality descriptors, they were then divided into groups of similar descriptors. Descriptors were grouped together on the basis of referring to related aspects (e.g. 'bigoted' and 'intolerant') or of referring to diametrically opposing aspects (e.g. 'tolerant' and 'intolerant'). In this the descriptors were combined in order to identify personality dimensions.

For example, one of the personality dimensions identified was Authoritarian / Liberal. This was made up of the 7 descriptors listed in Table 2.2.

Table 2.2 Descriptors Comprising the Authoritarian / Liberal Personality Dimension

Authoritarian Descriptors	Liberal Descriptors
Intolerant	Tolerant
Nationalistic	
'Shaper'	
Control freak	
Manipulative	
Overbearing	

Another dimension was Kind / Unkind. This was made up of the 18 descriptors listed in Table 2.3.

Table 2.3 Descriptors Comprising the Kind / Unkind Personality Dimension

Kind Descriptors	Unkind Descriptors
Sweet	Selfish
Compassionate	Thoughtless
Thoughtful	Cruel
Hospitable	Evil
Helpful	Tight-fisted
Caring	
Generous	
Supportive	
Devoted	
Considerate	
Thoughtful	
Loving	
Kind*	
Soft	

*Note that 'kind' appears as a descriptor as well as a dimension name. This was an example of a case where it was felt that a particular descriptor seemed to provide a good summary name for all the others in the group as a whole.

Overall, 17 personality dimensions were identified. These represented groups of between 2 and 43 descriptors each. Detailed information about the descriptors in each dimension was kept in the original workshop notes. Sadly, however, these have since been destroyed and the final report, which contains the 17 dimensions but not all of the descriptors which comprise them, is the only remaining record of the workshop. In the next section, these personality dimensions are listed, along with summary descriptions of each. The summary descriptions are based on the [lost] descriptors that were assigned to each personality dimension.

2.5 PERSONALITY DIMENSIONS

Kind/Unkind

Kind people are generous, caring, loving and compassionate. They are considerate of the needs of others and are supportive. Unkind people are selfish, uncaring and mean. They don't think of other's needs and can be cruel.

Honest/Dishonest

Honest people are straightforward and trustworthy. They do not tell lies or deceive others. Dishonest people are untrustworthy and hypocritical. They can be fake and deceitful.

Serious-minded/Light-hearted

Serious-minded people are strong, rational thinkers. They tend to take a professional approach to things and are competitive and goal-driven. Light-hearted people tend to be more emotion-based. They are joyful people with a sense of humour.

Bright/Dim

Bright people are gifted, creative and intelligent. They are full of original and imaginative ideas. Dim people are rather stupid and vague. They are losers with little to contribute.

Stable/Unstable

Stable people are self-confident, calm and mentally tough. Their moods are quite steady. Unstable people are insecure, touchy and temperamental. They are prone to mood swings.

Narcissistic/Humble

Narcissistic people have an exaggerated idea of their own importance. They are arrogant, egocentric and vain. Humble people underestimate their own importance. They are modest, meek and unassuming.

Flexible/Inflexible

Flexible people are spontaneous and unpredictable. They are inquisitive and like to use their initiative. They can find their way around difficulties, but may be weak-willed and unreliable. Inflexible people are systematic and organised. They are cautious people who like structured approaches. They can be determined but may be stubborn and obstinate.

Authoritarian/Liberal

Authoritarian people have conservative views and may be prejudiced. They enjoy having control over others. Liberal people are open-minded and tolerant of others. They prefer to allow others to have personal freedoms.

Value-driven/Non-value-driven

Value-driven people will act according to a set of principles in which they believe. They may be driven out of a sense of virtue or allegiance or by some other motive such as a 'work ethic'. Non-value-driven people will not act in accordance with any particular principles. They will act in order to try and exploit the situation in which they find themselves. This may sometimes mean reverting to the 'lowest common denominator'.

Extrovert/Introvert

Extrovert people are expressive and uninhibited. They enjoy the company of others. Sometimes they may be brash or vulgar. Introvert people are self-conscious and reserved. They may be rather shy. Sometimes they may be asocial or stand-offish.

Naive/Cynical

Naive people may be rather unsophisticated. They tend to assume that others are acting from the best intentions. Naive people can be easy to manipulate. Cynical people may be sophisticated; however, they tend to be rather negative in their view of the motivation of others. Cynical people are difficult to manipulate.

Excessive/Moderate

Excessive people tend to do things to extremes — whether it be work or play. They may be bon-viveurs, Epicureans, alcoholics or workaholics. Moderate people tend to do things to a 'sensible' degree. They live sensible but rather bland lifestyles.

Conformist/Rebel

Conformists are 'normal' people with respect for authority. They like to fit in and are unwilling to confront society's norms. Rebels are suspicious of authority and like to challenge it. They enjoy provoking argument and debate. Rebels may appear somewhat eccentric.

Energetic/Unenergetic

Energetic people are active, lively and enthusiastic. They are often busy and have a demeanour younger than their age. Unenergetic people are passive, lazy and unenthusiastic. Their demeanour is older than their age and they would rather others did the hard work.

Violent/Gentle

Violent people are insensitive and threatening. They can be destructive, dangerous and angry. Gentle people are sensitive and harmless. They tend to become melancholic rather than angry.

Complex/Simple

Complex people may have depth to their character. They are thoughtful people who may have rather 'mixed-up' and complicated lives. Simple people are down-to-earth and live straightforward lives. They may be rather shallow and vacuous.

Pessimistic/Optimistic

Pessimistic people tend to look at things negatively. They worry that things will go wrong. They may be defeatist. Optimistic people tend to look at things positively. They expect things to work out well. They may become complacent.

2.6 WORKSHOP PART 2: PRODUCT PERSONALITY ASSIGNMENT

Having identified 17 dimensions of personality, the next stage was to consider these in the context of products.

A questionnaire was compiled for rating product personality according to each of the 17 dimensions identified. This questionnaire is given in section 2.9. Each designer then individually rated 14 products on each of the personality dimensions. The products were in pairs:

- 2 irons;
- 2 men's shavers;
- 2 shaver bags;
- 2 epilators;
- 2 air cleaners;
- 2 hair dryers, and;
- 2 coffee makers.

These products are illustrated in the photographs (Figure 2.1 to 2.14).

Each designer also indicated which of each pair he or she preferred, and filled in the questionnaire one more time in order to give a self-rating of his or her own personalities.

Figure 2.1 Iron A

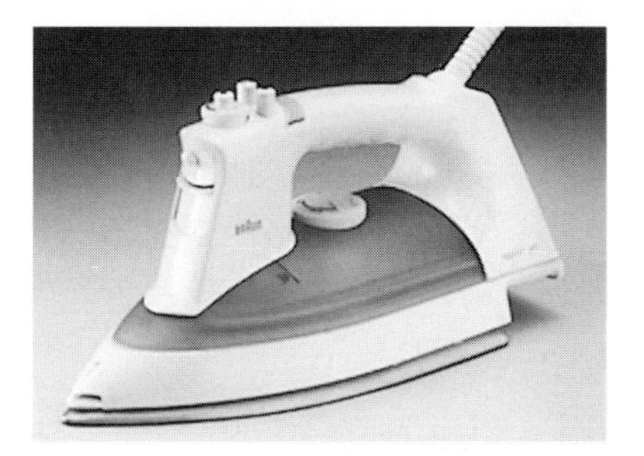

Figure 2.2 Iron B

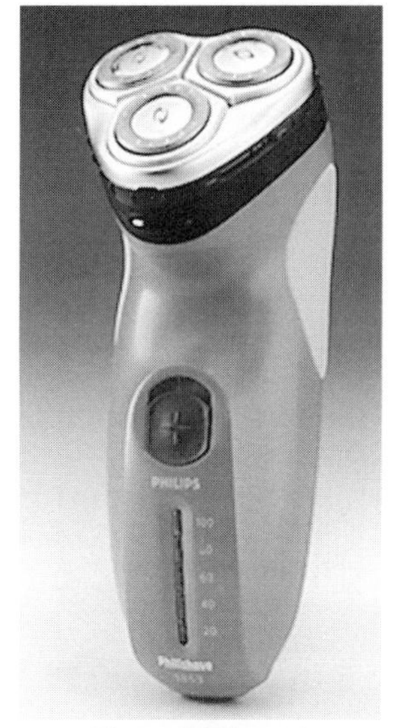

Figure 2.3 Shaver A

Figure 2.4 Shaver B

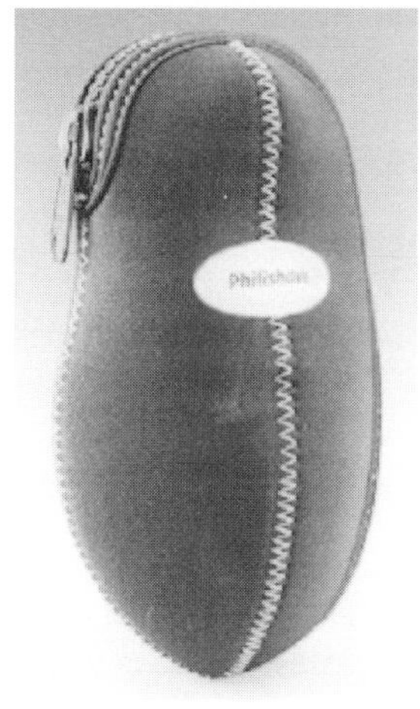

Figure 2.5 Shaver bag A

Figure 2.6 Shaver bag B

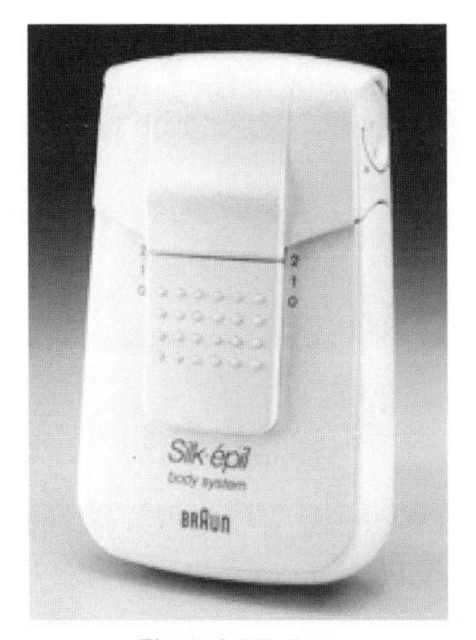

Figure 2.7 Epilator A

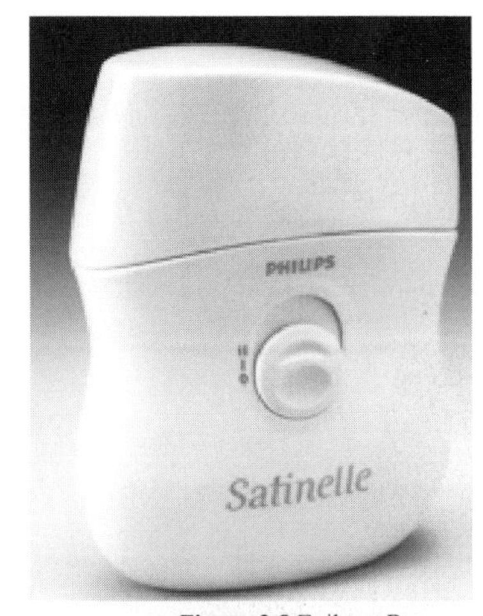

Figure 2.8 Epilator B

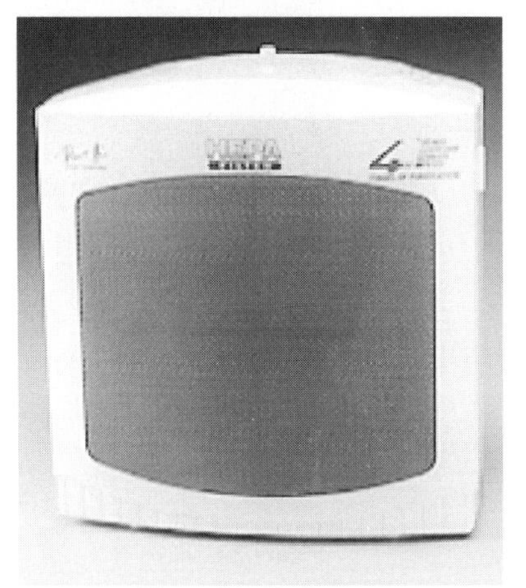

Figure 2.9 Air cleaner A

Figure 2.10 Air cleaner B

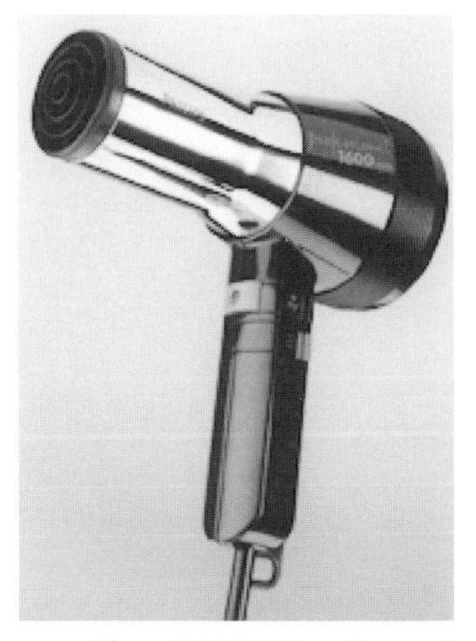

Figure 2.11 Hair dryer A

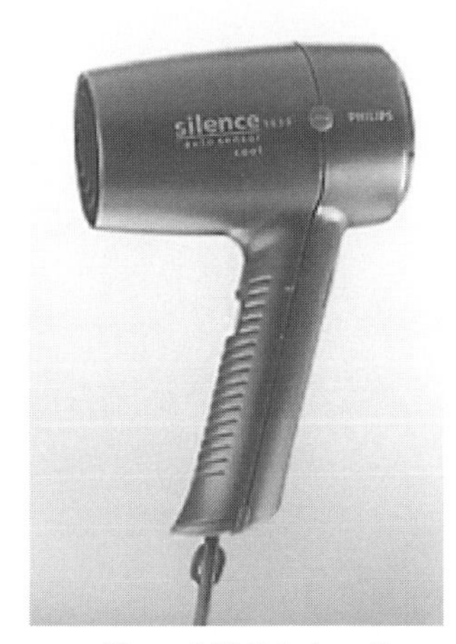

Figure 2.12 Hair dryer B

Figure 2.13 Coffee maker A

Figure 2.14 Coffee maker B

2.7 RESULTS

The personality profiles allocated to each of the products are given below. A figure is also given indicating how many of the designers preferred each product over the other in its pair.

IRON A

	1	2	3	4	5	
Kind	X					Unkind
Honest			X			Dishonest
Serious-minded				X		Light-hearted
Bright			X			Dim
Stable			X			Unstable
Narcissistic		X				Humble
Flexible		X				Inflexible
Authoritarian					X	Liberal
Value-driven			X			Non-value-driven
Extrovert	X					Introvert
Naive	X					Cynical
Excessive		X				Moderate
Conformist				X		Rebel
Energetic	X					Unenergetic
Violent			X			Gentle
Complex			X			Simple
Pessimistic				X		Optimistic

Number Preferring: 3 (75%)

IRON B

	1	2	3	4	5	
Kind			X			Unkind
Honest		X				Dishonest
Serious-minded			X			Light-hearted
Bright	X					Dim
Stable	X					Unstable
Narcissistic			X			Humble
Flexible			X			Inflexible
Authoritarian		X				Liberal
Value-driven	X					Non-value-driven
Extrovert			X			Introvert
Naive				X		Cynical
Excessive					X	Moderate
Conformist			X			Rebel
Energetic			X			Unenergetic
Violent			X			Gentle
Complex				X		Simple
Pessimistic		X				Optimistic

Number Preferring: 3 (75%)

SHAVER A

	1	2	3	4	5	
Kind		X				Unkind
Honest			X			Dishonest
Serious-minded					X	Light-hearted
Bright			X			Dim
Stable				X		Unstable
Narcissistic		X				Humble
Flexible		X				Inflexible
Authoritarian					X	Liberal
Value-driven			X			Non-value-driven
Extrovert	X					Introvert
Naive		X				Cynical
Excessive		X				Moderate
Conformist					X	Rebel
Energetic	X					Unenergetic
Violent		X				Gentle
Complex		X				Simple
Pessimistic					X	Optimistic

Number Preferring: 2 (50%)

SHAVER B

	1	2	3	4	5	
Kind			X			**Unkind**
Honest		X				**Dishonest**
Serious-minded		X				**Light-hearted**
Bright			X			**Dim**
Stable	X					**Unstable**
Narcissistic			X			**Humble**
Flexible			X			**Inflexible**
Authoritarian		X				**Liberal**
Value-driven		X				**Non-value-driven**
Extrovert				X		**Introvert**
Naive				X		**Cynical**
Excessive				X		**Moderate**
Conformist		X				**Rebel**
Energetic				X		**Unenergetic**
Violent				X		**Gentle**
Complex				X		**Simple**
Pessimistic		X				**Optimistic**

Number Preferring: 2 (50%)

SHAVER BAG A

	1	2	3	4	5	
Kind	X					**Unkind**
Honest			X			**Dishonest**
Serious-minded				X		**Light-hearted**
Bright		X				**Dim**
Stable				X		**Unstable**
Narcissistic	X					**Humble**
Flexible			X			**Inflexible**
Authoritarian				X		**Liberal**
Value-driven		X				**Non-value-driven**
Extrovert		X				**Introvert**
Naive			X			**Cynical**
Excessive		X				**Moderate**
Conformist					X	**Rebel**
Energetic	X					**Unenergetic**
Violent				X		**Gentle**
Complex			X			**Simple**
Pessimistic				X		**Optimistic**

Number Preferring: 4 (100%)

SHAVER BAG B

	1	2	3	4	5	
Kind				X		**Unkind**
Honest		X				**Dishonest**
Serious-minded	X					**Light-hearted**
Bright				X		**Dim**
Stable		X				**Unstable**
Narcissistic				X		**Humble**
Flexible				X		**Inflexible**
Authoritarian		X				**Liberal**
Value-driven				X		**Non-value-driven**
Extrovert					X	**Introvert**
Naive				X		**Cynical**
Excessive					X	**Moderate**
Conformist	X					**Rebel**
Energetic					X	**Unenergetic**
Violent			X			**Gentle**
Complex				X		**Simple**
Pessimistic			X			**Optimistic**

Number Preferring: 0 (0%)

EPILATOR A

Kind			X			Unkind
Honest	X					Dishonest
Serious-minded	X					Light-hearted
Bright		X				Dim
Stable	X					Unstable
Narcissistic			X			Humble
Flexible			X			Inflexible
Authoritarian		X				Liberal
Value-driven		X				Non-value-driven
Extrovert			X			Introvert
Naive				X		Cynical
Excessive			X			Moderate
Conformist	X					Rebel
Energetic		X				Unenergetic
Violent		X				Gentle
Complex			X			Simple
Pessimistic			X			Optimistic

Number Preferring: 2 (50%)

EPILATOR B

Kind	X					Unkind
Honest			X			Dishonest
Serious-minded				X		Light-hearted
Bright			X			Dim
Stable			X			Unstable
Narcissistic		X				Humble
Flexible			X			Inflexible
Authoritarian				X		Liberal
Value-driven			X			Non-value-driven
Extrovert		X				Introvert
Naive		X				Cynical
Excessive		X				Moderate
Conformist			X			Rebel
Energetic		X				Unenergetic
Violent				X		Gentle
Complex				X		Simple
Pessimistic				X		Optimistic

Number Preferring: 2 (50%)

AIR CLEANER A

	1	2	3	4	5	
Kind			X			Unkind
Honest		X				Dishonest
Serious-minded			X			Light-hearted
Bright				X		Dim
Stable			X			Unstable
Narcissistic				X		Humble
Flexible			X			Inflexible
Authoritarian				X		Liberal
Value-driven			X			Non-value-driven
Extrovert				X		Introvert
Naive		X				Cynical
Excessive				X		Moderate
Conformist	X					Rebel
Energetic				X		Unenergetic
Violent				X		Gentle
Complex			X			Simple
Pessimistic				X		Optimistic

Number Preferring: 2 (50%)

AIR CLEANER B

Kind				X		**Unkind**
Honest			X			**Dishonest**
Serious-minded	X					**Light-hearted**
Bright		X				**Dim**
Stable		X				**Unstable**
Narcissistic		X				**Humble**
Flexible			X			**Inflexible**
Authoritarian		X				**Liberal**
Value-driven		X				**Non-value-driven**
Extrovert		X				**Introvert**
Naive				X		**Cynical**
Excessive			X			**Moderate**
Conformist				X		**Rebel**
Energetic			X			**Unenergetic**
Violent		X				**Gentle**
Complex				X		**Simple**
Pessimistic		X				**Optimistic**

Number Preferring: 2 (50%)

HAIR DRYER A

Kind				X		Unkind
Honest			X			Dishonest
Serious-minded		X				Light-hearted
Bright		X				Dim
Stable			X			Unstable
Narcissistic		X				Humble
Flexible			X			Inflexible
Authoritarian	X					Liberal
Value-driven			X			Non-value-driven
Extrovert		X				Introvert
Naive				X		Cynical
Excessive		X				Moderate
Conformist		X				Rebel
Energetic			X			Unenergetic
Violent	X					Gentle
Complex		X				Simple
Pessimistic			X			Optimistic

Number Preferring: 1 (25%)

HAIR DRYER B

Kind			X			Unkind
Honest			X			Dishonest
Serious-minded			X			Light-hearted
Bright			X			Dim
Stable				X		Unstable
Narcissistic				X		Humble
Flexible				X		Inflexible
Authoritarian				X		Liberal
Value-driven			X			Non-value-driven
Extrovert				X		Introvert
Naive			X			Cynical
Excessive				X		Moderate
Conformist		X				Rebel
Energetic			X			Unenergetic
Violent			X			Gentle
Complex				X		Simple
Pessimistic				X		Optimistic

Number Preferring: 3 (75%)

COFFEE MAKER A

Kind			X			Unkind
Honest		X				Dishonest
Serious-minded	X					Light-hearted
Bright		X				Dim
Stable		X				Unstable
Narcissistic			X			Humble
Flexible			X			Inflexible
Authoritarian		X				Liberal
Value-driven		X				Non-value-driven
Extrovert		X				Introvert
Naive				X		Cynical
Excessive			X			Moderate
Conformist		X				Rebel
Energetic			X			Unenergetic
Violent		X				Gentle
Complex			X			Simple
Pessimistic		X				Optimistic

Number Preferring: 1 (25%)

COFFEE MAKER B

	1	2	3	4	5	
Kind		X				Unkind
Honest		X				Dishonest
Serious-minded			X			Light-hearted
Bright				X		Dim
Stable			X			Unstable
Narcissistic		X				Humble
Flexible			X			Inflexible
Authoritarian			X			Liberal
Value-driven			X			Non-value-driven
Extrovert		X				Introvert
Naive		X				Cynical
Excessive			X			Moderate
Conformist				X		Rebel
Energetic		X				Unenergetic
Violent					X	Gentle
Complex			X			Simple
Pessimistic					X	Optimistic

Number Preferring: 3 (75%)

2.8 ANALYSIS

After the workshop, the questionnaires were statistically analysed. The aim of this analysis was to answer the following questions:

- Is assigning personalities to products meaningful?
- Did the designers prefer products which they felt reflected their own personalities?
- Is it possible to characterise brands in terms of their personalities?

The next sections of the report will deal with each of these questions individually.

Is assigning personalities to products meaningful?

The first thing to consider is whether or not it is meaningful to designers to talk about products having personality. It could be that the whole concept of assigning personalities to products is meaningless, and that filling in a questionnaire to rate product personality amounts to little more than filling in rating scales in an arbitrary manner.

Analysis used

The analysis used to answer this question was based on comparing the ways that each designer had rated the different products. If product personality was a meaningful concept, then there should have been a degree of agreement between the designers about the personality characteristics of a particular product, even though they had rated the products independently. If they had all rated a particular product differently, this might mean that (at best) they all had different ideas about a product's personality, or (at worst) that they really were just making random assignments on the personality dimension scales.

Statistical tests of correlation and statistical significance were carried out to check this. These tests are described in section 2.10.

Results

The outcome of the tests gave strong support to the idea that the designers found that assigning personalities to the products was meaningful. This suggests that assigning personalities to products may be a meaningful concept for designers.

Conclusion

Assigning personalities to products and the general idea that products have personalities are meaningful approaches for designers.

Did designers show a preference for products which they felt reflected their own personalities?

A statistical analysis showed that the designers exhibited no systematic preference for products whose personalities they regarded as matching their own.

Conclusion

There was no evidence that the designers preferred products that matched their own perceived personalities.

Is it possible to characterise brands in terms of their personalities?

It may be the case that certain brands tend to be associated with particular product personalities. This can be checked by looking at different types of products manufactured by the same company and seeing if the perceived personalities of these products are closely matched.

There were three different Braun products in the sample. An analysis was performed on the ratings of these products to see if their personalities were significantly correlated with each other. If they were, then their common personality characteristics might be thought of as the 'Braun product personality'. If, on the other hand, the ratings of the Braun products were not correlated, it would not be meaningful to claim that there was a 'Braun personality'.

The Braun products included were:

- Iron B
- Epilator A
- Coffee maker A

THE BRAUN PRODUCTS

Iron B / Coffee maker A / Epilator A

A 't' test revealed that the mean correlation between the Braun products was statistically significantly higher than the correlation between the Braun products and the non-Braun products ($p < .05$). This indicates that there were indeed common perceived personality aspects across all three products. The most clearly defined aspects of the Braun personality were as follows:

- honest
- serious-minded
- bright
- stable
- authoritarian
- value-driven
- cynical
- conformist

Conclusions

Looking at products as personalities is one of the ways in which we can understand products as 'living objects' rather than merely as functional tools. The outcomes of the study indicated that the concept of products as personalities is meaningful to designers and that commonalties of perceived personalities can be seen across a number of products made by the same manufacturer. However, unlike in a previous study, no evidence was found to link preferences for particular products to the perceived degree of match between a product's personality and the personality of the designer.

A questionnaire was developed for rating product personality. It can be used free of charge, provided that the reference to this chapter is quoted. The next research question appears to be how to link the aesthetic features of products to their perceived personalities. Another challenge, as human factors helps in the creation of pleasurable products which are a genuine joy to own and use.

2.9 PRODUCT PERSONALITY ASSIGNMENT QUESTIONNAIRE

Instructions

Read through the descriptions of the personality dimensions as given below.

Imagine that the product you are rating were a 'living object'. Consider what sort of personality it would have with respect to each of these dimensions.

Mark the scales in order to indicate your opinions about the personality of the product that you are rating.

Kind/Unkind

Kind people are generous, caring, loving and compassionate. They are considerate of the needs of others and are supportive. Unkind people are selfish, uncaring and mean. They don't think of others needs and can be cruel.

Honest/Dishonest

Honest people are straightforward and trustworthy. They do not tell lies or deceive others. Dishonest people are untrustworthy and hypocritical. They can be fake and deceitful.

Serious-minded/Light-hearted

Serious-minded people are strong, rational thinkers. They tend to take a professional approach to things and are competitive and goal-driven. Light-hearted people tend to be more emotion-based. They are joyful people with a sense of humour.

Bright/Dim

Bright people are gifted, creative and intelligent. They are full of original and imaginative ideas. Dim people are rather stupid and vague. They are losers with little to contribute.

Stable/Unstable

Stable people are self-confident, calm and mentally tough. Their moods are quite steady. Unstable people are insecure, touchy and temperamental. They are prone to mood swings.

Narcissistic/Humble

Narcissistic people have an exaggerated idea of their own importance. They are arrogant, egocentric and vain. Humble people underestimate their own importance. They are modest, meek and unassuming.

Flexible/Inflexible

Flexible people are spontaneous and unpredictable. They are inquisitive people and like to use their initiative. They can find their way around difficulties, but may be weak-willed and unreliable. Inflexible people are systematic and organised. They are cautious people who like structured approaches. They can be determined but may be stubborn and obstinate.

Authoritarian/Liberal

Authoritarian people have conservative views and may be prejudiced. They enjoy having control over others. Liberal people are open-minded and tolerant of others. They prefer to allow others to have personal freedoms.

Value-driven/Non-value-driven

Value-driven people will act according to a set of principles in which they believe. They may be driven out of a sense of virtue or allegiance or by some other motive such as a 'work ethic'. Non-value-driven people will not act in accordance with any particular principles. They will act in order to try and exploit the situation in which they find themselves. This may sometimes mean reverting to the 'lowest common denominator'.

Extrovert/Introvert

Extrovert people are expressive and uninhibited. They enjoy the company of others. Sometimes they may be brash or vulgar. Introvert people are self-conscious and reserved. They may be rather shy. Sometimes they may be a-social or stand-offish.

Cynical/Naive

Cynical people may be sophisticated; however, they tend to be rather negative in their view of the motivation of others. Cynical people are difficult to manipulate. Naive people may be rather unsophisticated. They tend to assume that others are acting from the best intentions. Naive people can be easy to manipulate.

Excessive/Moderate

Excessive people tend to do things to extremes — whether it be work or play. They may be bon-viveurs, Epicureans, alcoholics or workaholics. Moderate people tend to do things to a 'sensible' degree. They live sensible but rather bland lifestyles.

Conformist/Rebel

Conformists are 'normal' people with respect for authority. They like to fit in and are unwilling to confront society's norms. Rebels are suspicious of authority and like to challenge it. They enjoy provoking argument and debate. Rebels may appear somewhat eccentric.

Energetic/Unenergetic

Energetic people are active, lively and enthusiastic. They are often busy and have a demeanour younger than their age. Unenergetic people are passive, lazy and unenthusiastic. Their demeanour is older than their age and they would rather others did the hard work.

Violent/Gentle

Violent people are insensitive and threatening. They can be destructive, dangerous and angry. Gentle people are sensitive and harmless. They tend to become melancholic rather than angry.

Complex/Simple

Complex people may have depth to their character. They are thoughtful people who may have rather 'mixed-up' and complicated lives. Simple people are down-to-earth and live straightforward lives. They may be rather shallow and vacuous.

Pessimistic/Optimistic

Pessimistic people tend to look at things negatively. They worry that things will go wrong. They may be defeatist. Optimistic people tend to look at things positively. They expect things to work out well. They may become complacent.

I rate this product as...

Kind						**Unkind**
Honest						**Dishonest**
Serious-minded						**Light-hearted**
Bright						**Dim**
Stable						**Unstable**
Narcissistic						**Humble**
Flexible						**Inflexible**
Authoritarian						**Liberal**
Value-driven						**Non-value-driven**
Extrovert						**Introvert**
Naive						**Cynical**
Excessive						**Moderate**
Conformist						**Rebel**
Energetic						**Unenergetic**
Violent						**Gentle**
Complex						**Simple**
Pessimistic						**Optimistic**

2.10 STATISTICAL ANALYSIS: MEANINGFULNESS OF PRODUCT PERSONALITY

A correlation analysis was used to check the extent to which the designers agreed on the characteristics of each product. If assigning personality to products is meaningful for designers, then the ratings for any particular product should correlate highly. So, if all four designers had given similar ratings to, for example, Shaver A, then assigning a personality would be meaningful in the context of this product. If, on the other hand, they had all rated Shaver A totally differently from each other, then the concept of product personality would not be meaningful.

We would, however, not expect the designers ratings for different products to correlate with each other — for example, there would be no reason to expect one designer's ratings for Coffee maker A to correlate with another's ratings for Coffee maker B or to expect a designer's rating for Hair dryer A to correlate with another's ratings for Hair dryer B.

To check for the meaningfulness of assigning personalities to products, the inter-rater correlations for the same product were compare with the inter-rater correlations for different products. If the correlations were significantly higher when rating the same product, this would indicate that personality assignment was meaningful. If they were not significantly higher, then this would indicate that it was not meaningful. 't' tests showed that same product correlations were significantly higher than different product correlations ($p=0.0004$). This indicated that the concept of product personality was meaningful to designers.

2.11 REFERENCES

Jordan P.W., 1997. Products as personalities. In *Contemporary ergonomics 1997.* Robertson, S. (Ed.), Taylor & Francis, London.

CHAPTER THREE

Beyond Usability, Computer Playfulness

JAN NOYES and RICHARD LITTLEDALE

Department of Experimental Psychology, University of Bristol,

Woodland Road, Bristol, BS8 1TN, UK

ABSTRACT

Since the launch of the first personal computer a little over 20 years ago in February 1978, the development of computing and associated facilities has been dramatic. This is in part fuelled by the growth of the Internet, but also by the increasing use of computers in our homes, schools and workplaces. As with the introduction of any new technology, the reaction to this ubiquitous use of computing has tended to be a negative one. For example, there has been a tendency in the literature to focus on the more negatively perceived aspects of computer use with books and articles on technophobia (Richards 1993; Mitchell 1994; Brosnan 1998) and the development of computer aversion (Meier 1988) and anxiety scales (Heinsssen, Glass and Knight 1987; Carlson and Wright 1993). In 1992, Webster and Martocchio noted that there has been little research on examining positive characteristics of human-computer interactions, and the aspect that they specifically mentioned was that of 'playfulness'. It is the intention of this chapter to consider playfulness within the context of usability, emotions and computer use.

3.1 PLAYFULNESS

There have been many attempts at defining 'playfulness', but this has not been easy because the word can be used in a number of different ways (Berlyne 1969). However, Liebermann (1977) viewed playfulness as a human trait, i.e. a stable characteristic of an individual that is relatively invariant to situational stimuli. Webster and Martocchio (1992, p. 202) put forward the following definition: '… a situation-specific individual characteristic represents a type of intellectual or cognitive playfulness. It describes an individual's tendency to interact spontaneously, inventively, and imaginatively with microcomputers.' These are represented in the following scales that they devised for assessing playfulness.

Table 3.1 Computer Playfulness Measure

Adjectives	Strongly Disagree	Disagree	Slightly Disagree	Neutral	Slightly Disagree	Agree	Strongly Agree
Spontaneous	1	2	3	4	5	6	7
Unimagin-ative	1	2	3	4	5	6	7
Flexible	1	2	3	4	5	6	7
Creative	1	2	3	4	5	6	7
Playful	1	2	3	4	5	6	7
Unoriginal	1	2	3	4	5	6	7
Uninventive	1	2	3	4	5	6	7

For each adjective, respondents had to circle the number that best matched a description of themselves when they interact with computers. Computer playfulness was calculated by averaging the responses to the seven items on the 7-point likert scales (taking into account reverse scoring) in order to obtain an overall index of playfulness. Webster and Martocchio stated that the measure showed good concurrent validity with positive correlations between playfulness and positive attitudes towards computers, and negative correlations with computer anxiety. They also found that the measure had good discriminant validity with the operational measures for computer attitudes and anxiety differing significantly in form and in content from the playfulness measure. Further, the measure showed good predictive validity and high test-retest reliability with a correlation of 0.85 ($p<0.001$).

One approach to viewing computer playfulness could be to see it as an attribute of usability. When considering the design of systems and products, the word 'usability' has entered the dictionaries of the 1990s. This has largely replaced the term 'user-friendly' that was popular in the 1980s, but 'frowned upon' by many professionals who regarded it as a subjective term with no recognised definition. This is in contrast to usability which was defined by the International Standards Organisation in 1994 as: 'The usability of a product is the degree to which specific users can achieve specific goals within a particular environment; effectively, efficiently, comfortably, and in an acceptable manner.' Hence, it can be seen that playfulness might fall under the umbrella of 'comfortably' in helping the user attain their goals. The latter is a subjective term in that its definition is open to interpretation.

The development and continuing refinement of the term 'usability' has always encompassed a subjective element. For example, one of the first definitions by Shackel (1981) included four components, one of which was attitude. Shackel (1986) defined attitude as the engendering of positive attitudes by the majority of users towards using the system, and suggested that its measurement should be within acceptable levels of tiredness, discomfort, frustration and personal effort. Booth (1989) in his definition of usability redefined attitude as 'likeability'. Likewise, definitions from Eason (1984) in his causal framework of usability included 'user reaction' as a dependent variable, and Nielsen (1993, pp. 193-4) suggested several measurements of usability, of which two were concerned with attitude. These were as follows:

- The proportion of user statements during the test that were positive versus critical towards the system.
- The number of times the user expresses clear frustration (or clear joy).

Therefore, it could be concluded that the subjective element, i.e. how people feel about using a system or product, is an important consideration in defining usability. More recently, Jordan and Kerr (1999) have suggested that pleasure is a key component in the usability of products. Taken to extremes, it could be argued that if people do not like or enjoy using a system or product, and provided there is a viable alternative, they will 'vote with their feet' and not use it. This must of course be the ultimate test of usability. It might be asserted that higher playfulness is likely to result in immediately enhanced subjective experiences such as positive mood. This would result in individuals more likely to experience psychological pleasure in their interactions with computers. This has implications for the design of products and systems, since it implies that more playful interactions will result in greater pleasure. Consequently, learning would be faster since those individuals who exhibited greater playfulness would be more likely to experiment and try out various aspects of the software (Liebermann 1977; Webster and Martocchio 1992).

3.2 PLAYFULNESS AND COMPUTER ATTITUDES

A great deal of research has been carried out on computer attitudes, and despite various controversies about actual definitions, it has long been recognised that the affective component is a major determinant in defining attitudes (Eagly and Chaiken 1993). How individuals feel about using computers in terms of their emotional response has been shown to affect the extent and value gained from their usage (Igbaria, Parasuraman and Pavri 1990).

If one considers playfulness as being (or at least, contributing to) an emotional state, this may pose a problem in that strong emotions are often perceived negatively because they are thought to impair rational decision-making. The stereotype of a person who is acting emotionally is that of someone who is behaving irrationally, perhaps inappropriately, and with poor judgement. Consequently, playfulness would have no role to play when considering control of some of the advanced technologies found in nuclear power plants, civil aircraft, and air traffic control. Indeed, a logical progression in this argument is that interactions resulting in high levels of playfulness are more suited to non-critical computer applications, e.g. those relating to the entertainment and leisure industries, where the effects of emotion are of no immediate consequence to anyone other than the individual involved. It might be concluded that there is a certain stigma attached to emotions when considered within the context of computer use. Further, this is exacerbated by the difference between humans and machines. Machines are perceived as being 'cold', computers are programmed to excel at deductive reasoning and logical decision-making, and emotions currently play no role in their operation. This difference may also be contributing to the negative stereotype of emotion in computing.

Where then does this leave playfulness in human-machine interactions? On the one hand, it is strongly linked to the negative connotations associated with emotions, while on the other, it has been viewed as a positive aspect of computer use (Webster and Martocchio 1992). In response to this, it may be that the negative view of emotion has overshadowed the positive one. Since emotion is a fundamental part of human communication, it will undoubtedly add to human-computer interactions. Picard (1997, p. 2) agrees with this stance: she argued that emotions in computing play an essential role 'in basic rational and intelligent behavior. Emotions not only contribute to a richer quality of interaction, but they directly impact a person's ability to interact in an intelligent way'. It can be seen that she was also questioning the extent to which emotions in computing result in the stereotypically irrational, unintelligent behaviour. In conclusion, the weight of the evidence suggests that playfulness can be viewed as a positive attribute in computer use, at least in certain circumstances. It is envisaged that those human-machine interactions that result in more playfulness will immediately provide more gratification for the individual in terms of enhancing mood and feelings of satisfaction (Csikszentmihalyi 1990).

3.3 PLAYFULNESS AND COGNITIVE STYLE

One affective aspect of computer attitudes that has been identified as a predictor of computer use is cognitive style (Mason and Mitroff 1973; Zmud 1979; Sage 1981; Parasuraman and Igbaria 1990). It was defined by Sage (1981, p. 642) as the way in which individuals formulate information or data for decision-making. Given that cognitive style is concerned with decision-making, an obvious progression is to consider the common metaphor that describes how individuals make decisions, i.e. the 'head versus heart' dichotomy. This expression is frequently used in the English language to refer to the thinking and feeling dichotomy, i.e. the head provides the rational approach to decision making whereas the heart provides the emotional response. Hence, cognitive style can be measured by assessing the extent to which individuals make decisions according to these two routes. One example of this polarisation can be seen in the personality test, the Myers-Briggs Type Indicator (MBTI). This test characterises personality in four axes: thinking (T), feeling (F), sensing (S) and iNtuition (N). The MBTI manual (Myers and McCaulley 1985, p. 15) described the four dimensions in the following way:

- T: (decision-making) relies on principle of cause and effect and tends to be impersonal;
- F: (decision-making) by weighing relative values;
- S: (gathering information by) perceptions available to the senses;
- N: (gathering information by) perceptions of possibilities, meanings and relationships.

It can be seen that thinking and feeling are at opposite ends of two axes. In other words, you are either a thinking or a feeling person. Likewise, sensing and intuition are at opposite ends of the other axes, but these dimensions are of less interest here. Support for using these two dimensions of the MBTI to reflect cognitive style can be found from Jones (1994) who investigated the relationship between MBTI types, computer attitude and use. Jones found no significant differences when using the dichotomous codes between MBTI types and computer attitudes. However, when the MBTI scores were interpreted as being continuous, it was found that more positive attitudes were associated with the T preference type*. Jones concluded that those individuals with this preference for logical problem solving were more likely to experiment with software packages. To a certain extent, this is in keeping with the stereotype that computer users are more likely to be 'thinking' rather than 'feeling' types. It can also be seen that some aspects of cognitive style as measured by the MBTI are mirrored in the playfulness measure described by Webster and Martocchio (1992). Further support for the MBTI was given by Picard (1997) and Kroeger and Thuesen (1992). Most students in technical programmes have been found to be biased towards the T axis; it is thought that these types of personalities

* Jones' reasons for using this approach to the MBTI was based on the work of Garden (1991), who stated that ignoring the 'middle ground' was likely to be problematic, and the work of McCrae and Costa (1989) who suggested that MBTI scores should be interpreted as being continuous.

place relatively little emphasis on emotions or feelings.

3.4 CASE STUDY: PLAYFULNESS, COMPUTER ATTITUDE AND COGNITIVE STYLE

An experimental study of the three dimensions has recently been completed. It was carried out in order to find out if thinking (T) types on the MBTI would score higher than feeling (F) types on playfulness; secondly, to see if playfulness scores were highly correlated with computer attitude, i.e. those individual exhibiting greater playfulness would have more positive attitudes towards using computers; thirdly, to study the pattern of responses on the MBTI dimensions to computer attitudes, i.e. relate attitudes to cognitive style.

40 volunteers (13 males, 27 females) attending a British Psychology Society Level 2 Occupational Psychology test filled in three questionnaire measures relating to playfulness, attitude and cognitive style. Participants' ages ranged from 19 to 56 (mean = 34.6); they are all psychology professionals, generally working in personnel departments and all using computers in their everyday lives. The following questionnaires were administered:

- Playfulness – this was measured using the scale given in Table 3.1.
- Computer attitude – this was tested via a modification of the Computer Attitude Measure (CAM) devised by Kay (1989), which is given below. The original questionnaire comprised four sections: demographic information, cognitive attitudes, affective attitudes, and behavioural attitudes. In this study, the first section on demographics was discarded, and the questions in the cognitive attitudes section were amended slightly. The CAM questionnaire was originally intended for university teachers, and the questions were not wholly appropriate for the sample of participants being tested here. For example, the question 'I would not need a computer in my classroom' was changed to 'I would not need a computer in my office'. In terms of reliability, Kay found relatively high alpha coefficients for the cognitive, affective and behavioural attitudes at 0.87, 0.89 and 0.94, respectively. The overall internal reliability coefficient was given as 0.94. However, it should be noted that these reliability scores only apply to Kay's population of student teachers, and it is doubtful whether they can be transferred across to the modified CAM questionnaire used here. A further problem relates to the fact that CAM was designed at least 10 years ago at a time when computer use was far less widespread than it is today. Consequently, some of the items, particularly in the cognitive scale, may seem inappropriate in respect of current computer technology.

3.5 COMPUTER ATTITUDE MEASURE

Section 1: Cognitive scale

Please state the level of your agreement with the following statements.

Note: Respondents had to state the level of their agreement using the same 7-point scale as shown in Table 3.1.

- Computers would help me be more creative.
- Computers would not significantly improve the quality of work for my colleagues.
- Computers would help make my work more interesting.
- It is important that I keep up with my computer innovations.
- I would not need a computer in the office.
- My colleagues' mental abilities would improve significantly by interacting with computers.
- Computers would make my colleagues lose valuable skills.
- Computers would help me be more positive.
- Computers would motivate my colleagues to do better work.
- Computers would make my life in the office more difficult.

Section 2: Affective scale

Please indicate, using the following 7-point scales, which phrase best describes your attitude towards computers for each of the 10 semantic differential scale items.

1. Unlikeable-Likeable

Extremely Unlikeable	Moderately Unlikeable	Slightly Unlikeable	Neither	Slightly Likeable	Moderately Likeable	Extremely Likable
1	2	3	4	5	6	7

2. Good-Bad

Extremely Good	Moderately Good	Slightly Good	Neither	Slightly Bad	Moderately Bad	Extremely Bad
1	2	3	4	5	6	7

3. Unhappy-Happy

Extremely Unhappy	Moderately Unhappy	Slightly Unhappy	Neither	Slightly Happy	Moderately Happy	Extremely Happy
1	2	3	4	5	6	7

4. Uncomfortable-Comfortable

Extremely Uncomfort-able	Moderately Uncomfort-able	Slightly Uncomfort-able	Neither	Slightly Comfort-able	Moderately Comfortable	Extremely Comfortable
1	2	3	4	5	6	7

5. Calm-Tense

Extremely Calm	Moderately Calm	Slightly Calm	Neither	Slightly Tense	Moderately Tense	Extremely Tense
1	2	3	4	5	6	7

6. Empty-Full

Extremely Empty	Moderately Empty	Slightly Empty	Neither	Slightly Full	Moderately Full	Extremely Full
1	2	3	4	5	6	7

7. Natural-Artificial

Extremely Natural	Moderately Natural	Slightly Natural	Neither	Slightly Artificial	Moderately Artificial	Extremely Artificial
1	2	3	4	5	6	7

8.Exciting-Dull

Extremely Exciting	Moderately Exciting	Slightly Exciting	Neither	Slightly Dull	Moderately Dull	Extremely Dull
1	2	3	4	5	6	7

9. Suffocating-Fresh

Extremely Suffocating	Moderately Suffocating	Slightly Suffocating	Neither	Slightly Fresh	Moderately Fresh	Extremely Fresh
1	2	3	4	5	6	7

10. Pleasant-Unpleasant

Extremely Pleasant	Moderately Pleasant	Slightly Pleasant	Neither	Slightly Unpleasant	Moderately Unpleasant	Extremely Unpleasant
1	2	3	4	5	6	7

Section 3: Behavioural scale

Please indicate using the following 7-point scale how likely you would be to perform the following actions.

Extremely Unlikely	Moderately Unlikely	Slightly Unlikely	Neither	Slightly Likely	Moderately Likely	Extremely Likely
1	2	3	4	5	6	7

- Use a word processor.
- Use a computer on a regular basis.
- Do a significant task on a computer.
- Buy or borrow computer software or hardware.
- Use a disc operating system.
- Investigate different types of software.
- Work with computer-aided instruction.
- Experiment with a new computer software package.
- Work with a computer graphics package.
- Use database software.

The CAM was scored by recoding the items so that high scores reflected 'positive' computer attitudes. The 10 items in each section were summed and averaged to create a score for each subscale, and these were averaged to produce an overall index of computer attitude.

Cognitive style – this was measured using the T and F dimensions on the MBTI. The MBTI shows good reliability of around 0.8 and good construct validity when compared to personality models such as the 'big five' (see McCrae and Costa 1989).

Did thinking (T) types score higher than feeling (F) types on the playfulness measure?

Analysis of the data using the usual interpretation of dichotomous codes on the MBTI indicated that the differences on the T-F and S-N dimensions did show a trend in that Ts and Ns scored higher on the playfulness measure (T=5.0204 and F=4.984). However, this difference was not significant (although it was found to be significant at the 1% level for the S-N dimensions, F=14.63, df 1,38, p=0.0005, with iNtuitive types scoring higher).

Were playfulness scores highly correlated with computer attitude?

It was found that there was a positive significant correlation between playfulness and attitude (r=0.64, p=0).

What was the relationship between computer attitudes and cognitive style?

Using the dichotomous code approach to interpreting the MBTI, there was found to be no significance to the pattern of results, although Ts scored higher than Fs and Ns scored higher than Ss. In contrast, analysis of the separate components of the computer attitude measure indicated significant differences between T and F on the affective scale (F=4.34, df=1,38, p=0.0440) and S and N on the behavioural scale (F=7.7531, df=1,38, p=0.0083). The latter was also found when analysis was carried out using continuous scores for the MBTI preferences.

The results of this study confirm that playfulness and computer attitude are correlated, i.e. those with higher scores on the modified CAM exhibited greater playfulness. This may not be surprising considering the similarity of some of the items on the two measure, i.e. if you have a positive attitude towards using computers, you would be more likely to respond favourably to some of the playfulness items.

In contrast, findings relating to cognitive style (as measured by the MBTI scores) were mixed. The original hypothesis that individuals with a preference for more logical problem solving in keeping with the stereotype of computer users would score higher on the playfulness measure was not upheld. However, participants with a preference for more intuitive modes of perception were found to report significantly more playfulness in their interactions with computers. This might indicate that Ss and Ns have a different approach to computer learning and hence may need to be trained in different ways. For example, Jones (1994) suggested that Ss might require a more hands-on approach whereas Ns require less external feedback in order to motivate themselves. Hence, Ns would benefit from a training and work environment that allowed them to explore and interact spontaneously with computers. It is also possible that this greater tendency to interact in this way would mean that Ns learn computer skills more quickly and would need less encouragement to use computers in the workplace.

Investigation of the three separate components of the modified CAM suggests that both the T-F and S-N dimensions are related to aspects of computer attitudes. On the T-F dimension, it appeared that those participants with a preference for logical problem solving exhibited more positive affective attitudes towards computers. Similarly, on the S-N dimension, those individuals with a preference for intuitive thinking showed more positive behavioural attitudes towards computers. It is not easy to explain these results within the context of the MBTI. Tentatively, it is suggested that logical thinkers are more likely to be objective in their assessment of computers, whereas those with a preference for problem solving with emphasis on the impact for other people may give more weight to negative subjective experiences. A possible link between intuitive perception and positive behavioural attitudes towards computers might be that more intuitive individuals see more possible uses for computers and therefore will use them more often. It should be noted that there is little evidence for these suggestions, and further work would be necessary before any explanation can be given with confidence.

In terms of methodological considerations, there may have been a number of problems relating to the sample used in this study. The data was collected from participants involved in the same type of job, i.e. personnel. One consequence of this could stem from the fact that personality can be a factor in an individual's choice of occupation (Bayne 1995). Hence, there may be some bias in terms of personality type although the MBTI data did not support this with a spread across the dimensions, e.g. T (n=21), F (n=19), S (n=16), N (n=24). Another possible bias in the sample emanates from the fact that all participants were professionals and likely to have some degree of experience with computers. The difficulty arises in that they would not all have the same amount of experience with computers. Zoltan and Chapanis (1982) showed that more experienced computer users are more likely to have positive computer attitudes when compared with less experienced users. Retrospectively, the study would have benefited from including the participant's level of experience as a covarying factor.

3.6 CONCLUSIONS

Playfulness as a concept associated with the use of machines is not new and indeed, Liebermann (1977) wrote a book on the topic over 20 years ago. What is new is the way we now need to turn our attention to considering the 'subjective aspects' of using machines given the rapid and increasing use of computers and other technologies. The human unlike machines is a fairly fixed entity; we cannot evolve to meet the demands of advancing technology. More importantly, we have emotions and this affective element will become more important when considering our use of computers. In the past, it may have been somewhat neglected as the emphasis was placed on getting the hardware and software to work. In human to human social interactions, emotions are of vital importance, and attention has now begun to turn to machines that will recognise the user's affective states, e.g. facial and vocal expressions, and eventually have their own set of emotions. The latter may sound fanciful, but to a certain degree we already anthropomorphise and attribute emotions to inanimate objects. This was clearly demonstrated in studies described by Reeves and Nass (1996). It could be argued that it is only when computers can recognise the frustration and irritation of the users that usability will truly be realised. The irony as pointed out by Picard (1997, p. 248) is that often people feel like dummies in front of the computer, when in fact the reverse is true. She expresses this graphically through the following scenario: 'Behind the technical façade is the fact that the computer simply does not care what its user thinks or feels; it does not speed up when he (sic) is bored, slow down when he is confused, or try to do things differently when he is frustrated with its current *modus operanti*. Today's computers are far from being human-centred systems; they cannot even see if they have upset their most valuable customer.' It should be noted that Picard is not advocating that designers build emotions into every piece of technology. There will be some devices that will never need to be affective; similarly, there are living organisms in the world that function perfectly well without emotions.

To summarise, although this chapter has focused on one particular emotion that of playfulness with reference to the user rather than the technology, it has begun to lay the foundations for measuring playfulness *per se*. Once we can understand more about the components of emotion and pleasure in human-machine interactions, we can then begin to design products and systems that are more usable for humans. This must be one of the major challenges for designers in the next millennium.

3.7 ACKNOWLEDGEMENTS

Our thanks go to Oxford Psychologists Press for their help in supplying the MBTI data.

3.8 REFERENCES

Bayne, R., 1995. The Myers-Briggs Type Indicator: A critical review and practical guide. Chapman & Hall, London.

Berlyne, D.E., 1969. Laughter, humor and play. In *The handbook of social psychology*. Lindzey, G. and Aronson, E. (Eds), Addison-Wesley, New York.

Booth, P.A., 1989. An introduction to human-computer interaction. LEA, Hove.

Brosnan, M., 1998. *Technophobia: The psychological impact of information technology*. Routledge, London.

Carlson, R. and Wright, D., 1993. Computer anxiety and communication apprehension: relationship and introductory college course effects. In *Journal of Educational Computing Research*, **9**, 3, pp. 329-338.

Csikszentmihalyi, M., 1990. *The psychology of optimal experience*. Harper & Row, New York.

Eagly, A.H. and Chaiken, S., 1993. *The psychology of attitudes*. Harcourt Brace Jovanovitch, San Diego, CA.

Eason, K.D., 1984. Towards the experimental study of usability. In *Behaviour and Information Technology*, **3**, 2, pp. 133-143.

Garden, A.-M., 1991. Unresolved issues with the Myers-Briggs Type Indicator. In *Journal of Psychological Type*, **22**, pp. 3-14.

Heinsssen, R.K. Jr., Glass, C.R. and Knight, L.A., 1987. Assessing computer anxiety: Development and validation of the computer anxiety rating scale. In *Computers in Human Behavior*, **3**, pp. 49-59.

Igbaria, M., Parasuraman, S. and Pavri, F., 1990. A path analytic study of the determinants of computer usage. In *Journal of Management Systems*, **2**, 2, pp. 1-14.

International Standards Organisation, 1994. Guidance on usability. In *International Standard* (Draft) ISO 9241- Part 11, 24 pp.

Jones, W., 1994. Computer use and cognitive style. In *Journal of Research on Computing in Education*, **26**, p. 514.

Jordan, P.W. and Kerr, K.C., 1999. Pleasure, usability and telephones. In *Interface technology: The leading edge*. Noyes, J.M. and Cook, M. Research Studies Press Ltd, Baldock, Herts, pp. 229-243.

Kay, R.H., 1989. A practical and theoretical approach to assessing computer attitudes: The Computer Attitude Measure (CAM). In *Journal of Research on Computing in Education*, **21**, 4, pp. 457-463.

Kroeger, O. and Thuesen, J.M., 1992. Type talk at work. Delacorte Press, Bantam Doubleday Dell Publishing Group, Inc., New York.

Liebermann, J.N., 1977. *Playfulness*. Academic Press, New York.

Mason, R.O. and Mitroff, L.L., 1973. A programme for research on management

information systems. In *Management Science*, **19**, pp. 475-487.
McCrae, R.O. and Costa, P.T. Jr., 1989. Reinterpretating the Myers-Briggs Type Indicator from the perspective of the five-factor model of personality. In *Journal of Personality*, **57**, pp. 17-40.
Meier, S.T., 1988. Predicting individual differences in performance on computer-administered tests and tasks: Development of the computer aversion scale. In *Computers in Human Behavior*, **4**, pp. 175-187.
Mitchell, S., 1994. Technophiles and technophobes. In *American Demographic*, February Issue, pp. 36-42.
Myers, I.B. and McCaulley, M.H., 1985. Manual: A guide to the development and use of the Myers-Briggs Type Indicator. Consulting Psychologists Press, Palo Alto, CA.
Nielsen, J., 1993. *Usability engineering*. Academic Press, London.
Parasuraman, S. and Igbaria, M., 1990. An examination of gender differences in the determinants of computer anxiety and attitudes toward microcomputers among managers. In *International Journal of Man-Machine Studies*, **32**, pp. 327-340.
Picard, R.W., 1997. Affective computing. MIT Press Cambridge, MA.
Reeves, B. and Nass, C., 1996. *The media equation*. Centre for the Study of Language and Information, Stanford University.
Richards, B., 1993. Technophobia and technophilia. In *British Journal of Psychotherapy*, **10**, pp. 188-195.
Sage, A.P., 1981. Behavioural and organisational considerations in the design of information systems and processes for planning and decision support. In *IEEE Transactions on Systems, Man and Cybernetics*, SMC-**11**, 9, pp. 640-678.
Shackel, B., 1981. The concept of usability. In *Proceedings of IBM Software and Information Usability Symposium*. Ploughkeepsie, New York, pp. 1-30.
Shackel, B., 1986. Ergonomics in design for usability. In *People and computers: Designing for usability, proceedings of the 2nd BCS HCI conference*. Harrison, M.D. and Monk, A.F. (Eds), University Press, Cambridge.
Webster, J. and Martocchio, J.J., 1992. Microcomputer playfulness: Development of a measure with workplace implications. In *MIS Quarterly*, **16**, pp. 201-226.
Zmud, R.W., 1979. Individual differences and MIS success: A review of the empirical literature. In *Management Science*, **25**, 10, pp. 966-979.
Zoltan, E. and Chapanis, A., 1982. What do professional persons think about computers? In *Behaviour and Information Technology*, **1**, 1, pp. 55-68.

CHAPTER FOUR

The Basis of Product Emotions

P.M.A. DESMET and P.P.M. HEKKERT

Delft University of Technology, Department of Industrial Design,

Jaffalaan 9, 2628 BX Delft, The Netherlands

ABSTRACT

This paper introduces a conceptual model for the process underlying emotional responses that result from the perception of consumer products. The model distinguishes different kinds of emotions on the basis of eliciting conditions. It is based on the presumption that all emotional reactions result from an appraisal process, in which the individual appraises the product as favouring or harming one or several of his concerns. In this process of appraisal, the personal concern gives the stimulus emotional relevance.

The model describes the various ways in which products can act as emotional stimuli, and the matching concerns that can either correspond or collide with these stimuli. Products can act as stimuli in three different ways: the product as such, the product (or designer) as an agent, and the products as a promise for future usage or ownership. The corresponding concerns that are addressed are respectively: attitudes, standards and goals.

By revealing the cognitive basis of product emotions, the model can be used to explain the nature and, often, mixed character of product emotions. The paper illustrates a possible application of the model in a tool for designers.

4.1 INTRODUCTION

After being neglected for many years, a sudden interest in product affect has emerged. The affective side of product experience has become a 'hot topic', which is probably best illustrated by this conference and similar events over the past few years (e.g. Overbeeke and Hekkert 1999). A difficulty with affective concepts such as pleasure and emotion is that they are probably as intangible as they are appealing. Although some interesting and promising studies have been reported, the research field is still short of conceptual clarity and therefore lacks consensus on what the actual subject of study should be. In fact, both the concepts of pleasure and emotion are somewhat undifferentiated; they are used as collective nouns for all kinds of affective phenomena. Design literature tends to refer to these when studying anything that is so-called intangible, non-functional, non-rational or, for that matter, non-cognitive. Some of the reported studies involve 'experiential needs' (Holbrook 1982), 'affective responses' (Derbaix and Pham 1991), 'emotional benefits' (Desmet, Tax and Overbeeke 2000), 'customer delight' (Burns, Barrett and Evans 2000) and 'pleasure' (Jordan and Servaes 1995). Naturally, it is inherent in any newly emerging research field that the emulsion has not even started to crystallise. On the other hand, an adequate definition of the subject of study would probably facilitate fruitful discussions between researchers.

In our view, a model of product emotions can help to get a grip on the concept of product pleasure and emotions. This paper introduces such a structured model that distinguishes different kinds of product emotions on the basis of their eliciting conditions. The model adheres to the cognitive (functional) view on emotions and finds its roots in a structure developed by Ortony, Clore and Collins (1988). A first step in developing the model is to clarify the relationship between the concepts of pleasure and emotion.

4.2 PLEASURE AND EMOTIONS

The concept of affect refers to a large variety of psychological states such as emotions, feelings, moods, sentiments and passions. Each of these affective states varies in duration, impact and eliciting conditions. Of these states, emotions are most relevant for product experience because only they imply a one-to-one relationship between the affective state and a particular object: one is afraid of something, angry at someone, happy about something, and so on (Frijda 1986). The other affective states, such as feelings and moods, do not involve a specific object. For example, a moody person will find it difficult to pinpoint the exact cause of his mood. In the study of affective reactions to products, the object, i.e. the product, is the starting-point. Consequently, the model of product affect presented in this paper focuses specifically on emotions.

The place of pleasure in emotions is debatable. Both the propositions, that pleasure is an emotion and that it is not, are defensible. On the one hand, pleasure is an emotion if it is merely used as an equivalent of 'fun' or 'enjoyment'. In this connotation pleasure is included in many of the taxonomies of emotions found in literature (e.g. Russell and Lanius 1984). On the other hand, this view of pleasure seems to be rather narrow for the current application. Design research literature refers to pleasure as a product benefit that exceeds just proper functioning. In other words, pleasure is an emotional benefit that supplements product functionality. In this sense, pleasure covers all pleasant emotional reactions, of which the experience of fun is just one example. Valence (a bipolar ranging from pleasant to unpleasant) is a dimension frequently discovered in scaling procedures of emotion terms. If pleasure is regarded as a dimension of emotions, it can be used to describe emotions, but it is not an emotion as such. This notion befits everyday experience: one never feels pleasant as such. One feels happy, cheerful, surprised, inspired, etc. Although each of these emotions might be pleasant, that does not make pleasantness an emotion. Therefore, in the light of this paper, pleasure is defined as any pleasant emotional response elicited by product design.

It might seem difficult, if not impossible, to find general relationships between product appearance and emotional responses because emotions are essentially personal. Nevertheless, although people differ in their emotional responses to products, general rules can be identified in the underlying process of emotion eliciting. A view that distinguishes such general rules is the cognitive view on emotions.

4.3 THEORY OF EMOTION

The cognitive, functionalist position on emotions posits that emotions serve an adaptive purpose. In this view, emotions are considered the mechanisms that signal when events are favourable or harmful to one's concerns. This implies that in each emotion-eliciting stimulus some concern can be identified. These concerns are more or less stable preferences for certain states of the world; they are our personal motives in life (Frijda

1986). Examples of human concerns are concerns for respect, safety and self-esteem. For instance, we are all concerned to be treated with the respect we believe we deserve. When a person receives a degrading comment from a colleague, he will probably find this event conflicting with this concern for respect. Consequently, this person will experience a negative emotion such as shame or anger.

The preceding example illustrates that the linking of the stimulus to the concern precedes the actual emotional response. This process of 'signalling the emotional relevancy of an event' is most commonly conceptualised as 'a process of appraisal' (e.g. Arnold 1960; Frijda 1986). Appraisal theories assert that it is not events per se that determine emotional responses, but evaluations and interpretations of events. Because emotions are intentional and essentially involve concerns, they seem to require an explanation that invokes these concerns. Moreover, as this paper focuses on emotions specifically elicited by products, the explanation should also include eliciting conditions. A good starting point for a model of product emotions is the model developed by Ortony, Clore and Collins (1988) because it particularly focuses on this relationship between different types of concerns and the eliciting conditions.

4.4 MODEL OF PRODUCT EMOTIONS

According to the cognitive model of Ortony, Clore and Collins (1988) there are three major aspects of the world we can focus on: events, agents and objects. We focus on events (e.g. a football match) for their consequences (e.g. a loss by your favourite team), we focus on agents (e.g. a dog) for their actions (e.g. barking at you), and we focus on objects (e.g. a painting) because we are interested in certain of their properties *as such* (e.g. its composition). Central to Ortony, Clore and Collins' view is the position that emotions are valenced reactions to one of these perspectives on the world. Based on this division they developed a structure of emotion types that are logically related to one of these three aspects. In focusing on product emotions, at first sight it may seem tempting to restrict ourselves to the third class 'products as objects'. Our major claim is, however, that all three perspectives are relevant when products are simply perceived with one of our senses, i.e. without requiring physical interaction with the product. What we will present here is an adjusted version of the original model, in which those elements are adopted that cover emotions that may result from product perception. This adapted model is presented in Figure 4.1. Figures 4.2 to 4.4 each show examples of one of the three classes of product emotions.

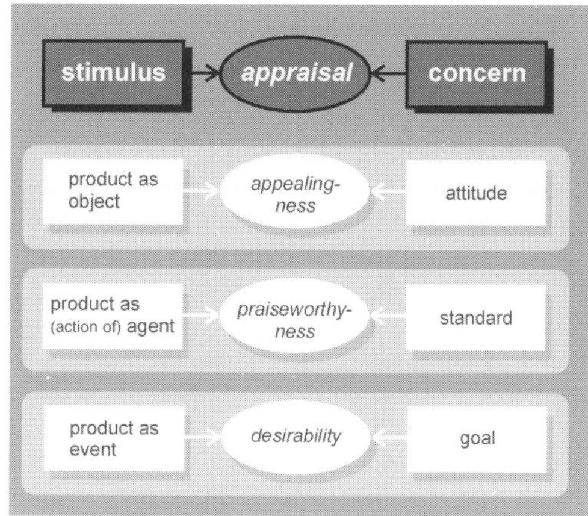

Figure 4.1 Model of product emotions

4.5 PRODUCTS AS OBJECTS

To start with the most obvious and simple branch of the structure in Figure 4.1, products are objects. Like all objects, products or aspects of products can be viewed in terms of their *appeal*. Products are simply liked or disliked for their appearance, for the way they look. The emotional reactions are basically unstructured and comprise attraction emotions such as love, attracted-to, disgust, and boredom.

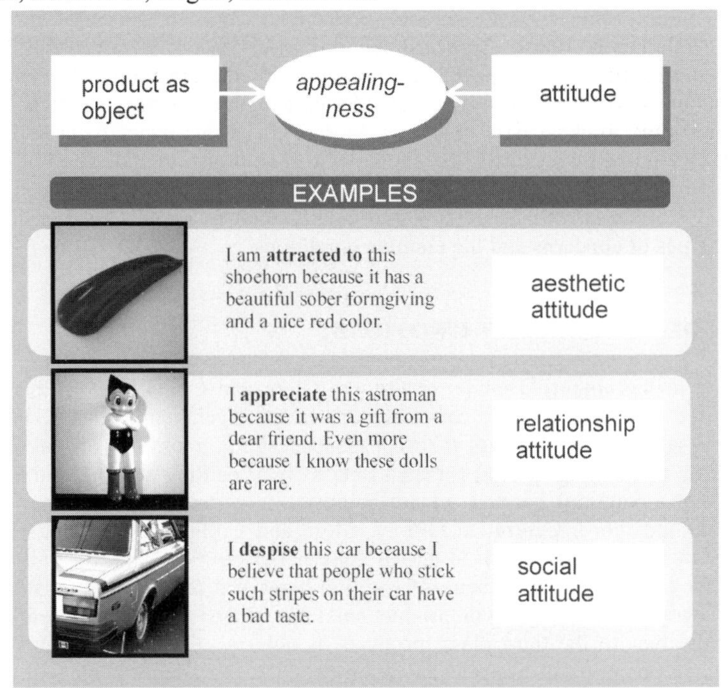

Figure 4.2 Examples of prodructs construed as objects

As argued in the previous section, emotions arise because products touch upon our personal concerns. What concerns are addressed by a product as such? Ortony, Clore and Collins (1988) call this type of concerns 'attitudes' or, as a special case, tastes. Some people have developed an attitude, a dispositional liking, for American cars, others simply have a taste for Italian design. When a product corresponds with such an attitude or aesthetic concern, it is appraised as appealing (see Figure 4.2, 'shoehorn').

In the previous examples, the appealingness concerns characteristics of the product itself, such as its size, shape or particular details. As a result, a dispositional liking for a certain type will be generalisable to most or all members of the product class. Sometimes, however, the dispositional (dis)liking is restricted to one specific product. In those cases the liking results from previous usage or ownership of that particular exemplar. One can have a dispositional liking for a ring because it was a gift from someone special, or for a particular backpack because one has travelled with it to many different countries (see Figure 4.2, 'astroman'). In these cases, the attitude is saturated with personal meaning due to the significant personal experience one had with the product. Such attitudes are not shared with others and are not applicable to other exemplars of the product class.

In some special cases, it is not the product itself that acts as the object of appraisal but other 'things', such as a situation, institution or person the product refers to through associations. A product can be associated with a particular user group, such as German cars, skateboards and antimacassars; a sunshade often refers to a summer holiday, and ties can refer to the business world. In those cases, it is often not the object itself that is (dis)liked but rather the user group or institution it is associated with (see Figure 4.2, 'striping').

4.6 PRODUCTS AS AGENTS

Agents are things that cause or contribute to events. Ortony *et al.* (1988) indicate that agents can also be inanimate objects such as products, and they give the example – familiar to all of us – of a car that is blamed for its malfunctioning. Although this type of product construal is dependent on actual product use this class remains relevant to our focus on product perception.

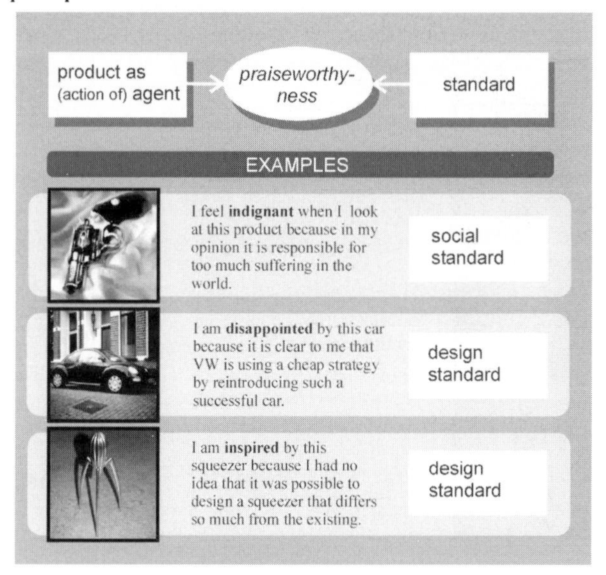

Figure 4.3 Examples of products construed as agents

First, products can be treated as agents with respect to the presumed impact they have or can have on people or society. One can, for instance, blame mobile telephones for the disturbance they cause in public spaces such as train compartments (see Figure 4.3, 'gun'). Secondly, products are often seen as the result of a design process, in which case the designer or company is the construed agent. While looking at a product one can praise the originality of its design(er) or blame the designer for blatant copying (see Figure 4.3, 'new Beetle'; 'lemon squeezer'). Seen in both ways, the *praiseworthiness* (or blameworthiness) of the 'actions' of the product or the work of the designer – through the design of the product – is the basis of our evaluation. This can lead to attribution emotions such as admiration, appreciation, contempt, and disappointment.

Agents or, in our adapted model, products or designers are appraised as praiseworthy in terms of standards. Standards are our beliefs, norms or conventions of how we think things ought to be. Some people might have a norm that one must be able to read quietly in a public room. Others might adhere to the norm that designers or industries should always strive for original product solutions. We approve of things that comply with such standards and disapprove of things that conflict with them.

4.7 PRODUCTS AS EVENTS

The third branch of the emotion model seems the least relevant to product emotions. In the model of Ortony, Clore and Collins, the aspect of the world we are focusing on here is the consequence of an event. Products are not events, but we nevertheless believe that an important class of product emotions falls into this category. These emotions result from the inclination of people to anticipate future use or possession of a product once they have seen it. The (foreseen) use or possession has become the event and the predicted consequences cause the emotion.

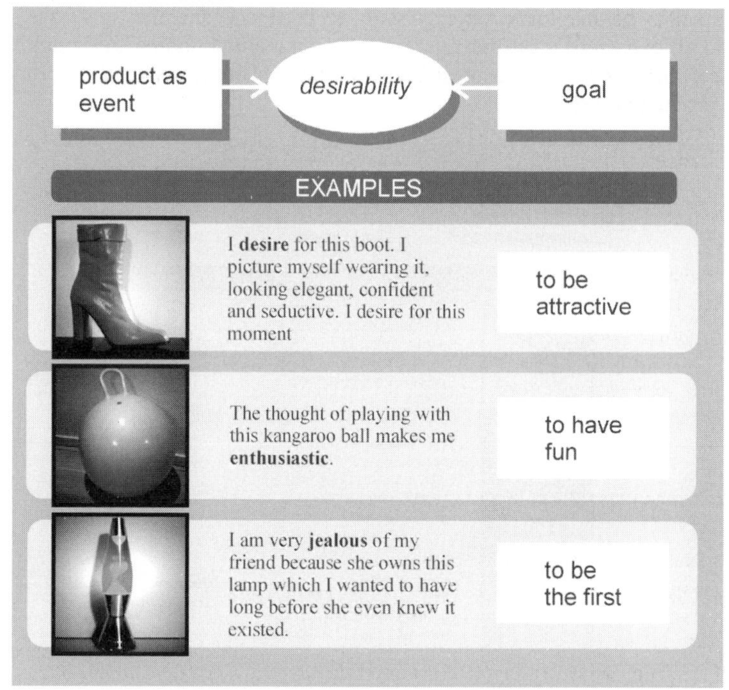

Figure 4.4 Examples of products construed as events

When we see a Ferrari, we can think of the status that owning and driving it will give us, and this fantasy may elicit an emotion like desire (see Figure 4.4, 'boot'; 'kangaroo ball'). The judged *desirability* (or undesirability) of the anticipated result of owning or using a product is the main criterion for our evaluation in this category.

This branch also has a distinction in that we can anticipate the consequences of product use/possession for ourselves and for others. In the first case, consequences can lead to anticipated event-based emotions such as desire or aversion. The latter (and relatively rare) case could play a role when we are looking for a gift for someone, or when we see a product as the possession of someone else. Emotions such as happy-for, jealousy and *Schadenfreude*, can be the result (see Figure 4.4, 'lamp').

The relevant concerns for this type of emotions are goals. Goals are "things" we want to see happen. One of our goals might be to acquire the status that can be attained by impressing other people with an expensive car. When someone else drives by with an even more expensive car, this goal is not reached. We appraise products as desirable when we anticipate that they will facilitate our goals, they are undesirable when they interfere with our goals.

4.8 IMPLICATIONS OF THE MODEL

The presented model of product emotions has some important implications for our understanding of the relationship between products and emotions. First, the model reveals that it is possible for a product to elicit several emotions simultaneously. Not only do we have the various concerns, i.e. goals, standards and attitudes; at the same time we also have a multitude of each one of them. As a result, one single product may be desired because we expect it will facilitate one of our goals, will be admired for the designer's achievement, will arouse contempt for its bad impact on the environment, or will be loved for its beautiful design. You might recognise the mixed emotions you have towards a particular car from this description. In measuring the emotional response to products, this possible co-occurrence of several different emotions must be acknowledged (see Desmet, Hekkert and Jacobs 1999).

Secondly, the model shows that there is often not a one-to-one relationship between the appearance of a product and the emotional responses it elicits. Emotions are not elicited by product characteristics as such, but by construals based on these characteristics. By definition, these construals are highly personal. Therefore, searching for general rules in a stimulus-response manner is a fruitless approach. The model of product emotions cannot support the designer with general rules concerning the relationship between product appearance and emotional responses. Nevertheless, on the level of eliciting conditions, the model does reveal some general patterns. In our view, understanding these patterns can be of great value to designers. Based on this notion, we are currently developing a tool – the Emotion Navigator – to assist designers in grasping the emotional potency of their designs.

4.9 THE EMOTION NAVIGATOR

The Navigator is an anecdotal database of some 250 photos of products that elicit emotions (note that the photos in Figure 4.2 to 4.4 are drawn from this database). These photos are made by 28 subjects who – for each photo – noted in a logbook which emotion was elicited and why it was elicited. The Navigator is structured in accordance with the model of product emotions and visualised in an open-ended manner that aims to be inviting and attractive. The interface shows photos, elicited emotions, addressed concerns, and the eliciting conditions described by the subjects.

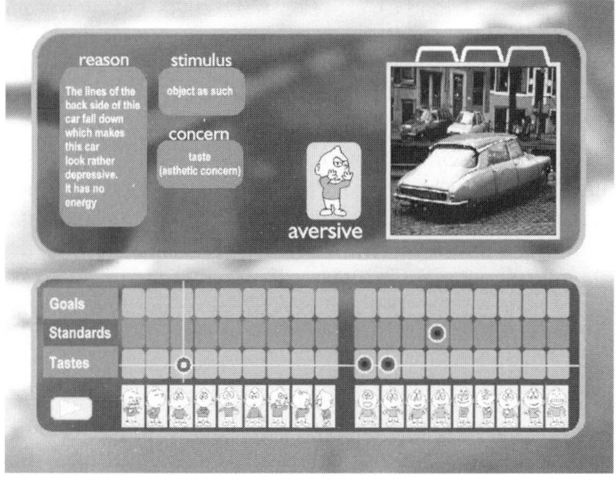

Figure 4.5 Interface of the Emotion Navigator

One can browse through the database by clicking on any of these four elements. For example, if an 'emotion' adjective is clicked, the Navigator will display examples of products that elicit that particular emotion. A second possibility is to click on a 'concern' in order to see products that elicit an emotion because they touch on this specific concern. If a product photo is clicked, the Navigator will display the (sometimes conflicting) emotional responses of other subjects to this specific product, and their particular construals. These are only a few of the browsing possibilities currently being explored in the development of the Navigator. A working prototype is planned for release in 2002.

The goal of the Navigator is not to be prescriptive but to offer food for thought and inspiration. In our view, the visualisation in a multitude of real life examples supports this goal by making the general patterns of emotions more tangible for designers. Browsing through the database can make the designer aware of the different ways in which products elicit emotions. In this sense, the Navigator can be a tool to support designers in developing a personal design vision that incorporates the emotional concerns of the users.

4.10 ACKNOWLEDGEMENTS

This research was funded by Mitsubishi Motor R&D, Europe GmbH, Trebur, Germany. We would like to thank Jan Jacobs and Kees Overbeeke, Delft University, for their constructive discussions. We would also like to thank the participants of the photo project for their useful research input.

4.11 REFERENCES

Arnold, M.B., 1960. *Emotion and personality*. Columbia University Press, New York.

Burns, A., Barrett, R. and Evans, P., 2000. Delighting customers through empathic design. In *Manuscript submitted for publication*, Lulea University.

Derbaix, C.M. and Pham, M.T., 1991. Affective reactions to consumption situations: a pilot investigation. In *Journal of Economic Psychology*, **12**, pp. 325-355.

Desmet, P.M.A., Hekkert, P., and Jacobs, J.J., 1999. When a car makes you smile; development and application of an instrument to measure product emotions. In *Advances of Consumer Research, vol. XXVII*. Hoch, Stephen J. and Meyer, Robert J. (Eds), in press.

Desmet, P.M.A, Tax, S.J.E.T. and Overbeeke, C.J., 2000. Designing products with added emotional value; development and application of a 'research through design' approach. In *Manuscript submitted for publication*, Delft University of Technology, Delft.

Frijda, N.H., 1986. *The emotions*. University Press Cambridge, Cambridge.

Holbrook, M.B., 1982. The experiential aspects of consumption: consumer fantasies, feelings and fun. In *Journal of Consumer Research*, **9**, pp. 132-140.

Jordan, P.W. and Servaes, M., 1995. Pleasure in product use: beyond usability. In *Contemporary Ergonomics 1995*. Robertson, S. (Ed.), Taylor and Francis, London.

Ortony, A., Clore, G.L. and Collins, A., 1988. *The cognitive structure of emotions*. Cambridge University Press, Cambridge.

Overbeeke, C.J. and Hekkert, P., 1999. *Proceedings of the first international conference on Design and Emotion*. Overbeeke, C.J. and Hekkert, P. (Eds), Department of Industrial Design, Delft.

Russell, J.A. and Lanius, U.F., 1984. Adaption level and the affective appraisal of environments. In *Journal of Environmental Psychology*, **4**, pp. 119-135.

CHAPTER FIVE

Product Appearance and Consumer Pleasure

MARIËLLE CREUSEN and DIRK SNELDERS*

Delft University of Technology, Department of Industrial Design Engineering, Section Product Management and Innovation, Jaffalaan 9, 2628 BX Delft, The Netherlands

5.1 INTRODUCTION

In consumer research, a number of methods are used to determine the consumer's preference for the functional possibilities of a product. Ergonomic research in the form of user tasks can also be performed to determine appropiate designs for consumer-product interaction which will maximise task efficiency and minimise discomfort in product use. Far less is known, however, about the ways in which the more subjective perceptions and preferences of consumers, that concern hedonic product value or pleasure, can be investigated. This is most notably the case for aesthetic pleasure, derived from the appearance of a product. How does product appearance appeal to consumers and how does it affect consumer evaluations of the product? Taking hedonic product value into account will help to attune the appearance of a product to the wishes of the consumer, and thus maximise their pleasure in product possession and use.

We will now take a close look at hedonic consumer judgments and the way in which they are formed. On the basis of this we will discuss possible avenues for taking into account the hedonic product value in product design.

5.2 THE ROLE OF PLEASURE IN JUDGMENTS ABOUT PRODUCTS

A number of studies show that a hedonic product experience or pleasure can be characterised by the way in which the product is represented by consumers in their judgment of the product. In a first study, Snelders (1995) validates a measure, originally developed by Ratchford (1987), of self-reported emotionality in the consumer's choice of certain products. Using two questionnaires, Snelders shows that pleasure is one aspect of a more general involvement with products, and that it is independent of the consumer's rational involvement with products. In the two questionnaires (N=405, N=624), three scales were developed that were combinations of self-reported buying criteria. An emotional, hedonic scale contained items about 1) the emotionality of the buying criteria and the decision itself, 2) the extent to which products allowed consumers to express their

*Mariëlle Creusen and Dirk Snelders are assistant professors of marketing at the School of Industrial Design Engineering, Delft University of Technology. Authors contributed equally to this paper.

personality, and 3) the intuitiveness of the decision.
Next to this hedonic scale, a rational scale was developed, containing items about 1) whether the criteria were logically applied, 2) whether they were based on objective product functions, and 3) whether the criteria were concrete product attributes.

Finally, a general involvement scale was created, containing items about 1) the importance of the product, 2) the amount of time and effort spent on buying it, and 3) the perception of risk in terms of what one will lose if the wrong product is chosen. Unlike Ratchford (1987), Snelders (1995) correlated these scales and found that the hedonic and rational scale were not correlated, while they both correlated with the general involvement scale. This means that consumers think of pleasure as a separate product value, unrelated to objective product functions, but just as important to them. If either the hedonic or the rational product value increases, then so does the general involvement that people feel with the product. Thus, hedonic and rational product values should be seen as complementary, as two independent aspects of a more general involvement that consumers have with products.

These findings suggest that pleasure is not simply the end result of rational deliberation. It can also point to another way in which pleasure seekers interact with products. This is demonstrated in a second study by Snelders, Creusen and Schoormans (2000). Synthetic clock radios and thermos flasks were created which systematically varied on a number of product attributes. By offering respondents a specific fraction of all the possible combinations, and by asking them to rate each combination on their degree of preference, the main effect of every attribute on the consumer's preference could be calculated (Green and Srinivasan 1978; 1990). The idea was that people who scored high on the hedonic scale would prefer products that contained attributes which described aspects of the form of the product, as it presents itself to the consumer; people who scored high on the rational scale would prefer products that contained attributes which describe how the product can be operated. The synthetic clock radios and thermos flasks both varied on a number of attributes that described aspects of the form of the product (global form, colour and form detail), and various product functions. These attributes were selected in such a way that they could be represented textually (e.g. 'clock radio has the form of a cylinder in a horizontal position') as well as pictorially (e.g. a drawing of a battery to show that a clock radio is protected against power cuts). This was to ensure that the contribution of each type of attribute (form or function) was not dependent on presentation format (pictorial or textual). The results show that people who score high on the hedonic scale are indeed more interested in the attributes that describe the form of the product, irrespective of whether these aspects are represented textually or pictorially.

Are people who pay more attention to form not thinking thoroughly about the consequences of the product form? A third study by Snelders and Schoormans (2000) shows that this is not the case. In a series of structured interviews about telephones and clock radios, 72 consumers were asked to provide arguments about why they valued certain attributes in the product (37 consumers provided 221 arguments for telephones and 35 consumers provided 217 arguments for clock radios). A substantial number of the arguments were about the pleasure the product could provide: 20%, both for telephones and clock radios, were about pleasure. This made pleasure the second most mentioned type of argument. Arguments about ease of use were mentioned most (54% for telephone and 43% for clock radio), but arguments about the finished quality or the price of the product were mentioned considerably less. It was also found that arguments about pleasure differed from the other arguments because, in two thirds of all cases, they were based on abstract attributes of the product* (i.e. qualifications such as modern design,

* As opposed to concrete attributes, which are attributes that one can easily point out in a product, like size, number of buttons, etc.

simple, automatic, tuneful sound, etc.): 67% for telephone and 66% for clock radios. Arguments about other product aspects were based much less on abstract attributes: 24% for telephones and 28% for clock radios. In a limited number of the arguments abstract attributes were presented as the result of more concrete attributes (e.g., 'a small number of buttons makes the telephone simpler'): 7% for telephones and 8% for clock radios. However, these arguments were typically about ease of use – not pleasure. Thus, pleasure-based factors find their way into the arguments that consumers construct about products, and these are characterised by the fact that they are based on abstract product attributes.

How should we think about these abstract product attributes, which are not inferred from more concrete product attributes? We do not want to imply that abstract product attributes do not exist through some physical aspect of the product. But not all physical aspects of the product are distinct and separate product parts with clear boundaries. If one thinks of concrete attributes, like a button or a specific colour, one usually thinks of distinct product parts that have a specific quality. However, if one thinks of abstract attributes, like elegance, simplicity or safety, then this is only the case to the extent that we can connect the abstract attribute with more concrete aspects of the product. For example, car safety may be connected to a number of concrete attributes like airbags, safety belts and big bumpers, all distinct product parts. But car safety may also reflect on product aspects that are not very distinct, like some aspects of the sound of a door, or the robust look of the car. These aspects are not present in the product in a distinct way, and consumers may not be able to describe them in very concrete terms, but they may still be valued by consumers and as such establish an indistinct sense of product quality. Thus, it may well be the case that abstract attributes describe product qualities that are indistinct. Different, more indistinct product qualities may be targeted by pleasure seeking consumers. This possibility is explored in the next section.

5.3 THE PERCEPTION OF PLEASURE IN PRODUCTS

Some studies have investigated what information consumers infer from looking at a product, thus from the product appearance (see Creusen 1998). It was assessed whether consumers inferred this information by using analytic information processing (by regarding specific characteristics or product parts), or by using holistic information processing (by regarding the overall impression).

Two methods were used to assess whether consumers use holistic or analytic information processing in making a specific product judgment. The first method uses scales to determine the way in which a product choice is made. The scales include items that indicate holistic information processing and items that indicate analytic information processing. The four 'holistic items', for example, are about whether the choice is based on the impression that the product gives, and on the product as a whole. The five 'analytic items' are about whether the subject compares the alternatives on several points, evaluates product characteristics separately, and considers them in detail. The second method of assessment involves interviewing subjects about the way in which they made their choice, on what information they based each choice and how that information was derived. A record was made of whether subjects mentioned specific characteristics or product parts (which would indicate analytic information processing), or said that it was a general or overall impression (indicating holistic information processing).

Four studies were performed, in which subjects had to make a choice from a limited number of product alternatives (two or three). Products used were telephone, thermos flask, desk lamp, coffee maker, clock radio, hairdryer, and answering machine. After making a choice, subjects filled in the type of processing scales and were then interviewed about the reasons for that choice, as described previously.

It was found that about half of the subjects in each study mentioned choice reasons that concerned hedonic (aesthetic) product value. Most of these hedonic judgments are made using holistic information processing. The results of both ways of assessing the type of processing (see above) point to this. Only a few hedonic judgments are based on a specific characteristic, such as a specific line or part of the product. However, colour is often mentioned as a basis for an hedonic judgment, although subjects often said that this played a role in the whole impression.

About one-third to one-half of the subjects in each study derived utilitarian information from a product's appearance*. Utilitarian judgments concern attributes of a more rational kind, like functionalities and possibilities, ease-in-use and durability /quality. These were mostly made using analytic information processing, based on details and specific parts (e.g. whether the display is in an easy-to-view place, or whether buttons are large enough). However, several utilitarian judgments were also based on the overall impression of the product appearance (between 7 and 15% of the subjects in each study). For example, the product gave a solid or durable impression or looked simple (not a complex shape, from which consumers infer that operation will also be simple). But most judgments that are made by using holistic information processing concern product pleasure, i.e. hedonic product value.

In these studies, some factors that influence the extensiveness of product evaluation before making a choice were investigated. These factors were the amount of knowledge about the product and the amount of time that subjects spent looking at the product alternatives. It was found that holistic information processing played a role in product evaluations, regardless of how extensively subjects evaluated the product. Analytic information processing was used to evaluate utilitarian attributes, and was used even more when the evaluation was more extensive. This means that holistic information processing plays a big role in consumer evaluation based on a product's appearance, and this is also the case when consumers have enough time and product knowledge to process detailed information. Thus, detailed processing comes in addition to global processing, and not in its place.

5.4 CONCLUSIONS AND IMPLICATIONS

It was found that many consumers derive pleasure from the form or appearance of a product. Furthermore, it was found that a large part of these judgments was made with the use of holistic information processing (Creusen 1998), and that they were based on abstract attributes (Snelders and Schoormans 2000). This points to the same conclusion, namely that it would be difficult to pin down specific characteristics that engender a certain holistic impression, as the relation with concrete attributes is not straightforward. This, in turn, may lead to problems of subjective interpretation.

How would abstract attributes, based on holistic impressions, make a consumer response more subjective? Why are they less clear when they are communicated to the consumer researcher? An answer to this question can be found in context availability theory (Schwanenflugel and Shoben 1983; Schwanenflugel, Akin and Luh 1992), where it is shown that concrete attributes can be applied to more restricted sets of situations. Since the situations to which words are applicable are learned during an individual's personal development, they must start off as a very idiosyncratic bunch of situations. Words, however, put social restrictions on our understanding of the world (after Vygotsky

* In the fourth study textual information about functionalities was presented next to the product itself. More than 70 % of the subjects mentioned utilitarian reasons, of which 40 % were based on the product appearance.

1986/1934). The more restrictions that a word puts on the situations to which it is applicable, the less room there is left for the subjectivity of the meaning of the word. According to Fish (1994), social restrictions lead to 'forceful interpretive acts', whose effects are 'the production of a situation in which for all competent members of a community the utterance of certain words will be understood in an absolutely uniform way' (p. 301). Context availability theory argues that the meaning of concrete words is the most restrained in this sense. Since this will lead to a more uniform way of understanding, we can regard concrete words as less subjective. In some instances, however, we may use rules to relate abstract words to more concrete words, and this will make the abstract words as objective as the concrete ones. However, Snelders and Schoormans (2000) have shown that this rarely happens in responses that are about pleasure and that, in most cases, an abstract word describes an aspect of a product that cannot be caught by elements in the product for which we have concrete words. This leaves room for abstract words to communicate a subjective understanding of the world around us, based on a more holistic impression of the product.

Since pleasure-based human factors are based mostly on holistic impressions, and since these are communicated mostly by abstract attributes, designing the appearance of a product so that it will give a certain impression to consumers is not a straightforward task. How then can we help designers to attune the appearance of a product to consumer wishes and perceptions? How can we assess which product appearance will appeal to (groups of) consumers and best suit the value consumers want from the product?

A possible way to assess whether a certain design/appearance engenders the 'wanted' perceptions (that suit the target group, usage situation, price, etc.) is to show the design to consumers and assess which perceptions it generates and whether it communicates the wanted product value (like ease-in-use). This can be done by asking people about associations with the design, or letting them indicate the extent to which a design communicates certain impressions (like cheap/expensive or business-like). Such research is often done for packages of food products. In this way it can be assessed whether the design/appearance engenders the correct or wanted perceptions. It is, however, not so clear from this how a design can be adapted in order to better suit the desired proposition.

Another possibility is to show a limited number of product concepts to consumers during the product development process, so that the most promising product appearance can be identified. In addition, the reactions of consumers to these concepts could be gathered and could give guidelines to designers as to how to further improve the appearance. There are, however, some difficulties with this approach.

Firstly, direct comparisons between different concepts may direct attention away from the more holistic and indistinct qualities of the product (Wyer and Srull 1989; Stapel, Koomen and Van der Pligt 1997). According to Wyer and Srull 'An abstract trait, concept or attribute with no clear object boundaries lacks the distinctness to be used as a comparison standard' (p. 134). Because of this, it will be mostly concrete information that is contrasted when consumers are asked to make a trade-off between concepts. However, there may be effective ways of solving this problem. For example, concepts could be generated that do not clearly delineate distinct product parts (e.g. rough sketches or pictures with details omitted). In a concept test, such concepts would be presented first and explicit information about distinct product parts would be provided only later. Another possibility, when comparing judgments about different concepts, is to vary designs between subjects so that each subject evaluates only one concept.

Secondly, it is difficult for consumers to envision how the change of a certain characteristic will change the total impression of a product. Consumers will have to see the change in order to be able to say whether or not it is an improvement in their opinion. This means that, in order to gather valid reactions about product appearance, consumers should be able to see many possibilities or, alternatively, to 'try out' several changes. Consumers could be shown a large number of pictures of an existing product and probed as to why certain of them are valued and others not. This would give the designer some feeling for the appearance that suits the specific target group, and what consumers infer from several aspects of the design. A computerised method which enables consumers to 'try out' several changes in product appearances is the 'Interactive Concept Test', developed by Loosschilder (1998). Here consumers can construct their 'ideal' product appearance from a limited set of options (e.g. form characteristics, material, colour), which gives the designer an idea of valued appearances. The advantage of this method lies in its efficiency in describing the variation in taste for larger samples. However, not much qualitative information can be gathered in this way that explains why consumers value a certain appearance.

Thirdly, it may be difficult for consumers to say concretely exactly what should be changed in a product concept in order to increase its hedonic value. Because a large percentage of consumer judgements that are based on the product appearance are formed by holistic information processing (see Creusen 1998), it is difficult for consumers to pin down the characteristics that influence their judgments. This does not mean that consumers have nothing to say about their holistic impressions. As the research by Snelders and Schoormans (2000) shows, consumers talk a lot about product appearance and the pleasure that it provides, albeit in very abstract terms that are almost never connected to more concrete product parts. It may therefore be worthwhile to find methods that help in the interpretation of these abstract attributes. An example of this may be the method employed by Snelders and Schoormans (2000), where consumers are asked why the abstract attributes are important to them. This can provide reasons (e.g. 'modern furniture gives me a sense of order, which structures my life, which is healthy for me'), which in turn can be used to interpret how a product should incorporate the abstract attribute (e.g. create modern chairs with clear orderly patterns). Another method is proposed by Durgee (1985) and Snelders and Schoormans (1999), who argue that opposites may clarify the meaning of abstract attributes. For example, the opposite of modern may be old-fashioned, but it may also be post-modern or classical or kitsch. Depending on the opposite, we can come to very different interpretations of what it implies for a product to be modern. A last method has to do with the possibility that problems of subjective interpretation arise because the consumer and the consumer researcher are from different communities, and a 'right' interpretation is seen as a test of competence for the researcher who breaks into the consumer community (Fish 1994). The solution for this would lie in an ethnographic study into the local meaning of certain products for members of that community (e.g. Hammersly and Atkinson 1983). As such, the design process can even be seen as an interventionist program to advance certain interests of the consumers in the local community (Thomas 1993).

5.5 REFERENCES

Creusen, M.E.H., 1998. *Product appearance and consumer choice.* Delft University of Technology (dissertation), Delft.

Durgee, J.F., 1985. Depth-interview techniques for creative advertising. In *Journal of Advertising Research* (December/January), pp. 29-37.

Fish, S., 1994. *There's no such thing as free speech...and it's a good thing too.* Oxford University Press, Oxford.

Green, P.E. and Srinivasan, V., 1978. Conjoint analysis in consumer research: Issues and outlook. In *Journal of Consumer Research*, **5**, pp. 103-123.

Green, P.E. and Srinivasan, V., 1990. Conjoint analysis in marketing: New developments

with implications for research and practice. In *Journal of Marketing*, **54**, 10, pp. 3-19.
Hammersly, M. and Atkinson, P., 1983. *Ethnography: Principles in practice*. Routledge, London.
Loosschilder, G.H., 1998. *The interactive concept test: Analyzing consumer preferences for product design*. Delft University of Technology (dissertation), Delft.
Ratchford, B.T., 1987. New insights about the FCB grid. In *Journal of Advertising Research*, **27**, 5, pp. 24-38.
Schwanenflugel, P.J., Akin, C. and Luh, W.M., 1992. Context availability and the recall of abstract and concrete words. In *Memory and Cognition*, **20**, 1, pp. 96-104.
Schwanenflugel, P.J. and Shoben, E.J., 1983. Differential context effects in the comprehension of abstract and concrete verbal materials. In *Journal of Experimental Psychology: Learning, Memory, and Cognition*, **9** (January), pp. 82-102.
Snelders, H.M.J.J. (Dirk), 1995. *Subjectivity in the consumer's judgement of products*. Delft University of Technology (dissertation), Delft.
Snelders, D., Creusen, M.E.H. and Schoormans, J.P.L., 2000. *The effect of emotionality on the relative importance of product form and function*. Delft University of Technology, Delft.
Snelders, D. and Schoormans, J.P.L., 1999. Are abstract product qualifications more subjective? A test of the relation between abstraction level and opposite naming. In *Advances in Consumer Research*, **26**, E.J. Arnould and L.M. Scott (Eds). UT: Association for Consumer Research, Provo.
Snelders, D. and Schoormans, J.P.L., 2000. *The relation between concrete and abstract product attributes*. Delft University of Technology (working paper), Delft.
Stapel, D.A., Koomen, W. and Van der Pligt, J., 1997. Categories of category accessibility: The impact of trait versus exemplar priming on person judgments. In *Journal of Experimental Social Psychology*, **33**, pp. 44-76.
Thomas, J., 1993. Doing critical ethnography. In *Qualitative Research Methods Series*, **26**, CA, Sage Newbury Park.
Vygotsky, L., 1986. *Thought and language* (trans. A. Kozulin). MA: M.I.T. Press (originally published 1934 as Myshlenie i rech. Moskow: State Social-Economic Publishing), Cambridge.
Wyer, R.S. and Srull, T.K., 1989. *Memory and social cognition in its social context*. Erlbaum, Hillsdale, NJ.

CHAPTER SIX

Product Design for Consumer Taste

MIRJA KÄLVIÄINEN

The Kuopio Academy of Crafts and Design,

PL 98, 70101 Kuopio, Finland

6.1 INTRODUCTION

This paper argues that in the current aestheticisation of everyday life, product taste is a significant factor incorporating embodied aesthetic experience, identity building and social display. Taste is a preference arising from the consumer's value-based capacity to make distinctions between physical objects and to get pleasure from them. Understanding the background to consumer taste is necessary for the critical and insightful interpretation of so complex a phenomenon. Useful theories can be found in consumer research, marketing research, psychology, anthropology, semantics, the cognitive sciences, the sociology of taste, aesthetics and usability studies. On the basis of these different approaches, three categories of the practical aspects of consumer taste can be identified. These categories combine the boundaries and possibilities of the consumers' use situation with their subjective aspirations and social interaction.

This chapter looks at the methods that would yield visual and tactile information for the meaningful astheticisation of future products. By using reflexive methods of analysis, connected to perception, meaning making and preferences in objects, the above mentioned aspects of taste can be investigated in practical design projects. This calls for interactive and observational methods such as distinction, associations, narratives and anthropological analysis. The advantages of these approaches can be combined in a comprehensive visual translation of the aspects of taste through 'staging' a view of the consumer's choice of objects.

6.2 TASTE AND AESTHETIC PLEASURE AS A CONSUMPTION EXPERIENCE

Product taste is the human capacity to make distinctions between physical objects and to favour some of them. This capacity is based on our beliefs and attitudes, and on values that are dependent upon our culture and history. To make these judgements requires symbolic capital and forms of communication that are particular to certain social groups at certain times. Taste offers this structured language comprising a set of suitable behavioural habits, and permitted limits of variation.

All this is presented and expressed by the selection of products with which individuals surround themselves for their self-identity and social interaction. For a designer, taste functions as a translation from lifestyle preferences and orientation to products, so that certain consumer groups display products that correspond with their orientation or demonstrate, via exclusion, what does not correspond with it.

Taste becomes evident in the compositions of the symbolic and communicative object features, which allow sensual appropriation. This subjective pleasurable experience is at the same time in many ways highly socially shared. In a BBC documentary about people's tastes in their home furnishing strictly aesthetic matters were infrequently referred to. Instead a complex mix of public and private concerns governed people's (Barker 1992: 3). Something might be aesthetically pleasing for a consumer in the pure, distanced sense but does not appeal to their taste because it does not fit their orientations and life-world.

The idea of experience, consuming wants, passions and pleasure, is one current theme in discussions on how users relate to products (Margolin 1997; Jantzen and Ostergaard 1998: 126-128). To capture the essence of the consumer's experience thoroughly the values the consumer perceives as important must be defined. This definition must be authentic and consistent in order to connect on an intuitive level with the consumer's experience of the defining object (Ashcraft and Slattery 1996). Instead of utilitarian satisfaction design is becoming more a communicator of the existential values and meanings embedded in the consumer experiences.

The quality of experiences has been explained through the consumer lifestyles that help to make sense of what people do, and what doing it means for them and others. These responses and choices are concerned with ethical and aesthetic significance – ways of living that are fundamental to a sense of identity (Chaney 1996: 4-12). On a deeper level consumers' life worlds and life goals explain their emotional experiences, their practical knowledge, and their intuitive understanding of cultural way of life (Thompson 1998: 130-136).

The aestheticisation of everyday life, individuality, and the rise of the inner orientated, self-searching consumer are evident in many contemporary lifestyle descriptions. Although the life worlds are expressed through appearances, they focus on authentic concerns of identity expression (Dant 1999: 30; Chaney 1996: 29, 122; Interview: Patrick Lalonde-Dade, planning director, Synergy Consulting, London, 19th of April 1999). Individualistic self-search raises questions for individual customisation of deep experiences in products.

Experiential design allows users to feel good about themselves by fitting the product to their identity and lifestyle (Crozier 1994: 4-5). Taste in products is about these intuitively and socially interpreted qualities such as emotional bonds, familiarity, aspirations, dreams, sentimentality, embodied aesthetics, general 'feel' and personality.

6.3 THE ASPECTS OF TASTE IN HUMAN-PRODUCT RELATIONSHIP

In this study a model has been formulated identifying the important aspects of consumer taste for potential application to design through marketing, consumer, psychological, usability and anthropological research. The various elements contributing to consumer taste are grouped into three categories, each with a different emphasis.

These categories cover contextual, aspirational and social areas of the use situation. It is necessary to consider all three of them in order to fully understand a phenomenon as complex as taste. This especially concerns deep structures, which can only be grasped by combining the boundaries and possibilities of objective points of view with subjective and social aspects (Kälviäinen 1996; Bordieu and Waquant 1992: 3-11, 228-231). Taste is all of these.

6.4 THE OBJECTIVE FRAMEWORK

First of all it is necessary to set the boundaries of possibilities by the objective framework of use. The context for the product use encompasses the environment, the location, place and region, the objective social situation and recognised routines and ritual of the whole use experience, where the product belongs. Furthermore, the new product serves as a component of the whole ensemble of products in this environment. The fit and impact of the new product on its use situation and environment must be evaluated.

Moreover, organisational issues in life are relevant in the use situation. There is the question of general behavioural patterns, which include, for example, whether consumption will be publicly conspicuous or private and what is the division of time, labour, decisions and space in the use situation. These all affect consumers' choice of product. Taste researchers have even shown that in a home environment these patterns have a stronger effect on product choice than demographic features (Douglas 1996: 95-104).

The role of the product is especially determined by the consumer's organisation of life. The product might be highlighted or just supporting the use experience. Also a functional object can achieve symbolically meaningful roles due to these organisational questions. For working mothers balancing through time-scarcity, a seemingly mundane item of household technology, the microwave oven, emerges as a 'heroic' figure capable of smoothing the emotional stress (Thompson 1998: 142).

Gender effects people's choices. In English households it is possible to observe a discussion about the feminine and masculine balance. Males often decide on the living room and the females about the bedroom. The masculine concept is about functional and plain and the feminine about cosy, comfortable, and even 'too pretty and nice' atmosphere (Signs of the Times).

Product involvement or personal relevance is often determined by the organisational questions in life. It increases with the object's closeness to the individual's central value system (O'Donohoe and Tynan 1997: 221). In the evaluation of central or culturally valid products the consumer expertise level is usually higher (Schneider and Rodgers 1996: 249-253; Jones, 1991: 11). Even a quite insignificant product has to fit into to the more significant product environment that the user has.

Some of the most interesting issues in the organisational questions of product use are time and place. The time of year or the time of the day might be influential in defining the context for product use and marked with different products. Many goods are sold on the basis of speed or mobility or explicitly in terms of how much time they can save us, but connectedness to the past and history is important for our present and future. Set against the contemporary freedom of place and high speed of things there are coping strategies to treasure nostalgic products with particular symbolic value of roots and generous amount of time.

The objective framework also incorporates the traditional consumer categorisation on the basis of demographic features. These features introduce the age, sex, education level, occupation, income and institutional bonds of the target user group. They present the user's objective resources and restrictions such as income, education, family structure and social background. Price is a restricting issue even when you have taste for something (ceramics focus group). A rented house sets restrictions on your interior design. What kind of home you can afford decides what suits it (lamps focus group).

6.5 THE MAKING OF MEANINGS

A designer should be interested in the aspirations and meanings the consumer is trying to build with the help of the product. Orientation in life and its visual display grow naturally through the consumer aspirations that also function as important glue for the future product use.

Consumers' constructions of meaning in product use rely on their capacity for symbolic thought and coding, which, in turn, is determined by the individual's cultural capital. Cultural capital refers to the education, knowledge, previous experience and embodied competence, which enable us to discriminate within the world of goods. These resources affect the nature of individual interpretation. They can influence both the intensity in using time and the depth in understanding certain products.

Acquiring and expressing identity through consumption has become a normal way of relating to products. Identity is created, maintained and reconstructed by an accumulation of knowledge, codes, values, stores of experience and guiding principles of action. It encompasses both personal and social characteristics of people as understood by themselves and others. The material possessions and aspirations for new possession that the person has support these things. The meanings attached to possessions reflect even different developmental stages of the self (Dittmar 1992: 115-116).

Self-image and product-image congruence is an important question. The way to express a pleasurable product relationship is to exclaim: 'This is just like me!' In identity building, products present gender, self-reflexivity, mysterious self-qualities, uniqueness, high quality, status aspirations, metaphors of self, self humour or spirituality (Signs of the Times 1992).

Identity building is also a process of self-fashioning. The body is not only a vehicle of emotional or sensorial pleasure, for example through the haptic system of touch, but an expression of self. Embodiment of experiences is about the physicality of meanings, where material things function as extensions of bodies, shelter for bodies or display bodies. Gender identity, for example, concerns different rules of body language, products and behaviour.

In individual self-identity building through embodied experience with the objects the question of authenticity is at stake. In a society where it is possible to reproduce everything the real origin and uniqueness become valuable. Rarity in a product is transferable to its owner and product uniqueness is used in the search for individuality. This distinction is achieved through acquiring outrageous, rare or unique things and environments (Signs of the Times 1992).

Stimulation seeking is connected with design tastes. People with high optimum stimulation levels or high sensory innovativeness may prefer novel, irregular, or unconventional and sensuous designs that offer greater arousal. Sources of arousal in certain physical properties of environmental stimuli include ambiguity, complexity, novelty and potential for surprise (Bloch 1995: 23; Crozier 1994: 62). To produce pleasurable products is to find the optimal arousal level. A high degree of novelty can be perceived as distressing and threatening, whereas a low degree of novelty can be repetitive and unchallenging (Bianchi 1998a: 2-4; Bianchi 1998b: 74-76). People might show readiness for stimulus, change, novelty, freedom or risk. Alternatively, they might have a desire for familiarity, peace, tradition and security. All these aspirations can manifest in products.

People also employ material objects to compensate for perceived inadequacies in their concepts of self: they engage in 'symbolic self-completion'. So a business student who had a weak symbolic basis for a business career, would tend to display the obvious symbols of a successful business career (Dittmar 1992: 101-103). Religious, economic or erotic desire can be displaced to an object. One way that an object's social value is overdetermined is through the demonstration of excess capability that suggests a latent property to deliver human qualities. A car can have an excess of power that cannot actually be used on the road but a powerful car makes the driver powerful (Dant 1999: 40-58).

The designer has to remember that people often have ideals of future symbolic capital. These ideas can be called meta-preferences, something whose use is unknown in the present yet possibly applicable in the future (Taste and Taste Formation 1997: 117-130). There is a willingness to change oneself through certain future profits. These changes often come from the user's capacity for self-criticism or reflexivity, and they demand changes of values. They may form a basis for new product ideas.

6.6 THE NETWORK OF INFLUENCES

Possessions play a profound role in differentiation from others, comparison with others and integration into social groups (Dittmar 1992: 92). Consumer's expectations and the way in which it is possible for them to appropriate products are moulded by social influence. Conversely, new products have an influence on the social environment and interaction within it.

Products help us to make sense of our social surroundings, locating us in the complex societal map of roles, groups and subcultures. Each of us has some sort of placement in society, even though it is in constant flux and disposed to negotiation. History and hierarchies are connected with this social placement. Social distinction grows from differences and status acquired through symbolic meaning. Many of the product meanings are derived from their association with social roles (Solomon 1983: 320). Each role is associated with a collection of products and activities. A Rolex watch, a Brooks Brothers suit, New Balance running shoes, a Sony Walkman, and a BMW automobile bear no relation to each other via function or form, but many consumers might nonetheless group them as a symbolic whole associated with a certain role (Solomon and Assael 1987: 192). People are concerned with how they think they are perceived by their product choices. Aspirations towards certain reference groups are expressed by choosing goods for oneself, which are congruent with those selected by typical reference group members (Dittmar 1992: 99-100). Taste is about possessing what someone else possesses.

Novice role players are more likely to rely more on stereotypical product exemplars of the role. This stereotype information is based on mass media, advertising stimuli, and experience that reinforce commonly accepted product-to-role associations (Solomon and Assael 1987: 195-197). Preferences arise through aspiration or dislike for some stereotype. We look at others, at least on first sight, through stereotypes. You can very easily tell from the product who it would belong to: a white porcelain cup with flowers and golden stripes belongs to an old lady in a wooden house with high rooms and nostalgic atmosphere (ceramics focus group). A designer should remember to go further in deeper structures, because stereotypes are easily misleading and aspirationally value-laden.

A large element in the network of influences is derived from social interaction with inbuilt rules. There can be respect or formal distance, as well as intimacy. Social adaptation, restrictions, pressures, competition, orientation, identification, and classification all affect the way we carry out the social interaction that occurs in the use of the product. They form patterns of activity through the symbolic income or profit. Underneath there is also a general value-basis that forms the ideology and norms accepted inside certain social networks. These restrict the types of discourse and products possible within the network.

Social interaction changes taste and these changes must be accepted through this interaction. Others must perceive a high status symbol as valuable and rare, and it becomes an everyday object when enough people have acquired it. The acceptance of changes is closely connected with the freedom and creativity evident in the social group a consumer most closely identifies with. This user's taste community concerns the user's social world: influential persons, possibilities, boundaries, limitations, rules, social codes and the changes occurring within all these. It also forms the nearest audience for the product.

In contemporary consumer culture, fashion is often described as collective taste, or even universal taste, a framework for the sharing of experience. Fashion changes, changing the boundaries of acceptable taste. Fashion works in every field of our society and influences different kinds of products. It creates shared expectations for a product's visual appearance. In considering the collective sharing of taste from a wider perspective, the forces of general social change over time must also be seen as an influential frame.

6.7 THE PLEASURABLE OBJECT PERCEPTION AND MEANINGS

In addition to the previous discussion about aspects of taste to formulate suitable research methods for taste orientated design, it is necessary to consider relevant issues of object perception and meanings. These emphasise the need for visual and other sensual research.

The judgements we make reflect our mental map which is used unconsciously to organise and understand sensory experiences. This map is made up of iconic imagery that is not easy to describe in words. Most human meaning is shared nonverbally through our intuitive emotional understanding that involves interdependencies among cognitive, physiological, expressive, and phenomenological components. Verbal information scales cannot grasp all these components of emotional experience (Holbrook and Hirschman 1993: 20; Belk 1998: 294-295; Zaltman 1997: 427-428). Visual and other sensual information helps designers to be empathetic through emotional understanding.

Our mental colour image includes memories, symbols such as good and evil, soft and hard, cool and warm, exhilarated and quiet feelings. Differences in interpretation occur also through cultures, gender, age, experience, and individual character (Hsiao 1995: 191). This complicated set of influences is difficult to describe in words and colour is just one factor in a product. The same words can be understood in many different visual ways and they may carry very different value-based meanings for different people. So it is possible to talk in words about a certain visualisation with quite different understanding about it and to talk with different words about the same thing as you can appreciate it for different reasons. It is essential to acquire information from the consumers about the images and atmosphere that they see as corresponding to the words describing their preferences.

Human experience is inherently multisensory through the process of synesthesia that enables us to perceive an experience in one sensory mode via another sensory mode. Our sense of touch can feel texture without the hand ever coming into physical contact. Some visuals summon us to draw from our tactile memory in order to grasp the feel and thus the meaning. In addition to the visual there are other sensory modes such as flavour, scent, sound, vibration, touch and even intuition. The synesthesia principle multiplies the possibilities of our sensing and experiencing. (Belk 1998: 290-291; Zaltman 1997: 296; Lupien 1995: 223-225). The useful approaches of product experience will have to consider the kinaesthetic and sensual experience. Pleasurable products are chosen through touching the objects and feeling them, their weight and the forms in you hand (ceramics focus group). They are chosen through feeling and trying the look against your own body.

Categorisation and prototypicality influence our object perception and judgements about taste fit. Categorisation is the mental activity of grouping like things together in mental conceptual categories. The boundaries for certain categories are fuzzy and dependent on context. Also, experiences and familiarity and times of encounter with the species of the category produce differences in categorisation. The conceptual representation of a certain category is lodged in prototypes that combine the characteristics of the typical members of the category. Each object has also a typical spatial layout with sizes and shapes (Roth and Bruce 1995: 19-54, 73-95, 111-130). The acceptable distance from typical features and layout forms the possible range of variation in a category member. The designer has to have knowledge about how far from the prototypical picture of the product he or she can drive the design. The originality can be pleasurable or it can prevent the categorisation of the product to the suitable group.

If a product is categorised on first impression as suitable for meeting the consumer needs it is likely to go through further consideration. Even though a product might be functionally ideal, it might fail by conveying the wrong associations (Shackleton 1996: 35). Product categorisation defines suitable user roles, gender associations, age associations, and stereotypes for the product.

The preference through typicality arises because a more typical object in one's own perception is easier to process. It demands expertise, involvement and higher attention to become interested enough to engage with an unfamiliar object. Research in categorisation suggests that consumers prefer goods that have moderate incongruity with respect to existing products. Then distinctiveness is high enough to warrant further processing, yet the product can still be categorised with relative success (Bloch 1995: 20; Crozier 1994: 69-70). Yet there are conflicting and differing results in research about typicality and preferences. The relationship between preference and typicality differs between product groups (Shackleton, Jung-Wan and Kazuo 1999). The consumer segment group in the context of the product's real life situation also affects typicality preferences.

There are two approaches to product-related beliefs: holistic visual perceptions of the product form or linear processing of one design element at a time. It is possible to assume that both occur. The product may first be perceived as a whole. In further processing individual elements may become prominent (Bloch 1995: 19). Categorisation is one example of how our brain organises the visual input into perceptual wholes. These are known as *Gestalts*.

According to the *Gestalt* position the whole has an organisation that has its own dynamic properties that cannot be reduced to the parts. To perceive *Gestalts* pleasurably we have a tendency to regularise, smooth out and simplify irregular and complex figures and to close up open shapes. Continuity, coherence in style makes it easier to understand an object as a composite (Crozier 1994: 41-49; Roth and Bruce 1995: 96-102). Also we need consistency in product compositions. If the product has the forms of modern design but does not achieve the same image in materials, weight and finish, it feels fake. It gives an uneasy feeling about the product as if everything would not be at the right place (lamps focus group).

The perception of the whole serves a useful purpose in revealing the whole flavour of the product with the emotions and experiences it evokes. The designer needs information about this general 'feel' in the middle ground between the detailed product features and its brand image. The general feel and flavour can be represented in different visual compositions that are either acceptable or not.

Style discrimination is an essential part of aesthetic enjoyment. Style differentiates products through distinctive quality or form, which is a composition of colour, shape, line, pattern in visual style, sound and scent. Style causes intellectual and emotional associations because it is a multisensory device easily triggering our emotional understanding (Schmitt and Simonson 1999: 84-85). It is thus important for our embodied experience. Different bipolar style traits such as romanticism/classicism even form part of the personality variables (Bloch 1995: 23). Style deals in surface impressions, yet it forms a corridor between the world of things and human consciousness. Style and meaning are always interrelated (Ewen 1990: 43). Style combines characteristic features both of what is said and how it is said.

In style distinction particularly certain people with design acumen have quicker sensory connections and exhibit more sophisticated preferences than those with little design acumen. Also, some consumers favour visual over verbal processing and these people attend more closely to visual design elements and have clearer preferences in making product choices than do those low in visualising. A person with expertise learns what to look for in certain product styles, and what the important determinants of attractiveness are. Levels of design acumen and experience for the target segment are important guides for design (Bloch 1995: 22). There is also no point in producing marginal variations which are not perceived as different.

Style can be the distinguishable factor even when the category and typicality of the product is the same. Style can also change the typicality factor. Style conveys mood and thus stretches even over product category limits. It enables comparisons of product styles to other product types (Signs of the Times 1992). As people have a tendency to posess different products that are of the same style, stylistic expression and interpretation makes stereotyped collections of products for certain roles possible.

We interpret intangible product meanings with the help of associations, 'Stimulus Chaining' that occurs in human cognition. It involves pairing of a secondary stimulus with a discriminative stimulus, resulting in the secondary stimulus becoming a condition reinforced in its own right. In this way, visual, stylistic attributes can become strongly associated with other product features. Low, sleek cars might be imagined to be fast. (Shackleton 1996: 15). Through association the owner of this car is seen as owning the same interpreted quality.

Products are used for interpreting the symbolism of collective dreams displayed in lifestyles. In their lifestyles people have experience systems such as eating or dressing. We use products to carry meaning through these systems governed by rules and context. Latent rules facilitate sign production, sign combinations and interpretative responses. (Mick 1986: 197). While these rules exist in the social language of goods, the object world is still expressive of complex social and individual realities (Jones 1991: 56-61). In the practises of everyday life meaning is mobile and ever changing as the possibilities of interpretation are complex and require structures that interlink with each other.

Conflicting expression and interpretations exist in products because our lives are filled with conflicting meanings and choices. There is the constant discussion about new versus old, exclusive versus common, expensive versus cheap, authentic versus artificial, sacred versus profane, and handcrafted versus mass-produced. At the level of meanings we strive to have roots with past spiritual connection with nature while we live in the busy cities among large social and technological changes (Barker 1992; consumer focus groups). Consumers build their product choices around individuality, subjectivity, privacy and closeness. At the same time they have to cope with their social lives, public display and demands for objectivity. Sensuousness, pleasure and emotion are important in product experiences but products must also be usable and reasonable.

Meaning is interpreted both from denotative and connotative messages in the product. In denotation an object conveys information about its functions and what it stands for. Connotation refers to an aesthetic dimension, which conveys a subjective impression and emotion about the product. There is the possibility of conflict in these different messages: the shape of the object can indicate its use but also adds an aesthetic dimension, which may contribute to a redefinition of the object function (Heilbrunn 1998: 196). As the nature of our understanding is already conflicting and complex this offers possibilities for openness in the meaning for the product. But there has to be enough consistency to give clues for understanding.

A designer can exploit two principles, similarity and continuity, when encoding meanings in a product (Mick 1986: 202). An iconic sign relates to its object by imitation or resemblance (whiteness is cleanness), an indexical sign relates to its object by some correspondence or fact, by causal relationship (newness as in opposition to the traces of use). A symbolic sign relates to its object in a conventional, knowledge bound manner (cultural habit can convey status). These categories are not exclusive, since a sign may function in all three capacities (Mick 1986: 199; Vihma 1995: 93-101, Appendix 1). Objects are capable of carrying many of these different signs and multiple layers of meanings at the same time.

Metaphor and metonymy provide a foundation for much of our understanding and communication of meaning in everyday life. Metaphor is structured through the similarity relationships that helps us to adapt to new situations by conveying an unfamiliar or abstract domain in terms of one that is familiar. Potential source domains share a similarity in function, structure, property, or sensory impression along any of the five senses (Spiggle 1998: 160). When a product expresses a quality like lightness, sorrow, energy, flexibility, cleanness and nobility it is conceived as having features similar to those found in domains that normally possess these qualities (Vihma 1995: 69-71). Metaphors enable us to build a pleasurable product character as a relatively coherent set of characteristics and attributes. These may cut across different functions, situations and value systems such as esthetical, technical and ethical in the product (Janlert and Stolterman 1997).

The unconscious knowledge elicited by metaphors is rooted in physiologically based image schema. These are patterns that arise from our bodily movements and manipulation or perceptual interaction with objects. Thus, positions such as 'down' associates with inactivity, 'upright' with activity, 'over' ideals and 'under' our reality. The physiological aspect of metaphors involves the transfer of experience from one sense to another (Zaltman 1997: 425- 426; Lupien 1995: 219). So the embodied experience involving our emotions is extremely relevant also in the interpretation of products symbolic meanings.

Metonymy works by associating things in the same whole. 'The crown' can represent the whole state. The objects may be causally, culturally, temporally, spatially, physically, or structurally contiguous. A part can stand for a whole or the whole can stand for a part from a set of elements linked by rules (Spiggle 1998: 160-161). Understanding through metonymy occurs in clothing ensembles where we can know that the men's shoes are combined to a suit and even to an office. This understanding arises from prototypicality and style in objects, their combination rules and stereotyped uses.

Narrative is an even wider composite explaining our understanding. It is crucial to how we make sense of the world, keep coherence, map reality, and even explore conflicts and contradictions. It allows people to give form and meaning to experience by providing a goal, a beginning, a middle, an end, and a central theme. People think about incoming information as if they were trying to form stories in order to understand them (Escalas 1998: 254-259; Thompson 1998: 130-137). Our emotions have different plots that are matched to current situations. Dramatic moments in a story awake more emotion than a steady one. Narratives make emotions meaningful by placing them in the context of an individual's personal history and future orientated goals (Escalas 1998: 256-258).

Emotional experiences with products arise often via object animation. Under appropriate circumstances the richness of information associated with objects interacts with spectators in such a way that they can respond to the objects as though they were alive. Inanimate objects are attributed human characteristics so they can be understood and it is possible to empathise with them. Empathy is a dynamic process that takes place between the object and spectator. The spectator projects his own felt life into the object and becomes fused with it (Crozier and Greenhalgh 1992: 70-74). Objects can thus even become part of our identity.

Taste preferences are formed on the basis of aesthetic pleasure and socially determined preferences. They rely on form and content that are inherently combined in a product. Perception and semiosis are interrelated and it is difficult to divide them. Taste is not about singular, swiftly changing experiences but about more permanent relationships to products that we want to purchase and possess. The pleasurable fit for our life is judged from the object perception and decoding through the aspects of taste described above.

6.8 TASTE INFORMATION ACQUISITION FOR DESIGN

The three categories of aspects of taste overlap each other, because many issues in the human-product relationship are difficult to discuss without referring to the whole complexity of inner life and social interaction. In the initial phase of the design process it should be decided what seem to be the most important aspects to start with if the purpose of the designed product is kept in mind. Of course this demands preliminary study of the use situation. This study has to be extremely critical and reflexive to avoid starting the design process with the designer's own value-based views of the use.

In the design task, it is economical to use set methods that provide complex product perception information and suit the holistic and sensual nature of design. The actual methods chosen depend on the stage of the design process, on the consumer segment, and on the use of the product. The need for information acquisition about taste is evident in the early stages of the design process, when the design starts to take on concept and form. In the proceeding stage the design must gain accuracy in form of language and tactile qualities that produce the right feel and meanings for the product. The whole character and the feel of the product have to be tested in their fit to the consumer's life and their perceptions and understanding.

Communication between the designer and the consumer raises many problems. Is it a question of doing research on real life products or about the mind images of the consumer? Can we use information gathering methods where customers are not limited by the questions put to them about meaning making? Does the stimulation material come from the consumer or from the researcher? How is it possible to get sensual feel responses from the consumer without using words? What is the capability of the consumer to imagine the final product from sketches, material samples or concept presentations? How could sketches and prototypes be made to improve communication between designer and consumer? The three-dimensional and tactile feel should be considered as major elements in the stimulus and communication. Environmental and contextual aspects also have to be considered.

Research should not diminish the value of designer's creativity and intuition. It can provide valuable input to the creative process in ideas and inspiration and a means to test the emerging designs. When designers are taking influences and inspirational material from the media, this source supports stereotypes and the values of creative media people, not nuances, insight and intuition about genuine consumer values and perceptions. The methods used can easily have an affect on how the creative process in design continues. To use a method that seeks for the narratives in object use can guide the product design to be based on narrative thinking, omitting other inspirational factors.

The essential aim of the research is to interpret data into future products. To picture the future of the consumer it is especially important to get information about the consumer's ideal products, dreams, lifegoals and interactions. All methods should be projecting the possible consumer futures.

Certain techniques for gathering information and analysing consumer preferences have already been proved to inform design. Participatory design projects, consumer panels, interviews, contrary pair distinction, analysis of product meanings and personality, cultural matrixes or maps, observation and visual documentation have been used. These techniques as such, however, are not necessarily enough to produce focused knowledge about taste.

Concentrating both on the aspects of taste and pleasurable product perceptions would provide more focus and depth. With the help of them this chapter now evaluates some approaches in terms of their usefulness for design purposes. These approaches overlap because many of them are underpinned by similar cognitive and perception activities. They are used in some form by market researchers, even more in cultural consumer research.

6.9 THE METHODS OF DISTINCTION AND CATEGORISATION

In distinction methods, the idea is that consumers evaluate products through many different attributes. The way people categorise things is a form of distinction of whole products through their perception of family resemblance. Distinction and categorisation techniques are useful for obtaining information about product features, product hierarchies and even holistic perceptions. Distinction methods that use very detailed analysis of the product parts or just certain distinctive features of the product omit the general feel. Both the distinction and categorisation methods often omit the environment of use.

Still, these methods can be used widely to inform design. Design acumen and experience level can be studied through nuances in differentiation. Perception of typicality can be studied through abnormal-normal distinction. Distinction and categorisation could also be used to determine for whom or for what purposes the product seems suitable. The participant might be asked to differentiate the alternatives by positioning them in a certain use context and environment.

There are many quantitative ways of producing information about product distinction and categorisation. Complex systems of distinction can be studied on a large participant scale through quantitative analysis. Computer-based distinction study tools for product development can be based on visual or attribute materials (Chen and Owen 1997; Shackleton 1996; Tovey 1997; Bloch 1995: 26).

Attribute distinction, however, has often been done verbally, which emphasises a cognitive approach to objects. There has been an argument that it is therefore more suitable to the study of utilitarian product features than for aesthetic elements (Bloch 1995: 26). Compared with other attributes, the aesthetic of a product is more difficult to isolate as it relates to a unified Gestalt rather than to sum of individual elements. Attribute distinction can study profoundly the attributes for the product without capturing the general 'feel'. A product with other attributes than the ideal ones can correspond better to the general 'feel' than a product with the right attributes.

Because the starting point for these methods rests with people's capacity to distinguish and categorise objects, material 'cues' should be available to assist this process. For design purposes this can be images, sketches, products, product qualities, prototypes, general feel samples, colours, materials, alternative design solutions or products. Larger compositions, such as different environments, could also be used. Embodied experience involving all the senses and the phenomenon of synesthesia should be covered with the material 'cues'. The problem with existing alternatives such as products is that they can help to obtain information about product qualities and make comparisons with competing products, but they do not necessarily yield any insight into the consumer's mental images of the future. The important question is how freely the consumer is allowed to choose the materials forming the basis for the distinction, and the categories structuring the distinction and grouping.

Distinction studies are often based on distinction via contrary pairs that define attributes or style. In a semantic differential scale, a series of bipolar adjectives that anchor either end of a set of numbers is used. The respondent evaluates the concept according to the various dimensions. It must be certain that the pair for distinction really do exclude each other and that no third or fourth possibilities exist in the consumer mental map to define the same 'feel' of difference. Problems in the bipolar scaling occur when products have multidimensional or contradictory qualities.

Distinction and categorisation tell how the consumer sees differences between artefacts. It can also describe the pleasurable choices. The question of taste goes further in trying to determine what is suitable for the consumer's orientation. The research has to determine what things make sense in relation to the consumer's life and identity. This calls for a semiotic analysis that involves a correspondence between stimuli and pleasurable meaning.

6.10 PROJECTIVE TECHNIQUES

Projective techniques are based on the appropriation of associations and the narrative nature of peoples' understanding. They seem promising in that they approach the mental images in unconscious level and intuitive assessment. By using projective techniques, it is possible to obtain information about how consumers construct meanings, and how they build a whole experience for the product. They can be used to show the connection between our meaning making and pleasure.

Projective techniques permit a wide variety of subject responses, are highly multidimensional, and elicit rich data. The stimulus material presented by the projective test is ambiguous, interpreters of the test depend upon holistic analysis, and there are no correct or incorrect responses. In principle, projective techniques allow creative 'make do' methods where different sensuous mind images of the consumer are produced in tangible materials.

Product image associations that combine values, function and aspirations can be useful in gaining knowledge about taste. By associations, different types of attributes, features and user identities or roles can be connected to the same mental map. Pictures, colours, materials, sentence completion, fine art, music or products can provide stimulation for associations. Instead of just concentrating on the associations raised by the stimulus, it is important to determine what associations a pleasurable product would invoke.

Mood- or storyboards of visual images can be used and are easily made by the consumers. Also mind mapping is another good way of building and presenting associative results. Mind maps can be made both in words and visuals or other 'sense' materials, in order to represent the associative connections attached to a product feel.

In capturing the feel of the product it is possible to use complex communication artefacts or sensual experiences, such as fine art, music, food taste, smell or touch or metaphoric impression to capture and transfer the flavour of the desirable product without being constrained by words. The problem lies in the interpreting this feel into products. The product should embody the consumer's understanding of a 'feel', and not the designer's. By using associative methods with the consumer, the designer could insightfully use a work of art that feels right for the consumer as a starting point for design.

The narrative nature of object meanings and actions can be used as a tool that reveals motivation and goals of action that are important for future taste. For instance 'ideal world' stories tell about our dreams and aspirations. Also preferable themes of life can be studied though narratives.

The consumer's identity aspirations can also be revealed through story telling. An interesting story might be based on what the consumer would like to be and what these characters would like to do. Personification of objects gets people to describe the emotional character of the product naturally, in a human like and personal feature-empathised way. The consumer's own personification and that of the products could be matched in design work.

Here again, it is important to think about the starting point for the narrative. The consumer's own experience as a user can be beneficial as stimulus, in the form of pictures or films of the consumer's use situation. The starting point could also be an unfinished story about an imagined future context of use. A narrative can be told about the use of the object and the consumer's relationship to it. The narrative created by the object itself can also be considered. The designer can analyse how these two narratives can be built to fit one another.

A narrative can be constructed by combining smaller associations, or even single word associations. A professional visualiser can be used to allow the immediate and interactively controlled visualisation of the user's ideas of an ideal product (Pavesi and Sommer 1996). It is not even necessary to put the stories into words, as visual storyboards, such as psychodrawing, tell a pictorial story about the product's use.

Interpretations of associations, narratives and product characters and the fit these have for our lifestyles and goals, define the preference for the object. By researching these meanings visually and sensually the designer could use a complex set of dreams, life goals and associations in the design instead of simple signs.

6.11 ANTHROPOLOGICAL VIEW

Anthropology is a study of patterns of meaning, values, perspectives and worldviews that also exist in material objects. It provides the context and the story of human experience from a cultural-systems approach (Tso 1999: 70). Anthropology uses ethnographic research that articulates layers of meaning underlying a behaviour constellation such as cooking. These layers can well correspond to the categories of taste described earlier. Multiple sources of data are considered useful in ethnographic practice to access the different realms of experience (Arnould 1998: 86-91).

In taste studies the triangulation in information acquisition ensures rich information feedback from the different frameworks for taste. The objective framework encourages interviews, observation, photography, and the use of video. The making of meanings can be studied through identity and aspiration narratives or by searching for products that correspond the consumer in their character. An interaction narrative, observation, interviews, photography, and the use of video can map the network of influences.

The task of the ethnographer is to give a detailed account of the arrangement of experience. The usefulness of this approach lies in the capacity to describe the structure of life and the setting for this life (Arnould 1998: 86-89). This is a good source of information for the designer who is trying to form future arrangements for experience, but in the creative inspirational sense it can also be restricting.

Ethnography for design tries to make sense of how other people make sense (Salvador, Bell and Anderson 1999: 35). This approach tries to grasp the guiding theme or the guiding values of behaviours. Behaviours that informants interpret and experience as unique may, when observed, can represent cultural regularities (Arnould 1998: 86-103). The important issue is to capture those especially charged moments in human life, which are pregnant with meaning and are relevant for the product use and users.

Visual and product based observation research from the ethnographic tradition has not been used for design as much as it could have been. An example is socio-semantic analysis developed by Harold Riggins (Riggins, 1994: 101; Riggins 1990: 341-367). Riggins emphasises the complex relationship between the self and domestic objects and shows how this can be analysed through observation of these objects.

The selective display of domestic objects establishes and reinforces personal identity. It forms the living stage of social interaction where messages about the self are presented to the public. Variety in interpretation is a result of the same sort of artefact being put to varied uses and settings by the different consumers (Riggins 1990: 342-343).

Riggins suggests different categories for analyses. There are categories connected with use, such as active (touching/moving) and passive objects (contemplation). There can be normal or original and alien or unintended use by the creative consumer. The role and meaning of objects is important in this analysis. Status objects exist via object's cost or apparent cost. Esteem objects relate to achievements. Occupational objects connect with people's careers. Objects can be role connecting, distancing or misidentifying devices. Indigenous objects and exotic objects provide information about cosmopolitanism. Collective objects can be national symbols or signs of membership. Stigma objects are often hidden because of their negative associations.

People have a lot of time indicators in their possessions indicating time and periods of time, even the future. The size and proportions of objects conveys typicality or atypicality in meanings. The way objects have been made, by hand or machine, connote value. The spatial positioning of objects is relevant for their meanings. Objects can work as social facilitators and can structure interaction. Objects are displayed in relation to others conveying the importance and even the dominance of certain persons.

Some constellations of analysis relate to the questions of general style and feeling such as status consistency and inconsistency. A democratic ethos can be conveyed by the display of objects chosen primarily for sentimental reasons or by treating objects of varying value in the same way. The degree of conformity to current tacit rules of interior decoration measures person's social integration or marginality. The general flavour summarises the intuitive impression of an inhabited room. The flavour can be interpreted on the basis of a general impression such as atmosphere or character. The general impression can be, for example cosy, conservative, impersonal, chaotic, controlled, formal, casual, deprived, bohemian, minimal, nostalgic or extravagant.

This sort of analysis shows how also visually observable aspects of a certain product and its use are relevant. This object and environment based analysis forms the setting for the subjective and socially defined experience. In the case of living room analysis, the home plan, social and resting functions, display or general impression to be felt and shown to others are important, and make up possible concepts of acceptable furniture or interior decoration.

6.12 CONCLUSION

Staging as the visualisation of taste

A comprehensive picture of the use situation requires an understanding of the consumers' visual world of experience. This consumer world could be embodied in a 'stage', constructed either in accordance with their desires, or as an extension of their current environment. This visual construction should be supported by information about the attitudes, values, aspirations and social actions for which this composition was made.

To understand taste it is essential that the idea of sets of objects can be applied to a research method. Images tied to singular possessions or particular brands are only a minor contribution to the evaluation of others and us. For this we use combinations of objects. Also an object can change its meaning drastically if different objects surround it.

Social interaction can be viewed in terms of impression management that helps both the impression maker and other participants to orient towards the person and the situation. Impression management is built disguised from the eyes of the audience, on the backstage, and then visibly performed in the front stage. The performance can be idealised with part of the image hidden on the backstage or it can depict a true picture of the person's life (Goffman 1990). Impression management is about the whole composition and structure of the products displayed by the consumer.

From a designer's point of view, it is beneficial to take a look at a certain stage with certain boundaries that the consumer has set, and at the aspirations and identity this structure supports. The metaphor of staging relates to the definition of taste as an understanding of what occurs in a certain stage where we are participating in the action. Staging as the visualisation of taste could therefore be concerned with building the visual composition that depicts the necessary things about the consumer's orientation, action and experiences.

The elements needed for building the staging are those employed by the set designer: line, shadow, colour, texture, and style suggest such contextual information as place, historical period, time of day, season, mood, and atmosphere. For these purposes metaphor is also deployed (Laurel 1993: 10). Especially important is the construction of the whole structure of the consumer's surroundings, consisting of space, highlighting, relations and product choices. Not only what has been displayed is of importance: the way of displaying and hiding has also to be considered.

However, set design is not the whole play – for that we also need representations of character and action (Laurel 1993: 10). Identity is played in the stage where the staging supports it. Also, the actors' characters are interpreted through the characters of the objects surrounding them. When the setting is made by a group of people, the question of task organisation has to be considered. Who makes the choice, or is it jointly made? What is the position and interaction of the different performers in the act? What are the different products required by these positions?

The metaphor of the stage might help us to look visually at the use situation from different perspectives. The stage can be orientating, changing and creative in action. There can be a backstage and a front stage or many levels of action in a use situation. The circular stage of 'theatre in the round' shows the different sides of the life, the public and private, in turns. Quite different things can go on in these different spheres of life. For design it is useful to see the different situations in which the product is involved.

It is also important to consider how the metaphor of staging helps in evaluating the different aspects of taste discussed earlier. An objective framework is the basis for the stage construction, because it creates a certain setting for the use situation. It considers those who have constructed the stage, and those who are going to use it. The role of the product, the way of using it, the role of the surrounding environment, the time and organisation of action are all taken into account. In looking at something that does not yet exist, objective resources and other boundaries in the use situation naturally set limits on possibilities.

In the staging method, the making of meanings becomes apparent in the display of identity, roles, values and aspirations. The environment should correspond to these visually. The meanings provided by the use environment consist of the meanings in the objects, the meanings of objects in relation to other objects, and the meaning of the whole structure of the setting. Aspirations and values have to be studied to ensure appropriate fit with the meanings in consumers' lives.

In looking at the social network of influences and activities, the interaction possibilities and the social distance that the setting offers are revealed. Social meaning is considered through desirable impression management in social interaction. This impression consists also of things that the consumer wants to keep at the back stage or hidden, but that have meaning in its construction. Social aspects call for studies of social networks of the target group and social interaction surrounding the use environment. The circle of people who form the inner taste community for the consumer's use situation define the display, social use and taste sharing of the environment.

To form a stage, a triangulation of information acquisition is necessary. In addition to interviews and discussions, various visual means can be used such as collage, sketches, photos and videos. It is also important to consider the active role of the consumer in forming and structuring the stage from the different pictorial, colour and tactile materials. Alternatively, future use situations could be built through the research, with the consumer being asked to choose the most pleasurable one.

Some of the methods described earlier can be applied to this visual composition of a use situation. Through principles of distinction and categorisation, the consumer can be offered building material from which to choose a suitable composition. Associations can be used to build appropriate connections of meanings in the staging composition. It is also possible to look at those narratives and cultural scripts that the staging is meant to support. The narrative explains how people need to act physically and emotionally in the setting. Through staging, narrative is changed to a visual story of the product use.

From the anthropological point of view, staging means the structural and meaningful organisation of things in the life setting. The use environment and the other objects surrounding it are depicted in a naturalistic way. The role of the designed object can be observed in the whole setting of objects. It can be supporting other more important ones or be in the centre of the consumer's use. An anthropological perspective also emphasises impression management, and the differences in behaviour in the public and private spheres.

The benefits of using the metaphor of staging in building the picture of taste are many. Staging depicts the action through a set of products. Often people do not use products on their own. In addition to their interaction with people, objects exist in interaction with other objects and products, which have roles in this interaction. Although staging is about the whole composition of the use situation, it still allows the categories of objects to be observed, for example their prototypicality. Furthermore, the general trend in the consumer's choice of objects and the general flavour of the product environment can be observed. New products can be adjusted by their character to fit this flavour and composition. Different people who use the same objects might have various stages, and through building them it is possible to see how the future product fits them.

At its best staging could mean that the future consumer lifestyle and the visual composition displaying it are presented together. This would be the most profitable taste research result for design purposes. The idea of staging combines two future-bound methods used in design. It forms a depiction of product use situations as in action-based scenarios and also grasps the embodied experience by an atmospheric, visual and tactile product environment as in fashion trend forecasting.

Design is about visions of the future. Through studying taste, it is possible to find solutions that are more pleasurable, arouse more passion and support more fully the aspirations and interaction of the consumer than previous products. At its best, design which tries to provide desirable future settings for the consumers' lives can raise the consumers' quality of life with a positively experiential fit to their life aspirations.

6.13 REFERENCES

Unpublished:

Interview:
Patrick Lalonde- Dade, planning director, Synergy Consultant, London 19.4.1999.
Focus groups:
Lamps Salford, 16.7.1999.
Ceramics and spectacles Kuopio 6.3.1999.

Published:

Arnould, E.J., 1998. Daring consumer-orientated ethnography. In *Representing consumers. Voices, views and visions.* Stern, B.B. (Ed.), Routledge, London, New York.
Ashcraft, D. and Slattery, L., 1996. 'Experiential design. Strategy and market share.' In *Design Management Journal*, **7**, 4.
Barker, N., 1992. *Signs of the times. A portrait of the nation's taste.* Photographs M. Parr, Barker, N.(Ed.), Cornerhouse Publications, Manchester.
Belk, R.W., 1998. Multimedia approaches to qualitative data. In *Representing consumers. Voices, views and visions.* Stern, B.B. (Ed.), Routledge, London, New York.
Bianchi, M., 1998a. Introduction. In *The active consumer: novelty and surprise in consumer choice*. Bianchi, M. (Ed.), Routledge, London.
Bianchi, M., 1998b. Taste for novelty and novel tastes. The role of human agency in consumption. In *The active consumer: novelty and surprise in consumer choice*. Bianchi, M. (Ed.), Routledge, London.
Bloch, P.H., 1995. Seeking the ideal form: Product design and consumer response. In *Journal of Marketing*, **59**, July 1995, pp. 16-29.
Bourdieu, P. and Wacquant, L.J.D., 1992. *An invitation to reflexive sociology.* Polity Press, Cambridge.
Chaney, D., 1996. *Lifestyles.* Routledge, London, New York.
Chen, K. and Owen, C.L., 1997. Form language and style description. In *Design Studies*, **18**, Elsevier Science Ltd, pp. 249-274.
Crozier, R. and Greenhalgh, P., 1992. The empathy principle: Towards a model for the psychology of art. In *Journal for the Theory of Social Behaviour*, **22**, 1.
Crozier, R., 1994. *Manufactured pleasures: psychological responses to design.* Manchester University Press, Manchester.
Dant, T., 1999. *Material culture in the social world. Values, activities, lifestyles.* Open University Press, Buckingham, Philadelphia.
Designing the Experience, 1996. *Strategies for developing consumer products and services.* In *Design Management Journal*, **7**, 4.
Dittmar, H., 1992. *The social psychology of material possessions. To have is to be.* Harvester Wheatsheaf, Hemel Hempstead.
Douglas, M., 1996. *Thought styles. Critical essays on good taste.* SAGE Publications, London, Thousand Oaks, New Delhi.
Escalas, J.E., 1998. Advertising narratives. In *Representing consumers. Voices, views and visions.* Stern, B.B. (Ed.), Routledge, London, New York.
Ewen, S., 1990. Marketing dreams. The political elements of style. *Consumption, identity, and style: marketing, meanings, and the packaging of pleasure*. Tomlison, A. (Ed.), Routledge, London, New York.

Goffman, E., 1990. *The presentation of self in everyday life*. Penguin Books Ltd., Harmondsworth.

Heilbrunn, B., 1998. In search of the lost aura. The object in the age of marketing romanticism. In *Romancing the Market*. Brown, S., Doherty, A.M., Clarke, B. (Eds), Routledge, London, New York.

Holbrook, M.B. and Hirschman, E.C., 1993. *The semiotics of consumption. Interpreting symbolic consumer behaviour in popular culture and works of art*. Mouton de Gruyter, Berlin, New York.

Hsiao, S.W., 1995. A systematic method for color planning in product design. In *COLOR Research and Application*, **20**, 3, John Wiley & Sons, Inc., NY.

Jones, P.L., 1991. *Taste today. The role of appreciation in consumerism and design.* Pergamon Press, Oxford, Seoul, New York, Tokyo.

Janlert, L-E. and Stolterman, E., 1997. The character of things. In *Design Studies*, **18**. Elsevier Science Ltd., pp. 297-314.

Jantzen, C. and Ostergaard, P., 1998. The Rationality of 'irrational' behaviour. George bataille on consumer extremities. In *Romancing the Market*. Brown, S. , Doherty, A. M., Clarke, B. (Eds), Routledge, London, New York.

Kälviäinen, M., 1996. *Esteettisiä käyttöesineitä ja henkisiä materiaaliteoksia. Hyvän tuotteen ammatillinen määrittely taidekäsityössä 1980-luvun Suomessa.* (Aesthetic functional products and spiritual material works of art. The professional definition of a 'good product' in the 1980's art craft in Finland). Taitemia 4. The Kuopio Academy of Crafts and Design, Kuopio.

Laurel, B., 1993. *Computers as theatre*. Addison-Wesley, Reading.

Lupien, J., 1995. Polysensoriality in plastic symbolic discourses. In *Advances in visual semiotics. The semiotic web 1992-93*. Sebeok, Thomas A. and Umiker-Sebeok, J. (Eds), Mouton de Gruyter, Berlin, New York.

Margolin, V., 1997. Getting to know the user. In *Design Studies*, **18**, Elsevier Science Ltd., pp. 227-236.

Mick, D.G., 1986. Consumer research and semiotics: Exploring the morphology of signs, symbols, and significance. In *Journal of Consumer Research*, **13**, September.

O'Donohoe, S. and Tynan, C., 1997. Beyond the semiotic strait-jacket. Everyday experiences of advertisement involvement. In *Consumer research: postcards from the edge*. Brownd, S. and Tyrley, D. (Eds), Routledge, London.

Pavesi, G. and Sommer C.M., 1996. Washing Machines: Users' perceptions and expectations as input into the professional design process. In *Proceedings of the Design Management Institute's Research and Education Forum 'Fostering Strategic Design Cultures'*. November 1996. The Design Management Institute, Barcelona.

Riggins, S.H., 1990. The power of things: The role of domestic objects in the presentation of self. In *Beyond Goffman. Studies on communication, institution, and social interaction*. Riggins, S.H. (Ed.), Mouton de Gruyter, Berlin, New York.

Riggins, S.H., 1994. Fieldwork in the living room: An autoethnographic essay. *The socialness of things. Essays on the socio-semiotics of objects.* Mouton de Gruyter, Berlin, New York.

Roth, I. and Bruce, V., 1995. *Perception and representation: current issues.* Open University Press, Buckingham.

Salvador, T., Bell, G. and Anderson, K., 1999. Design ethnography. In *Design Management Journal*, Fall 1999.

Schmitt, B.H. and Simonson, A., 1999. *Marketing aesthetics: the strategic management of brands, identity, and image.* The Free Press, New York.

Schneider, K.C. and Rodgers, W.C., 1996. An 'importance' subscale for the consumer involvement profile. In *Advances in Consumer Research*, **23**.

Shackleton, J.P., 1996. *The application of a 'prototype theory' framework to the modeling of product perception and the emergence of new product groups.* Graduate School of Science and Technology Chiba University.

Shackleton, J., Jung-Wan, H. and Kazuo, S., 1999. 'Typicality' and consumer preference - Prototype theory in product design. In *Design Cultures. Proceedings of the Third European Academy of Design Conference*, **2**, Sheffield Hallam University, 30th March - 1st April 1999. The European Academy of Design, Salford.

Signs of the Times. *A portrait of the nation's taste.* Television documentary. M. Parr original idea, research H. Goodman, series producer N. Parker. BBC MCMXCII.

Solomon, M.R., 1983. The role of products as social stimuli: A symbolic interactionism perspective. In *Journal of Consumer Research*, **10**, December.

Solomon, M.R. and Assael, H., 1987. The forest or the trees?: A Gestalt approach to symbolic consumption. In *Marketing and semiotics. New directions in the study of signs for sale.* Umiker-Sebeok, J. (Ed.), Mouton de Gruyter, Berlin, New York, Amsterdam.

Spiggle, S., 1998. Creating the frame and the narrative. In *Representing consumers. Voices, views and visions.* Stern, B.B. (Ed.), Routledge, London, New York.

Taste and Taste Formation Part II, 1997. *Cultural Economics: The Arts, the Heritage and the Media Industries*, **1**. Towse, R. (Ed.), The International Library of Critical Writings in Economics, Edward Elgar Publishing Limited, Cheltenham, Lyme.

Thompson, C.J., 1998. Living the texts of everyday life. In *Representing consumers. Voices, views and visions.* Stern, B.B. (Ed.), Routledge, London, New York.

Tovey, M., 1997. Styling and design: intuition and analysis in industrial design. In *Design Studies*, **18**, Elsevier Science Ltd., pp. 5-31.

Tso, J., 1999. Do you dig up dinosaur bones? Anthropology, business, and design. In *Design Management Journal*, Fall.

Vihma, S., 1995. *Products as representations - a semiotic aesthetic study of design products.* Publication series of the University of Art and Design UIAH A 14, University of Art and Design, Helsinki.

Zaltman, G., 1997. Rethinking market research: Putting people back. In *Journal of Marketing Research*, November, pp. 424-437.

CHAPTER SEVEN

Pleasure versus Efficiency in User Interfaces: Towards an Involvement Framework

ANTONELLA DE ANGELI, PAULA LYNCH and GRAHAM I. JOHNSON

Advanced Technology and Research, NCR - Self Service Strategic Solutions, Kingsway West, Dundee, DD2 3XX, UK

7.1 BEYOND USABILITY

The concept of usability is regarded by many as a milestone in the history of computer system design, specifically the design and evaluation of user interfaces. Since the term entered common usage in the early 1980's, the consideration of usability has greatly impacted on the way in which many interactive systems are developed. Usability compels designers to think from the very beginning about end-users. Therefore, it has contributed to the evolution, from the traditional top-down design approach (all requirements were specified in the planning phase and then developed by stepwise refinements) to a more iterative-design approach (evaluation and implementation are closely linked, user input at key stages, and ongoing requirements definition and system specification).

Despite the apparent current popularity of usability thinking, the acceptance of the user-centred metric has not been easy in a development world dominated by a previously unquestioned system-centred philosophy. Historically, the latter approach assumed that users could, and would, adapt to whatever was built. Training, support documentation and 'Help' functions were considered to be the most appropriate solution to serious interaction difficulties. Only within the usability framework has the end-user become the focus of the design process. The recognition that usability is a fundamental aspect of product quality as well as of marketing is a fairly recent reality. Nowadays, much usability research effort is devoted to the development and validation of cost-effective evaluation tools to encourage the integration of usability issues into design (e.g. Nielsen 1993; Johnson 1996; Jordan *et al.* 1996; De Angeli *et al.* 2000).

Currently, the value of the usability framework is so well acknowledged (ISO 1991) that proposing major revisions will no doubt sound outlandish to many usability and human factors specialists. A number of objections can be raised to the modification of a successful metric, which is reliable and capable of driving the design of effective systems. This chapter attempts to meet these objections, proposing an evolutionary perspective.

Our belief is that the need for re-examination and update of the general usability framework is prompted by the powerful combination of technological progress, consumer expectations and the evolution of novel applications. A new generation of interactive systems, utilising varied animated characters, sophisticated behaviours, embodied conversational agents and interfaces with 'personality', will revolutionise the way people interact with computers. enlarging the bandwidth of communication to fully include social and affective dimensions (Picard 1997; Cassell 2000; Cassell *et al.* 2000). These next generation systems will elicit different psychological reactions from users; hence, they need to be evaluated in a specific framework respecting their key characteristics.

Human-Computer Interaction (HCI) has many examples of successful frameworks and approaches that have been significantly modified as a result of the movement of the state of the art of technology. Consider, for instance, the GOMS model, a cognitive modelling tool to represent the procedural knowledge that a user must have in order to carry out specified tasks on a system (Card, Moran and Newell 1983). GOMS analyses apply well to situations in which users perform simple and linear goal-oriented tasks which they have previously mastered. Therefore, it is well suited to model text-editing tasks, but has difficulties in extending towards creative tasks, involving parallel activity, such as those supported by contemporary multimedia systems.

HCI is a dynamic discipline whose object of study is continuously evolving (De Angeli 1997). The interrelation between HCI and technology can be represented as a loop: HCI modifies technology, the new technology modifies HCI. Computers change, users change: As a result, the interaction is different and need to be investigated with different research tools. A constant update of the theoretical and methodological frameworks within HCI is required, as a consequence of this evolution of technology.

The chapter discusses our vision of the future with respect to personality-enriched interface developments, providing some initial thoughts on an involvement framework, and augmentation of a traditional usability engineering perspective. A brief review of the pertinent literature and the results of a preliminary survey of an exemplar financial interface embracing critical attributes of personality are presented in the following sections of this paper.

7.2 SOCIAL ARTIFACTS AND RELATIONSHIP TECHNOLOGIES

In recent years, the interaction bandwidth between users and computer-based systems has exponentially increased and the future is promising even more stimulating scenarios. Also, digital networks have moved far away from their original habitat: they have moved from academic and industrial settings to private and public spaces. In parallel, the Internet has evolved and the consensus is that the future impact of the Internet on all personal and business relationships will be tremendous.

In the Network Economy (Kelly 1998) the real business is in building relationships between customers and information providers. According to the Knowledge Lab vision (Lynch, Emmott and Johnson 1999), the future of interactive technologies is not centred on raw information or on Information Technologies (IT). Rather, the focus will be on *Relationship Technologies*. These are primarily aimed at gaining the attention of consumers, attracting them, understanding their needs, communicating with them, supporting them, and inducing them to use the services. The term was originally proposed by the NCR Knowledge Lab, to define existing and emerging technologies and models that enable, support and enhance relationships between customers and providers, or between groups of customers. Here, the definition is extended to include those direct relationships between users and (social) interfaces.

Historically, IT has been aimed at creating effective machines, encompassing all of the rationality, logic and abstract knowledge available to human beings. IT has typically produced cognitive artifacts: objects that store, manipulate and retrieve information (Norman 1991). Cognitive artifacts are artificial tools made by human beings to support representational functions. In contrast, relationship technologies are going far beyond efficiency, building social artifacts – agents that create and maintain meaningful relationships between users, groups of users and interfaces. These systems will massively impact on the focus of and approaches within HCI, introducing emotions, affect, attitudes and social intelligence, as well as enjoyment, pleasure and humour in the interaction with computers and via interfaces.

From the user's point of view the difference is enormous. Cognitive artifacts mainly support instrumental activities, from perception to problem solving or reasoning. These artifacts elaborate commands according to rational rules and are, therefore, predominantly passive objects, which can be completely under our control. A linear causal model can explain the interaction between them and users. On the other hand, to build lasting and meaningful relationships, social artifacts need to be active agents, capable of responding socially to users, appealing to emotional states, dynamic, showing a 'sense' of personality and attitude. These artifacts have the potential to enhance the anthropomorphic perception elicited by computers: they can easily be perceived as animate, aware, and active (De Angeli *et al.* 1999).

Little published research exists on whether the intentional introduction of social dimensions and emotions in HCI is a positive (or a negative). Until very recently, the prevailing notion was that computer systems must reduce learning times and task-performance times, as well as error rates. Peripheral attributes, such as playfulness or humour, were typically considered wasteful – something which could distract a user and cause them to take their work less seriously. Nevertheless, a new trend is emerging in both the Artificial Intelligence (AI) and HCI community (see e.g. Proceedings of UM'99 Workshop on Attitude, Personality and Emotions in User-Adapted Interaction; Proceedings of the 3rd I^{3} Annual Conference, Workshop on Affect in Interactions 1999). Researchers are starting to examine issues of personality (e.g. Lynch *et al.* 1999) and humour (e.g. Morkes, Kernal and Nass 1999) in interfaces and interaction. The latter team has demonstrated that humour can have similar effects, whether the source is a person or a computer system. In principle, humour can be used to improve the overall user experience, helping to relieve the tension that is often associated with the performance of a demanding task.

The following section of the paper describes how amusement and enjoyment can be introduced in a stereotypically serious context – that of a public self-service cash machine, an ATM (Automated Teller Machine). In this application, the purpose is to service consumer banking needs, normally considered a chore, and rarely (if ever) regarded as a pleasurable or enjoyable experience.

7.3 USER INTERFACES FOR BANKING

Background

Many of today's financial consumers visit their (physical) bank branch less than three times a year. For these consumers, interaction with their bank normally takes place remotely via technology (ATMs, telephones, and personal computers). Financial service providers recognise the intrinsic importance of user interfaces as their primary means of contact or relationship with customers, and are striving to further enhance the consumer experience, as a way to differentiate from competitors, and as a means of improving service quality. In recent years, research work in this area of financial user interfaces has focussed on new interaction scenarios and interfaces (e.g. Johnson 1995; Johnson 1996b; Johnson and Westwater 1996; Baber, Johnson and Cleaver 1997; Coventry and Johnson 1999). Research has also investigated evaluation and design methods for usability in self-service interface solutions (Johnson 1993; Johnson and Briggs 1994; Westwater and Johnson 1995).

The NCR Knowledge Lab is exploring many opportunities to make consumer experiences of e-commerce applications more engaging, personalised, and relevant (Lynch *et al.* 1999). We are investigating how interaction with financial user interfaces can be improved by introducing a sense of personality, attitude or character into the equation. As a result, we created a project to examine the potential of, and reactions to, a prototype (ATM) user interface which is enriched by a specific (type of) character: we affectionately called her 'Granny'.

'Granny': a financial interface with 'personality'

The following scenario describes how a virtual user interface (see Figure 7.1) could be employed, within the common context of an ATM user interface, to facilitate consumer interaction with a financial service provider in the near future:

Emma, a 22-year-old maths student visits an ATM, close to where she lives. A few weeks ago, she chose 'Granny' as her personal banking interface. There were many other 'agents' or personalities that she could have chosen; however, Granny looked like she could be a lot of fun. Once logged on, via a PIN, Granny greets Emma and displays her personalised options such as cash (withdrawal), her accounts and local information.

The overall layout, animations and icons on the screen reflect the general character of Granny – for example, clicking on the 'handbag' icon Emma can access her diary and shopping list. Granny is learning about Emma's preferences by overtly observing her transaction behaviour and by retrieving relevant personal details from the (financial provider's) data warehouse. When Emma requests £30 Granny reminds Emma that it's Bob's birthday next week. In the course of the transaction, Granny prints a receipt presenting Emma with her daily horoscope.

Granny has had a very brief life thus far, and has moved from a basic concept to a full-blown (interface) application within an ATM, which can dispense real cash very quickly. The concept came from a brainstorm session, which concluded with a role-play, two of the authors 'acting' the part of Emma and Granny. Sketched storyboards and then an interactive prototype swiftly followed this stage. The move from PC to a touch-screen ATM represented the final stage of her development.

Granny is an example of a social artifact that attempts to inject a specific personality or character into the interaction between a financial service provider and a consumer. In this context, personality is defined as a stable set of traits that determines the artifact interaction style, describes its character, and allows the end-user to (understand and) predict the artifact's general behaviour.

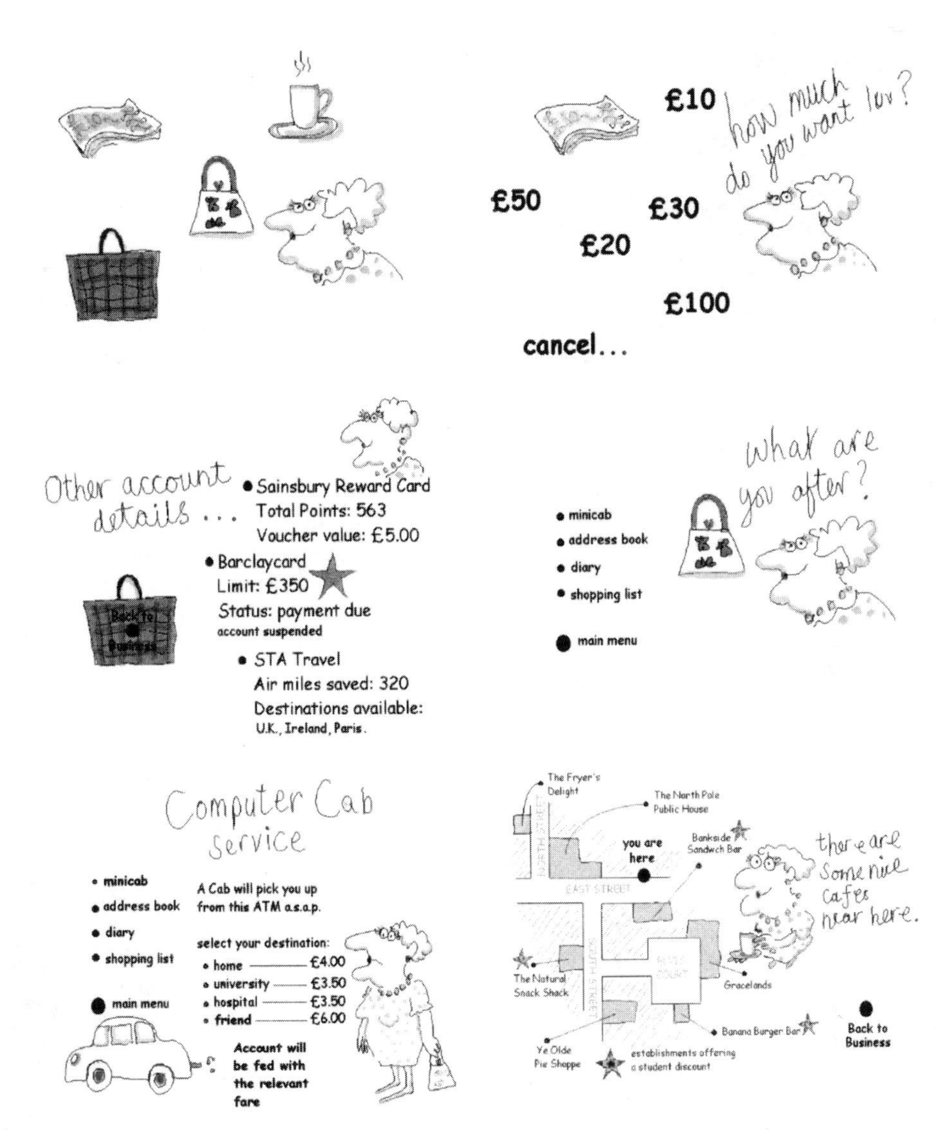

Figure 7.1 'Granny': A financial user interface with personality – example screens

7.4 EVALUATING SOCIAL ARTIFACTS

Despite the general importance attributed to the concept of personality in the development of intelligent (user interface) agents, little is known yet about what type(s) of personality should be implemented. The prevailing approach, typified by Granny, is deliberately anthropomorphic in nature, but more research is needed before effective synthetic personalities can be developed. We have begun to investigate the potential for personality profiles to be used for e-commerce social artifacts, specifically synthetic characters. Starting with the five-dimension OCEAN model of personality (see e.g. Costa and

McCrae 1992), we outlined the possibilities using the five dimensions: Neuroticism, Extraversion, Openness, Agreeableness, Conscientiousness. Each dimension contains a set of traits that tend to occur together and can be conceptualised as a continuum ranging from two opposite personalities. We are working towards a characterisation using this framework, based upon these five factors, the context and task/application in hand. It strikes us that one of the biggest challenges is in developing a framework that can *reliably* measure the effects that social artifacts have on consumer behaviour and attitudes. For example, how effective is Granny in building strong and lasting (versus amusing, casual, superficial) relationships with consumers.

ISO Standard 9241 defines usability as 'the extent to which a product can be used by specified users to achieve specified goals with *effectiveness*, *efficiency* and *satisfaction* in a specified context of use'. *Effectiveness* refers to the accuracy and completeness with which specified users achieve specified goals in particular environments. *Efficiency* refers to the resources expended in relation to the accuracy and completeness of goals achieved. *Satisfaction* refers to the comfort and the acceptability of the system for its users and other people affected by its use. Such a framework refers to computers as tools. Despite being a multidimensional concept, usability mainly relies on objective parameters describing task performance. Indicators of effectiveness include quality of solution and error rates; indicators of efficiency include task completion time and learning time. Satisfaction is the only parameter that directly addresses feelings, opinions and attitudes of the user. Nevertheless, this dimension has traditionally been treated as a measure of perceived effectiveness or efficiency. This approach, explicitly or implicitly, relies on the assumption that users are satisfied when they perform their task efficiently. Nevertheless, the correlation between the three major attributes of usability is still not fully understood, and in the case of complex tasks has been proved to be quite weak (Frøkjær, Hertzum and Hornbæk 2000). These findings breach the traditional framework suggesting that users could be satisfied by something other than efficiency.

Cognitive artifacts are designed to help perform tasks effectively: They need to be simple and efficient. Social artifacts are designed to establish relationships; they need, at some level, to be pleasurable and engaging. The major aim of social artifacts is to create and maintain relationships. For all these reasons, social artifacts do not just need to be *user-friendly*, they need to be *friendly to users,* responding socially to them and showing a 'sense' of personality, attitude and understanding.

Applying the media equation

Following the media equation paradigm, as stated by Reeves and Nass (1996), the framework to evaluate social artifacts could be directly imported by social psychology or by the psychology of personality. According to the media equation (*media* = *real life*) individuals' interactions with computers, television and new media are fundamentally social and natural, just like interactions in real life. This assumption implies that the same social rules guiding human-human communication are equally applied to HCI, even though such behaviour is not necessarily conscious. Research has demonstrated that people engage in polite, socially desirable behaviour when interacting with computers (Nass, Moon and Carney 1999). Moreover, users apply gender stereotypes to computers (Nass, Moon and Green 1997) and human personalities (Nass *et al.* 1995; Moon and Nass 1996; Nass and Lee 2000). Also, when a person is monitored, either by an animated agent or a human, the same psychological reactions are elicited (Reeves and Rickenberg 2000).

The media equation implies that computers should be evaluated by the same instruments used by psychologists to measure the strength of interpersonal relationships. Hence, the evaluation framework for social artifacts can be based on personality tests, attitude and motivation questionnaires. Moreover, the same social rules and norms that drive interpersonal interaction can explain user behaviour. In our opinion, the limit of the media equation can be identified in its generality: the media equation applies to everyone and to all media (De Angeli *et al.* 1999). In this view, social behaviour encompasses any exchange of meaning between users and computers. Such an assumption generates a paradox: every instrumental act (like pushing a key) is also a social act. Taking the media equation to extremes poses the problem of differentiating computers from users, instruments from agents, cognitive from social artifacts. Recently, the same authors have started criticising the initial equation, claiming that HCI and CMC (Computer-Mediated Communication) are not identical (Morkes, Kernal and Nass 1999).

Many empirical studies have demonstrated that face-to-face communication is not an adequate model to explain natural-language interaction (Bernsen, Dybkjær and Dybkjær 1998). Talking to a computer, people maintain a conversational framework, but tend to simplify the syntactic structure and to reduce utterance length, lexicon richness and pronoun usage. We can reasonably expect similar simplification effects affecting social attribution for synthetic characters. The challenge, however, is in understanding which key dimensions are used to evaluate social artifacts and in then translating them into a reliable evaluation framework. The next section describes a preliminary empirical study aimed at understanding how users actually perceive social artifacts, taking Granny as the example.

7.5 A PRELIMINARY STUDY OF 'GRANNY'

Our initial study was aimed at identifying the fundamental perceptions of three different financially related *targets*. These were:

- *Traditional ATM*, representing a typical example of cognitive artifact, an operative tool providing a familiar functionality of basic financial information and dispensing cash.
- *Granny*, as the social artifact prototype: 'she' has a clear personality and is able to adapt to the specific user and context where the interaction takes place. Further, this prototype introduces amusement and humour to the interaction.
- *Cashier* (a human teller), representing a simple control condition, to assess possible differences between social artifacts and human operators.

As our working hypothesis, we assumed that the general perceptions of, and mental representations encouraged by, social artifacts are essentially different from those concerning cognitive artifacts and human operators. Moreover, we expected that the overall reaction towards social artifacts would be generally positive.

Approach

A few days before the survey study, all participants were introduced to Granny during a formal demonstration at the NCR Knowledge Lab, in March of this year. The prototype was installed in a standard multimedia ATM that uses a touch-screen. The interaction started with inserting a bank card and ended with the provision of cash and a personalised receipt. Participants could not directly interact with Granny, but the speaker (demonstrator) showed all the functionality. The demonstration lasted for almost ten minutes, after which participants could ask questions.

Twenty-two Knowledge Lab employees were invited to participate in the study via an e-mail based survey form. The survey instrument itself was an Excel file, composed of three sheets (one for each target). Participants were asked to list six adjectives or short sentences that described each target. Respondents were explicitly instructed to follow the order of presentation of the sheets, switching to a new one only when the previous one was completely filled. To minimise any carry-over effects in the evaluation, the order of presentation (of the targets within the survey) was counterbalanced across participants.

Coding and analysis

Data were available from a sample of 16 respondents (5 F, 11 M; yrs mean = 29), yielding a corpus of 266 adjectives or short sentences (see Figure 7.2). Of these 90 described the cashier, 93 the ATM and 83 Granny. The corpus was initially cleaned up, deleting all the meaningless sentences (N=10). All of the remaining adjectives were then aggregated in semantic categories following the synonymous taxonomy, provided by WordNet 1.6,
http://www.cogsci.princeton.edu/.

The analysis suggested the existence of three superordinate *Evaluation Dimensions (ED)* along which the corpus could be differentiated. Each ED reflects a particular aspect of the targets, which were described in terms of:

- functional quality,
- aesthetic quality, and
- social quality.

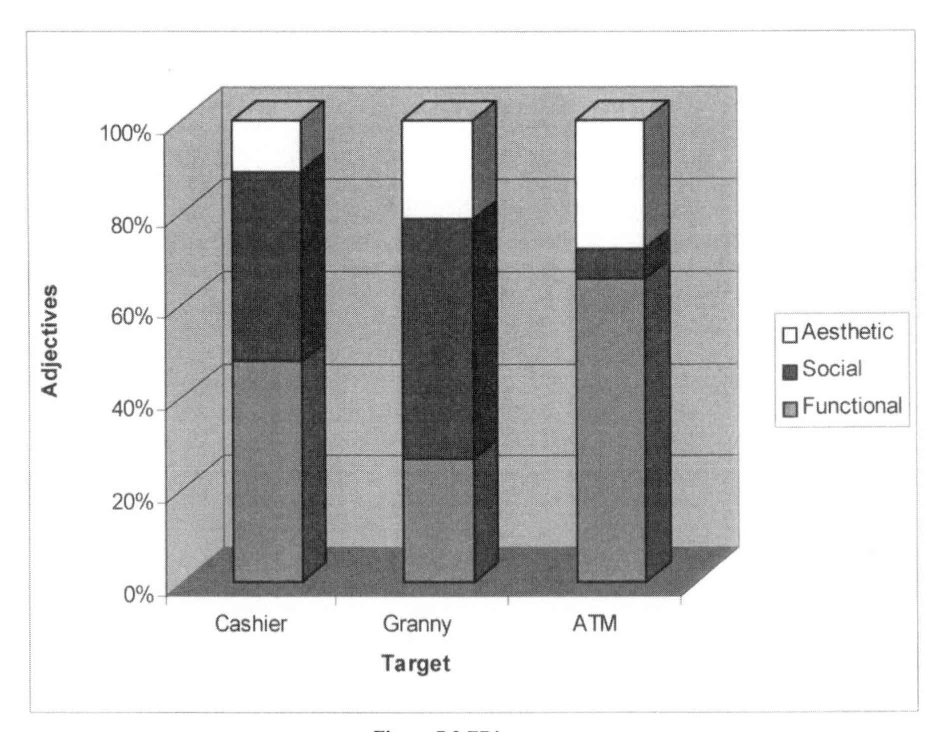

Figure 7.2 ED's

The *functional* ED provides an instrumental description of the target. Here, the ATM, the Cashier or Granny were portrayed as *tools* used by customers to perform financial tasks. This ED mainly reflects a functional perception describing how the target works with respect to the available functionality, the procedures to operate it, overall effectiveness, efficiency or usefulness. The ED category includes all those adjectives describing the target in terms of operative abilities related to problem solving or reasoning, interaction abilities related to the communication of task-related information, accessibility of the service, and novelty of the service. It is evident that the functional ED closely resembles the traditional usability concept: the target is described in terms of a tool, and in relation to the task to be executed. Example adjectives categorised in this ED were: 'reliable', 'service oriented', 'numerical', 'mechanical' and 'efficient'.

The *social* ED reflects a more complex representation of the target. Here, the functional aspect of task execution is disregarded and the stress is on the (personal) relationship between the customer and the target. The category provides a perception or mental representation of the targets as agents capable of actively shaping the relationship. The social ED includes all those adjectives describing personality traits (with the exception of intellectual capabilities related to task execution, previously included in the functional category) motivations, attitudes, feelings or emotions. Example adjectives classified within this dimension include 'chatty', 'engaging', 'personal' and 'smiling'.

The *aesthetic* ED reflects a mental model of the target in terms of its overall physical appearance and perceptual qualities. With this dimension, the stress is on the sensorial perception of the target, primarily on the 'look and feel'. The category includes descriptions of physical aspect, design or form of the target relating to visual, tactile and auditory perception. Example adjectives included 'colourless', 'ugly', 'pink', and 'grey'.

Each adjective or short description contained in the corpus was tabulated in one ED. Double scoring was conducted for 100% of the corpus and any discrepancy was resolved before further analyses were performed. The adjectives were then further classified according to their value into two categories: positive vs. negative quality. The double scoring had a reliability of .89. The adjectives on which a correspondence was missed were not included in the analyses, since their value was considered highly context-dependent (most of which referred to humour or amusement).

Results and discussion

The pattern of ED percentages as a function of the evaluation target is illustrated in Figure 7.2. It clearly emerges that the typology of the target strongly affects the elicited description, the perception or attributions made, and that there are clear differences between the three stimuli of Granny, ATM, and Cashier.

Let us first look at the key differences between the two 'machines' (traditional ATM and Granny). The ATM is mainly described in terms of functional qualities (over 65%) related to task execution, which may not come as too much of a surprise. In contrast, Granny is primarily described in terms of social qualities (over 50% of the adjectives provided for Granny), referring to the relationship between the consumer and the system via Granny. From an evaluation point of view, this difference is not trivial. Granny is a social stimulus; the ATM is mainly a cognitive one. It is fundamental to understand that the two stimuli elicit very different psychological reactions from the end-user or consumer. Social stimuli are more complex than cognitive ones: they are more likely to be two-way interactive agents, perceiving while they are perceived, changing because they are perceived, and involving the observer. Further, many important attributes of social stimuli are not directly observable (e.g. traits, intents, attitudes), and the accuracy of observations is difficult to determine. All these characteristics have to be taken into account when evaluating the user's reaction to social artifacts and will deeply complicate the evaluation framework.

As regards the comparison between the human cashier and Granny, a counterintuitive result emerged. Note the relative difference between the percentage of functional and social ED (which shows that for the cashier or teller there were 48% and 41% functional and social adjectives respectively). Surprisingly, many more (by relative proportion) social attributes were produced to describe a machine (in the form of Granny) than to describe a human being: 52% for Granny and 40% for the cashier. Such a gap can probably be attributed to a typical anchoring effect affecting the evaluation. The cashier is a *functional human* whose social characteristics are irrelevant if compared to important others, such as friends or relatives. We see the effect of people's perception of the *occupation* of bank cashier, rather than the description of Doris the teller at the local branch. Granny is (intentionally) a *social machine* whose functional characteristics are largely irrelevant if compared to stereotypical machines.

The percentages of positive adjectives in each ED as a function of the evaluation target are presented in Figure 7.3. It is clear that Granny received a very positive overall evaluation. The weakest aspect is the functional dimension. Most of the negative attributes were related to the time demand associated with the use of Granny as a functioning ATM. From a purely functional point of view, the ATM appears to be the best solution (in that, of the three targets, it has the most positive functional adjectives). However, it does not elicit any positive reaction along the other two dimensions.

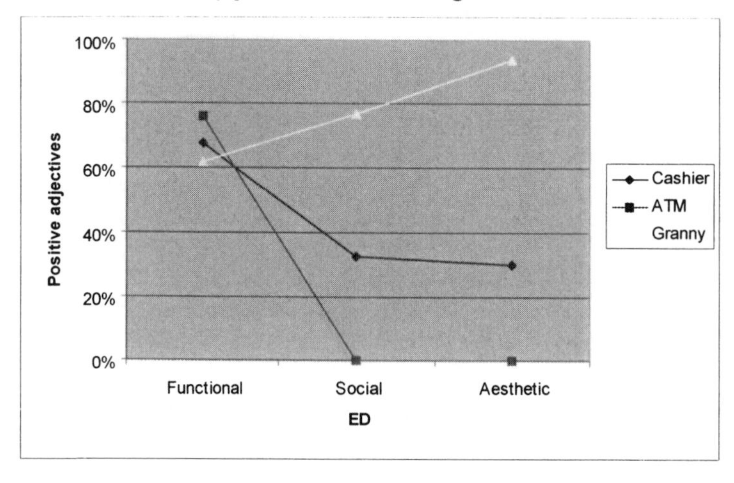

Figure 7.3 Percentage of positive adjectives in each ED as a function of evaluation target

The physical appearance of Granny was highly appreciated. With a very few exceptions, the sample evaluated it as appealing, attractive, and colourful. Also, the social dimension was considered positively, but it can still be improved. Most of the people evaluated Granny as friendly and entertaining, but someone pointed out that it could be over-friendly and silly.

The evaluation trend attributed to the human operator is somehow surprising. Only the functional dimension received a majority of positive evaluations. From a social point of view, human operators were often perceived as unfriendly, bored, depressed or sad. Again, we are most probably witnessing associations to do with the occupation per se with its stereotype of repetitive and mundane counter-based work.

Our initial study was simple in format and our results are preliminary, but they clearly demonstrate that social artifacts tend to elicit particular representations and attributions, which differentiate them from both cognitive artifacts and people (such as a cashier) performing exactly the same task set. Hence, the results offer some support for our basic premise that both the usability framework of ISO and conventional wisdom and the media equation are not fully adequate to provide a reliable framework for social artifacts evaluation.

7.6 TOWARDS THE INVOLVEMENT FRAMEWORK

Evaluating social artifacts is a complex task, since relationships, emotions and personality are, to a large extent, subjective and individual variables. Previous work has mainly focussed on anthropomorphic comparisons, assessing the naturalness of synthesised speech or of automatic talking-faces, as well as a general idea of user satisfaction (see, for instance, the evaluation session in Cassell *et al.* 2000). It is time now to develop an engineering-oriented evaluation framework, so as to ensure acceptability, comparability and utility of results. This section reports an initial proposal for the requirements for user interaction with Relationship Technologies. It is intended merely as a very preliminary contribution, essentially aimed at generating discussion and at laying a foundation for the involvement framework.

The concept of involvement refers to the strength and the quality of the relationship between a user — or a group of users — and a social artifact. Unlike usability, which has been mainly considered as a property of the interface, involvement is a relational property, generated by the encounter between two active agents: the user and the artifact. The difference is fundamental. Involvement is by definition a relative concept that strongly depends on the system's features, on the user's characteristics and on the level of familiarity between the two agents. Familiarity is a key concept and a clear component in the evaluation of social artifacts: It refers to the amount of personal knowledge about the partner which is available during the information exchange. Relationships unfold over time. Hence, familiarity is an evolving factor enhancing the control over the interaction: predicting the behaviour of a close friend is much easier than predicting the reaction of a stranger. Nevertheless, familiarity can have, as many realise, two opposite effects on the relationship: reinforcing it (attraction effect) or dissolving it (tedium effect).

Involvement is a multidimensional concept. Following the early results from our first study, we propose to divide it into the three dimensions referring to social, functional and aesthetic qualities. The relative weight of each dimension varies according to the task to be executed, the context of interaction and the nature, or personality, of the end-user. A simple representation of the involvement framework evaluating e-commerce social artifact is illustrated in Figure 7.4. The circles represent the dimensions and are labelled with the initial of the dimension to which they refer. The overlapping area stresses the blurred outlines between the dimensions. We assume that a number of mutual influences affect the social, aesthetic and functional qualities of any object. For this reason, user satisfaction is determined by the convergence of the perceived quality of each dimension.

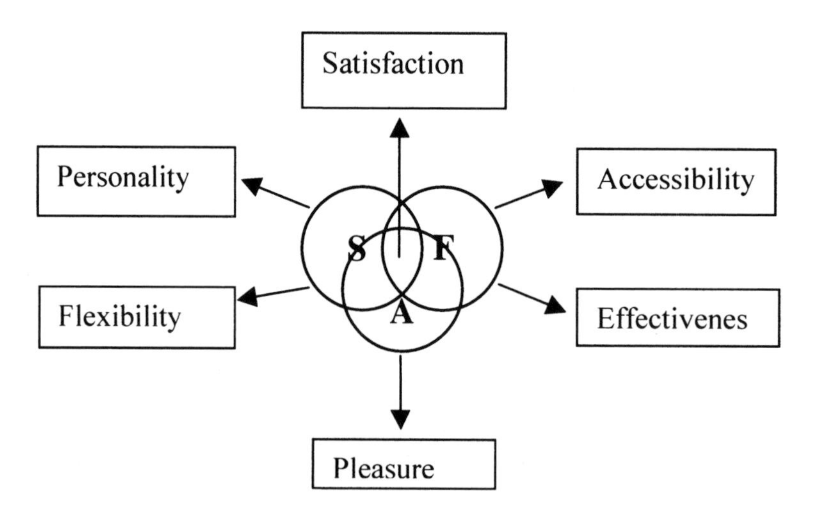

Figure 7.4 The involvement framework

Two basic functional qualities are considered relevant for successful social artifacts: Accessibility and Effectiveness. Accessibility refers to the effort required to interact with the social artifact, in terms of ease of communication and availability of the artifact via different technology channels mediating the same service(s). Natural Language (NL) appears to be a favourite interaction medium for social artifacts, since users can express their communicative intentions in a spontaneous way (Cassell *et al.* 2000). Moreover, NL interfaces have a tendency to elicit social representations (De Angeli *et al.* 1999).

Effectiveness reflects the key usability dimension. In this context, however, a number of indicators can be added to the traditional 'error rate'. The task of a social artifact is to provide a service whilst creating and maintaining a relationship with the user. Therefore, the 'return rate' and the 'amount of the bi-directional information flow' become fundamental variables to assess its overall success. The first refers to the number of times a customer interacts with the artifact. The second refers to the quantity and the quality of personal information disclosed by the user to the system, in addition to how many times the user acts on the information provided by the agent. The interaction time is a controversial variable. It refers to the temporal interval used to successfully exchange information with the system from the consumer's perspective. In the traditional usability model, time is often the major indicator of efficiency: the lower the time, the better the system. Social artifacts tend to invert this assumption: the higher the time, the better the system.

The major attribute underlying the aesthetic dimension is pleasure. It refers to the sense of enjoyment produced by all the aspects, and especially the physical aspect, of the interface. To maximise it, great care should be devoted to designing attractive and appealing interfaces. Moreover, the user should be encouraged to personalise the aspect of the social artifacts. Note that the aesthetic dimension overlaps with one of the basic social attributes, flexibility. Each form of successful communication is mediated by at least three factors: (a) the initial mental model of the conversational partner; (b) the evolution of this model over the course of communication; and (c) the ability to adapt to different partners. Hence, a social artifact must be capable of dynamically understanding the user and the context in which the interaction takes place, and in modifying its behaviour according to this knowledge.

A flexible system must be both adaptable (the user can directly personalise some features according to her/his preferences) and adaptive (the system is capable of modifying its behaviour according to a user model). Traditionally, adaptivity in human-computer interaction has been focused on cognitive factors such as the knowledge, plans, interests, and preferences of the user. Here, the concept is extended to include extra-rational factors such as the user's attitudes, personality, or emotional state. In this sense, flexibility implies social intelligence (i.e., knowledge and understanding of shared social values and norms) and emotional sensitivity (i.e., the ability to recognise the emotional state of the conversational partner). It is evident that huge technical challenges must still be overcome before computers will be capable of easily or reliably understanding user's moods and desires. Nevertheless, affective computing (Picard 1997) is growing at a tremendous pace, which is likely to shape the future of computing (Ball and Breese 2000).

In addition to being flexible, a social artifact must exhibit a personality (a stable set of traits defining its overall character). Providing the system with a personality can increase the user's control over the interaction, and help to optimise the learnability of the system. To fulfil these objectives, the artifact must exhibit a consistent and stable personality. Unexpected and unpredictable swings between different attitudes can disorient the users and create a strong sense of discomfort. The system's personality must be predictable, both over time and across channels, for the same task (e.g. using a mobile 'phone' or an ATM to access a financial account balance, or personal information service).

In conclusion, we have argued that the current adherence to and reliance upon the accepted usability framework is challenged when we encounter social artifacts. We have created an exemplar 'interface agent' in the form of Granny and undertaken an initial study, which reveals great differences in people's perceptions of such an artifact, when compared with other actors/agents providing the same service. Finally, as a consequence of our study and the existing consensus on social artifacts, we have attempted to sketch the foundations of a framework based upon the construct of involvement, describing its components, which we believe offers a pragmatic means of evaluation. The next stages for our work in this area will be in the creation and evaluation of further social artifacts, manipulating their characteristics, and conducting empirical studies, as we develop the thinking behind the involvement framework. Our paper has taken us from a consideration of 'efficiency' as a key element of accepted usability definitions, to 'pleasure' as a component of involvement, a brief tour hosted in part by Granny.

7.7 ACKNOWLEDGEMENT

We are most grateful to NCR Corp. for their support, our colleagues in the Knowledge Lab for their encouragement and help with this project, to Sandra Ensby for her impressive work on the Granny visualisation, to Govert de Vries (now at in2sports.com) for his early input and enthusiasm, and to NCR Self-Service Advanced Solutions for their assistance in creating the ATM-Granny prototype.

7.8 REFERENCES

Baber, C., Johnson, G. I. and Cleaver, D., 1997. Factors affecting users' choice of words in speech-based interaction with public technology. In *International Journal of Speech Technology*, **2**, 1, pp. 45-60.

Ball, G. and Breese, J., 2000. Emotion and personality in a conversational agent. In

Embodied conversational agents. Cassell, J., Sullivan, J., Prevost, S. and Churchill, E. (Eds), The MIT Press, Cambridge, Mass.

Bernsen, N.O., Dybkjær, H. and Dybkjær L., 1998. *Designing interactive speech systems*. Springer Verlag, London.

Card, S., Moran, T. and Newell, A., 1983. *The psychology of human-computer interaction*. Erlbaum, Hillsdale, NJ.

Cassell, J., 2000. Embodied conversational interface agents. In *Communications of the ACM*, **43**, 4, April 2000, pp. 70-78.

Cassell, J., Sullivan J., Prevost, S. and Churchill, E., 2000. *Embodied conversational agents*. The MIT Press, Cambridge, Mass.

Costa, P.T. and McCrae, R.R., 1992. *NEO PI-R: Professional Manual*. Odessa, FL: Psychological Assessment Resources.

Coventry, L. and Johnson, G.I., 1999. More than meets the eye: Usability and iris verification at the ATM interface. In *Human-Computer Interaction: Interact 99*, Proceedings, **2**. Brewster, S., Cawsey, A. and Cockton, G. (Eds), IOS Press/IFIP, pp. 151-156.

De Angeli, A., 1997. *Valutare i sistemi flessibili: un approccio globale alla HCI*. Ph.D. Thesis, Psychology Dept., University of Trieste.

De Angeli, A., Gerbino, W., Nodari, E. and Petrelli, D., 1999. From tools to friends: Where is the borderline? In *Proceedings of the UM'99 Workshop on Attitude, Personality and Emotions in User-Adapted Interaction*. Banff, Canada, June 23, pp. 1-10.

De Angeli, A., Matera, M., Costabile, F., Garzotto, F. and Paolini, P., 2000. Validating the SUE Inspection Techniques. In *Proceedings of AVI2000*, June 23-27, Palermo, Italy.

Frøkjær, E., Hertzum, M. and Hornbæk, K., 2000. Measuring usability: Are effectiveness, efficiency, and satisfaction really correlated. In *CHI 2000 Conference Proceedings*, ACM Press, NY, pp. 345-352.

ISO - International Standard Organisation, 1991. ISO 9241: *Ergonomics requirements for office work with visual display terminal (VDT)* - Parts 1-17.

Johnson, G.I., 1993. Spatial operational sequence diagrams in usability investigations. In *Contemporary Ergonomics 1993*. Lovesey, E.J. (Ed.), Taylor & Francis, London.

Johnson, G.I., 1995. Consumer reactions to voice at the ATM. In *Proceedings of the '95 Annual AT&T Human Factors & Behavioural Science Symposium*. AT&T Bell Labs, Holmdel, NJ, pp. 54-58.

Johnson, G.I., 1996. The usability checklist approach revisited. In *Usability evaluation in industry*. Jordan, P.W., Thomas, B., Weerdmeester, B.A. and McClelland, I.L. (Eds), Taylor & Francis, London.

Johnson, G.I., 1996b. Exploring novel banking user interfaces: Usability challenges in design and evaluation. Invited address, IEEE: *Interfaces – The leading edge*, C5 Group – Human Computer Interaction; Digest 96/126.

Johnson, G.I. and Briggs, P., 1994. Question-asking and verbal protocols – Qualitative methods of usability evaluation. In *Qualitative methods in organisation research: A practical guide*. Cassell, C. and Symon, G. (Eds), Sage Publications, London.

Johnson, G.I. and Westwater, M.G., 1996. Usability and self-service information technology: Cognitive engineering in product design and evaluation. In *AT&T Technical Journal*, Jan/Feb. AT&T, NJ, pp. 64-73.

Jordan, P.W., Thomas, B., Weerdmeester, B.A. and Clelland, I.L., 1996. *Usability evaluation in industry*. Jordan, P.W., Thomas, B., Weerdmeester, B.A. and McClelland, I.L. (Eds), Taylor & Francis, London.

Kelly, K., 1998. *New rules for the new economy*. Fourth Estate, London.

Lynch, P., De Vries, G., Johnson, G.I. and Waller, M., 1999. *Financial interfaces with personality*. Paper presented at the 3rd I^3 Conference, Affect in Interaction, October 20-

22nd, Sienna.
Lynch, P., Emmott, S.E. and Johnson, G.I., 1999. The NCR Knowledge Lab. In *Human-Computer Interaction: Interact 99, Proceedings*, **2**. Brewster, S., Cawsey, A. and Cockton, G. (Eds), IOS Press/IFIP, pp. 229-230.
Moon, Y. and Nass, C., 1996. How "real" are computer personalities? Psychological responses to personality types in human-computer interaction. In *Communication Research*, **23**, 6, pp. 651-674.
Morkes, J. Kernal, H.K. and Nass, C., 1999. Effects of humour in task-oriented human-computer interaction and computer-mediated communication: A direct test of SRCT theory. In *Human-Computer Interaction*, **14**, pp. 395-435.
Nass C., Moon Y., Fogg, J., Reeves, B. and Dryer, D.C., 1995. Can computer personalities be human personalities? In *International Journal of Human-Computer Studies*, **43**, pp. 223-239.
Nass, C., Moon, Y. and Carney, P., 1999. Are respondents polite to computers? Social desirability and direct responses to computers. In *Journal of Applied Social Psychology*, **29**, 5, pp. 1093-1110.
Nass, C., Moon, Y. and Green, N., 1997. Are computers gender-neutral? Gender stereotypic responses to computers. In *Journal of Applied Social Psychology*, **27**, 10, pp. 864-876.
Nass, C. and Lee, K.M., 2000. Does computer-generated speech manifest personality? An experimental test of similarity-attraction. In *CHI 2000 Conference Proceedings*, ACM Press, NY, pp. 329-336.
Nielsen, J., 1993. *Usability engineering*. Academic Press, San Diego, CA.
Norman, D.A., 1991. Cognitive artifacts. In *Designing Interaction: psychology at the Human-Computer Interface*. Carroll, J. M. (Ed.), University Press, Cambridge.
Picard, R., 1997. *Affective computing*. The MIT Press, Cambridge, Mass.
Reeves, B. and Rickenberg, R., 2000. The effects of animated character on anxiety, task performance, and evaluations of user interfaces. In *CHI 2000 Conference Proceedings*. ACM Press, NY, pp. 49-56.
Reeves, B. and Nass, C., 1996. *The media equation*. Cambridge University Press, New York.
Westwater, M.G. and Johnson, G.I., 1995. Comparing heuristic, user-centred and checklist-based evaluation approaches. In *Contemporary Ergonomics 1995*. Richardson, S. (Ed.), Taylor & Francis, London.
WordNet 1.6: available via http://www.cogsci.princeton.edu/

CHAPTER EIGHT

The Scenario of Sensory Encounter: Cultural Factors in Sensory-Aesthetic Experience

ALASTAIR S. MACDONALD

Course Leader, Glasgow School of Art, Product Design Engineering,

167 Renfrew Street, Glasgow G3 6RQ, Scotland, UK

'Man has no Body distinct from his Soul; for that called Body is a portion of Soul discerned by the five senses, the chief inlets of Soul in this age' (Blake c. 1790).

8.1 INTRODUCTION

This chapter discusses the cultural values of sensory phenomena. The concepts of 'aesthetic intelligence' and 'sensory encounter' are used to assist this discussion and our understanding of product preference.

8.2 AESTHETIC INTELLIGENCE AND THE SCENARIO OF SENSORY ENCOUNTER

Aesthetics

Although our culture is largely visually-orientated and our first encounter with an object is invariably visual, the original Greek *aisthetika* meant 'that which is perceptible through the senses', so an inclusive definition of aesthetics is concerned not just with visual form, colour, or texture, but with understanding and predicting the effects of information from all the senses on human perceptions and cognition. The sensations which are aroused from sight, touch, taste, smell, hearing, balance, movement and muscular effort, all help to form an aesthetic appreciation of an object or environment. Each sense is finely tuned: our senses are extremely discriminating and able to distinguish, often subconsciously, very subtle detail.

Human values and aesthetic intelligence

Our use of language reveals that when physical qualities are being described, the language also portrays human values. For instance, weighty language is that which carries authority, in distinct contrast to its opposite 'lightweight' – the sense that something has little substance. Physical qualities have somehow acquired the ability to express, through attributed value, a human orientation towards life '...our use of physical terms to exemplify ... values in our lives shows that they have a sensory value in our consciousness' (Fukasawa 1995). To help us discuss the idea that we possess an innate, (subconscious) ability to perceive 'human' values in products, it would be useful to acknowledge that we all possess what might be called 'aesthetic intelligence', i.e. the capacity for perceiving, comprehending and utilising a language of aesthetic sensibility (Macdonald 1999b).

The scenario of sensory encounter

To illustrate this concept of aesthetic intelligence it would be useful to develop a scenario of the sensory events which occur when one encounters an object, a 'scenario of sensory encounter', e.g. with a private car. On approaching the car one's initial impressions, formed visually, will lead to attraction, indifference or dislike, depending on one's own preferences for colour, form, and detail. As one opens the car door, one subconsciously judges its weight and quality by feel and sound: does the door-hinge feel secure, and does the door catch reveal a satisfyingly solid sound as it closes? One's tactile and auditory senses have come into play. As one slips into and sits in the car comfortably, or perhaps with some discomfort, one becomes aware of the smell of the material – of old patinated leather, or is it a more acrid smell of torn leatherette and other harsh plastics? As one drives away one senses its powerful acceleration, or perhaps a more sluggish response. Finally, there is the 'throat' of the exhaust, or maybe just a tinny hollow rattle. What has just occurred is a process of 'sensory encounter', a process in which value judgments are made, consciously or subconsciously, from information perceived through the senses. This example describes a hypothetical scenario for someone with full sensory faculties. How would it differ for someone with, say, a sight or hearing deficiency, or with mobility or agility problems? Their 'sensory encounter' would differ markedly from our own: already this suggests that the world might be experienced very differently for people with different sensory capabilities.

8.3 DREYFUSS' PARADIGM

In the field of human factors, one exemplary mapping of the human body's tolerance of physical agents is Dreyfuss' Environmental Tolerance Zone (ETZ) which set the paradigm for ergonomists in mapping the robustness of the body and its senses measured on a 'comfort-to-tolerance' scale. His diagram (Dreyfuss 1967) (Figure 8.1) shows a plan view of a schematic individual within two concentric circles delineating the comfort and tolerance zones for a range of environmental factors which include gases (oxygen, carbon dioxide and carbon monoxide), light, noise, vibration, shock waves, ultra-violet and atomic radiation, electricity, acceleration, pressure, temperature, humidity and heat loss. Dreyfuss describes the first circle as the bearable zone limit and the band between the two circles as indicating the zone from comfort to the tolerance limit '... outside this limit great discomfort or physiological harm is encountered'. Other factors not shown in Dreyfuss' diagram but which he states must be considered are infrared radiation, ultra-sonic vibrations, noxious gases, dust, pollen, chemicals and fungi.

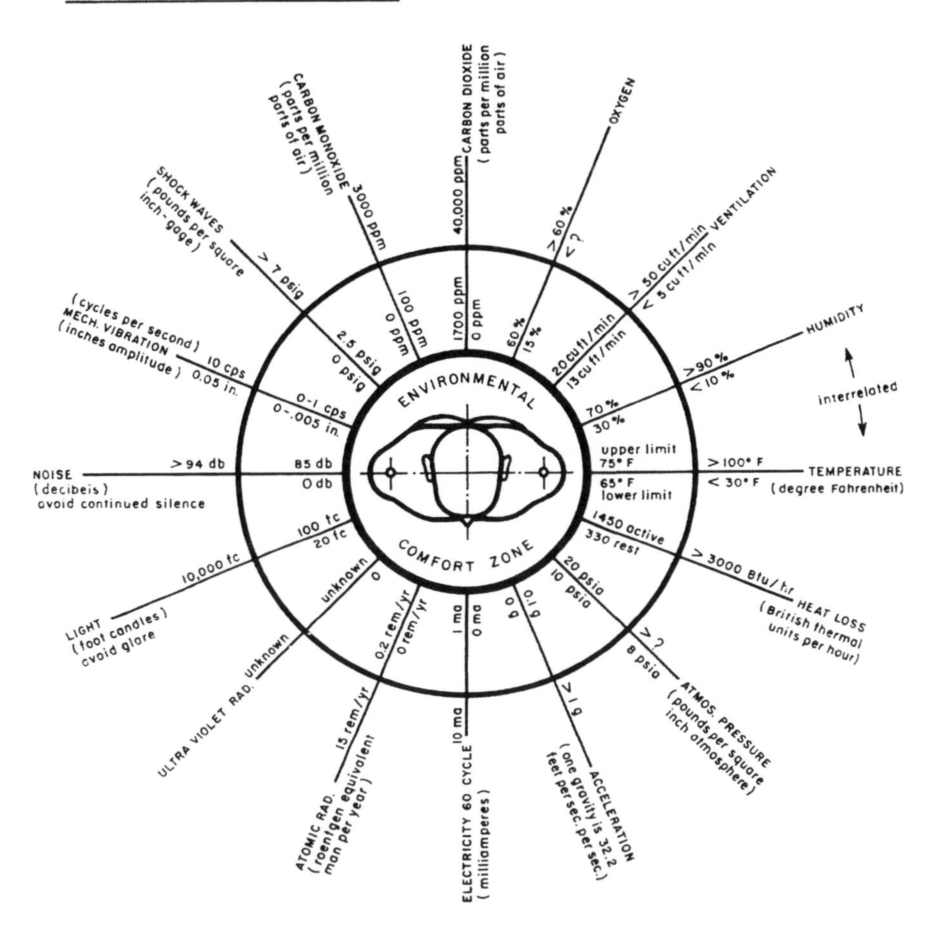

THE BAND BETWEEN THE CIRCLES INDICATES THE ZONE FROM COMFORT TO THE TOLERANCE LIMIT. OUTSIDE THIS LIMIT GREAT DISCOMFORT OR PHYSIOLOGICAL HARM IS ENCOUNTERED. OTHER FACTORS NOT SHOWN AND TO BE CONSIDERED ARE: INFRA-RED RADIATION, ULTRA-SONIC VIBRATIONS, NOXIOUS GASES, DUST, POLLEN, CHEMICALS & FUNGI.

Figure 8.1 Dreyfuss' Environmental Tolerance Zone
(reproduced courtesy Henry Dreyfuss Associates)

The ETZ crucially alerts us to environmental factors which may, at best, reduce our efficiency in the workplace, or at worst, cause us harm or even death. Dreyfuss' schematic model has been adopted, extended and refined, for instance by Finch and Stedmon (1998) who have developed a taxonomy of stressors in response to the complexities of stress in the operational military environment. There now exists a substantial body of knowledge and legislation, embedded in directives and standards to ensure optimum environmental conditions prevail that safeguard us from physical stress and harm.

With reference to the scenario of sensory encounter above involving the car, the level of noise from the exhaust, the temperature and humidity of the interior, the level of vibration as the car travels along the motorway, the glare of light entering the driver's field of vision or from the instrument panel would be easily embraced by Dreyfuss' model. But valuable though Dreyfuss' diagram has been, it is inadequate for explaining some of the territory revealed in the 'scenario of sensory encounter', at both a physical and at a cultural level. The environmental agents that Dreyfuss describes are purely physical phenomena for 'able' individuals. It is perhaps not surprising that these phenomena are ones we have learnt to describe and codify with relative ease: as biological beings we have evolved with many of these environmental agents over millions of years. Wengenroth (1999) argues that the codification of these phenomena is a process of reductionism '... reducing the complexity of an object by creating a simplified abstract representation'. This process has resulted in a narrowing or exclusion of the full complexity and subtlety of other perhaps 'softer' or more human values that one needs to embrace when considering human factors for products, and that accompany those phenomena we have already learned to encode.

8.4 DEVELOPING A RICHER MAPPING

The traditional model of human factors, according to Fulton's (1993) critique, is one which concentrates on harmful or efficiency-reducing factors. She states that this approach does not encompass some of the more positive factors which would provide enhancements for those working on tasks, within environments, at interfaces, or using products. Fulton says 'the human factors profession was actually developed specifically to respond to problems and deficiencies ... the new challenge is to make products and environments which do better that keep us from getting sick, damaged or irritated ... [they] can enhance the quality of our physiological experience'. The logical extension in one direction of Dreyfuss' scale of 'comfort' and 'tolerance' is towards 'pain' and, in the extreme, to 'death'. Jordan (1997) extends the scale in the opposite direction to encompass 'pleasure' as a positive factor and points the way towards other issues discussed below.

Mapping the senses

The work of the Canadian neurologist Penfield reveals, through his sensory 'homonculus', the proportional representation of sensations obtained from different parts of the body's surface onto the right cerebral cortex (Blakemore 1976). This sensory 'homonculus' reveals a peculiar make-up of the acuity of sensitivity to touch in different parts of the body, e.g. enlarged lips and fingertips (Figure 8.2). As a culture which presumes that vision is the largely dominant sense, it perhaps comes as a surprise that, in this form of human mapping, the mouth and the thumb appear so large until one discovers, for instance, that the fingers are one of the most sensitive parts of the surface of the skin, with up to 135 sensors per square centimetre at the fingertip.

Figure 8.2 Penfield's human homonculus and those of other species (adapted from Kandel, Schwartz and Jessell (1991))

Inclusive design considerations

The 'homonculi' for different species, e.g. rabbit, cat or monkey each reveal a particular sensory emphasis (Kandel, Schwartz and Jessell 1991). In humans, when one sense is for whatever reason absent or lost, the cortex compensates through allocating proportionally more of the cortex to another sense, or senses (Sacks 1995). If one is to develop a more inclusive approach to design, then one requires a model which describes and accommodates people's individual physical differences and needs. With particular reference to the senses, this inclusive approach is especially important when considering, e.g. the design of products for an ageing population, where the acuity of the senses change as one ages, or those with sight, hearing, or mobility impairments. Pirkl (1994) has assembled physiological data about the deterioration, with age, of each of the senses of vision, touch, hearing and balance. Loss of visual or auditory acuity, or the diminution of the sense of smell often accompany the ageing process. Pirkl's demographic charts (Hewer 1996) illustrate this process in a graphic manner, showing the translation of quantitative data (from the domain of the physiologist) into terms of reference which designers can use, i.e., design guidelines and product design specifications. For example, the fluid in the eyeball discolours with age, turning more yellow, affecting colour perception. Designers require to acknowledge this physiological change through their choice of contrast and colour, for instance in the design of graphics used by older people. This demographic tool is useful because it understands the needs of the end user of the information, i.e. the designer, and is written in an accessible way avoiding obscure terminology. Pirkl's demographic charts help to develop a more individual profile of sensory capability in the way that Dreyfuss' ETZ cannot.

But this only deals with some of the issues: we are not only physical and biological beings, but also socio-cultural beings, i.e. *human* beings. Given that as we have been cultural beings for only a relatively short part of our biological evolution (Baxter 1999) there is an argument that we have not yet been able to easily encode, e.g. cultural phenomena to the same extent as physical phenomena. What is also required is an accessible and useful way of articulating the human or cultural *value* of information perceived through the senses. It is this area of the *meanings* and *values* that we attribute to physical sensory phenomena that the remainder of the chapter will now address in more detail.

8.5 REVISITING THE SCENARIO OF SENSORY ENCOUNTER

In terms of the car scenario described earlier, it would be understating the case to say that the process of sensory encounter is very complex: Restak (1995) usefully describes a four level cognitive process involving the visual and association cortexes during the process of visual perception. To highlight what might be happening during this process of sensory encounter, it would be useful to rerun the scenario to explore each of the senses in turn and in some more detail.

Sight and empathy

The first sense normally to be engaged in an 'encounter' is sight. Our senses can be extremely discriminating in their ability to detect subtle detail. Seymour (1996) demonstrated just how finely tuned is our visual sense by showing three slides of a famous supermodel's face. In each successive photograph, the size of the nose was altered by only one millimetre. By the third photograph, the subtle but unmistakable change was clearly detectable. One makes similar judgments on viewing a product, where a process of personification is evident if one analyses the language used to discuss the physical features of a car: some language is distinctly human (anthropomorphic) or animal (zoomorphic) in reference. Perceived 'body tone' is important: bodywork details are referred to as 'lean', or 'perky'. The features, e.g. the shoulders, on some 4-wheel drive sports utility vehicles can look distinctly muscular, as if they have been overdeveloped in the gym. Similarly, on the front elevation, the configuration and shape of headlights, grills and bumpers help determine its characteristic 'face' and thereby one's emotional response to the vehicle (Macdonald 1999a). Smets (1989) explored the perceived 'age' of products by comparing, say, the morphology of a vehicle's shape with a young or old human or animal face. Glancey (2000) uses feline and other animal analogies, describing Jaguar's new F-Type as having '... a high waistline, tall wheels tucked into swooping, voluptuous bodywork, pert, boat-like tails, aggressively curved front wings that appear to be pawing their way through the air with the grace, pace and muscularity of a big cat'. While some cars' details will appeal to individuals at a purely personal level, others, such as the finned Cadillacs of the 50's, or the Citroen DS will undoubtedly reflect the spirit of the times in their general form and configuration. Visual features embody codes, manifest as styles, which act as shorthand for sets of social or cultural values.

Touch: shifts in cultural values

However, if one's visual sense is diminished or absent, through disability or the ageing process, then this 'sensory route' will be missing, other senses will compensate and the individual's world will be mapped in a different way. The next sense engaged in the car scenario was touch. In European culture, there has been a distinct correlation between weight and value, where we have become used to associating the weight of a particular type of object with a particular value (Macdonald 1999a).

A way to illustrate this is to develop a simple 'cultural matrix' to map material qualities against cultural values (Figure 8.3). Using the example of a personal stereo, and giving an audience a choice of a thin, heavy object, or a thick, light object; experiments have revealed a consistently strong preference for the thin, heavy personal stereo.

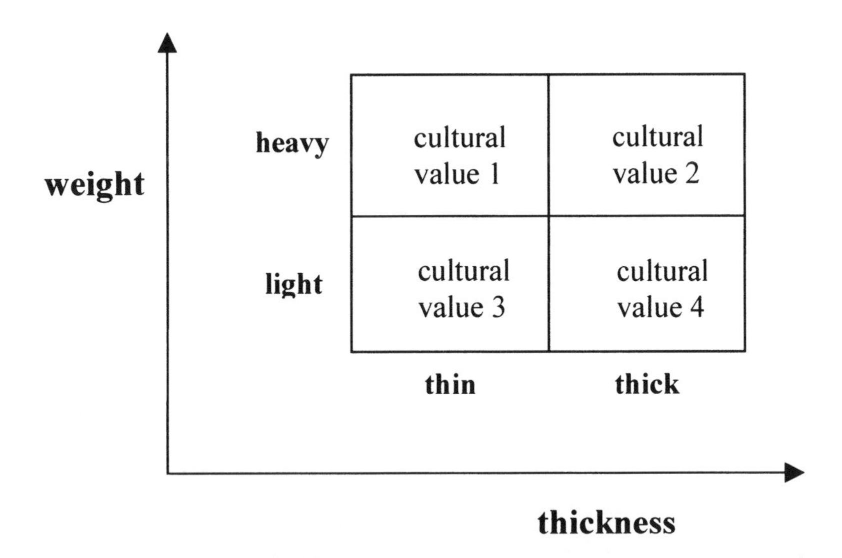

Figure 8.3 Cultural values and physical properties

Another illustration of this cultural disposition occurred when injection moulded telephones replaced those made from the heavier wood flour-filled phenolic resins: initially they were perceived to be too light and flimsy. To understand this preference, Manzini (1989) states that, '... in European culture ... the value of weight (and the equations that derive from it, such as 'weight' = quality, longevity, solidity, safety), is rarely rejected, and has certainly left its mark on the quality of our physical surroundings'. We had become used to associating the weight of that particular object with a particular value. Using this example it is possible to say that one can correlate particular material qualities with distinct cultural values, i.e. that a quality personal stereo is expected to have some weight and compactness (signalling a technologically advanced product). However, cultural values evolve, and whereas the value standard with reference to cars used to be 'weight = strength + safety', one's perceptions require reeducating when shifts in material performance and manufacturing technology allow, for example, one car manufacturer to pose the question '*how can light be strong?*' (Toyota 1999) and then to justify that its new model is strong due to a tough new body structure. This example (Figure 8.4) serves as an illustration that a new cultural norm may be in the process of being established.

Figure 8.4 Toyota Yaris advert
(reproduced courtesy Toyota (GB) PLC)

Another issue to consider in discussing cultural values is demographics: one generation may have its own set of associations and corresponding values for the material properties of products which are distinct from another (as in the telephone example above - bakelite models versus mobiles). The value standards for these older groups may evolve less slowly than younger, emerging groups. The differing levels of acuity which accompany the ageing process, such as loss of tactile, pressure and thermal sensitivity, also condition one's perception of a product. As Penfield's homonculus reveals, tactile qualities have been largely undervalued in design and this particular territory offers good opportunities for enhancing the design and attractiveness of products.

Hearing and smell: emotional cues

In the car scenario, sound was discussed – the sound of the door opening, or the sound of the exhaust. Recent research into the treatment of the hearing difficulty tinnitus has made great progress through recognition of the 'emotional label' associated with each and every sound we hear and learn the meaning of 'which may change from time to time according to how we feel in ourselves and the context in which we hear it' (Hazell 1999). This is due to the discovery that patterns of sound are detected by subconscious filters in the hearing pathways, and the conditioned response triggers activity outside the auditory system where there are large numbers of connections with the limbic system (concerned with emotion and learning). The nature of the design and engineering specification of a computer keyboard can result in either a pleasant or irritating ambient clatter of keys in the office environment, despite alternatives being ergonomically correct and possessing parity of functional performance. This may be due, in part, to personal preference for a sound, one's state of mind, or the fact that we are predisposed to some types of sounds over others. Kansei engineering recognises the importance of attractively tuning sound emissions in car exhausts (Nagamachi 1995), tapping into a collective value system of what a quality sports car should sound like. If a product relies heavily on sound cues, the value of this sensory route would be diminished, or totally absent, for those with various forms of hearing impairments which would affect sound intensity, pitch, and directional hearing.

In a similar vein, recent research at the Defence Evaluation and Research Agency has suggested that 'odour cues' might improve the recall of material learnt in a particular environment. 'It is believed that memory will be significantly improved if there is a match between the context in which it is learnt and the context in which it is to be retrieved' (DERA 1999). Smell is different from other senses as it goes straight into the brain's limbic system, which suggests that strong emotional connections are made with smells. There is a need to consider the importance of these emotional labels which occur during the process of perception and cognition of products. While, on the one hand, these may be more associated with an individuals' particular experiences rather than broadly held values, on the other there may be odours which are broadly attractive or repulsive through long evolutionary associations. In this context 'new car smell' may be a feature to be given some consideration. Physiological changes condition our sensory perceptions which in turn affect our emotional condition. With age one's sense of smell tends to become less acute: how would a designer of a supermarket tackle the issue of shopping as a sensory experience to accommodate older people?

Speed and acceleration: balance and movement

Moving away in the car, the rate of acceleration affects the adrenaline levels in the body, increasing alertness and speed of reaction, a residual 'fight or flight' effect. Involuntary movement caused by poor suspension, for example, will jolt the body and affect the ear's balance mechanism and equilibrium. Various modes of transportation can be unpleasant for some people due to the (poor) engineering specification upsetting the inner ear balancing mechanism. As elderly people experience a gradual decrease in their sensory functions, their sensory losses and changes in stature tend to lead to a loss of balance. In terms of design this may affect, for example, the height of the car seat, and in their homes the location of chairs and other unsecured objects in stable positions in room settings.

8.6 STRUCTURING SENSORY EXPERIENCE

As has been illustrated during the 'scenario of sensory encounter' and the discussion of each of the senses, there is a complex layering of relationships with the object, whereby physical qualities perceived by the senses are linked with value judgments and the emotions. Hofstede's (1991) model of culture, which discusses different layers of preference or cultural level, proves useful in our discussion. According to Hofstede, culture is, 'the collective programming of the mind which distinguishes the members of one group or category of people from another'. Hofstede's model of human mental programming suggests 'three levels of uniqueness': those values which are innate or universal, i.e. part of our human nature; those specific to an individual; and those specific to a group or category. He describes values as 'broad tendencies to prefer certain states of affairs over others'.

This author's interpretation of Hofstede's model suggests that the innate level response reflects how the way our brains and senses have evolved biologically influence the particular meanings and values we give to information perceived through our senses – a kind of evolutionary or ecological approach to sensory perception revealed by Penfield's 'homonculus'. The personal level response will be due to both one's physical make-up (young, old, able, disabled etc.) and a personal history of circumstances and experiences which might have accompanied sensory encounters in the past and how these have been 'emotionally cued' or 'emotionally labelled' by the individual. Finally, one's socio-cultural level response is due to factors which shape values shared with a broader sector of the population, the kind of value association which Manzini describes. Each and all of these complementary sets of responses requires to be considered as they combine to shape our value judgments when choosing or responding to products.

8.7 CONCLUSION

Dreyfuss' ETZ provides a valuable, though partial model of the impact of physical agents on our bodies and our senses. What has been lacking in human factors is a method which considers the sensory and emotional values associated with physical qualities. Penfield's 'homonculus' gives us a visual clue for the relative sensitivity of e.g. touch, but one needs to remember that Penfield's model has to be adapted for the ageing process, and for those with sensory impairments. Hofstede's model helps differentiate three levels of response which separate out inherited from learnt levels of response, and the concept of aesthetic intelligence is useful in raising the issue of tacit knowledge gained, or inherited, through our senses.

The use of the 'scenario of sensory encounter' as a method would help designers and ergonomists structure an approach which highlights the range and extent of the different sensorial and cultural profiles of different sectors of the population, to design as fully as possible for all the senses and for a broad range of capabilities. It would also remind them of the layers of response to be considered – the innate, the individual and the cultural. Use of this method would allow a far broader, more inclusive consideration of human factors with which to enhance the appropriateness and attractiveness of products, tasks, interfaces and environments.

8.8 ACKNOWLEDGEMENT

This paper was developed from *Sensory encounter: the codification of 'soft' qualities* (Macdonald 2000) produced for the Ergonomics Society Annual Conference, and published in Contemporary Ergonomics 2000 by Taylor & Francis Ltd, London.

8.9 REFERENCES

Baxter, S., 1999. *Deep design.* Duncan of Jordanston University, argued this point during his lecture. Glasgow School of Art, 2nd November, Glasgow.

Blake, W., 1975 (c. 1790). *The marriage of heaven and hell.* Oxford University Press,

London.
Blakemore, C., 1976. *Mechanics of the mind 79-80*. BBC Reith lectures 1976, Cambridge University Press, Cambridge.
DERA, 1999. Website Nov 99 http://www.deva.gov.uk/html/news/devanews/smell.htm
Dreyfuss, H., 1967. *The measure of man: human factors in design*. Whitney Library of Design, New York.
Finch, I.M. and Stedmon, A.W., 1998. 'The complexities of stress in the operational military environment'. In *Contemporary ergonomics 1998*. Hanson, M.A. (Ed.), Taylor & Francis, London, pp. 388-392.
Fulton, S., 1993. New human factors: physiology and design. In *American Center for Design Journal*, 7, 1. Chicago, Illinois.
Fukasawa, N., 1995. *The theory of hari in design*. Ideo, San Francisco, CA.
Glancey, J., 2000. Cat's whiskers. In *The Guardian G2*, 17 January 2000, pp. 8-9.
Hazell, J., 1999 *Tinnitus retraining therapy based on the Jastreboff model*. March 1999, http://www.ucl.ac.uk/~rmjp101/tin2.htm
Hewer, S., 1996. *Incorporating age-related issues into design courses*. Design Age pilot teaching pack. Hewer, S. (Ed.), Various authors Royal College of Art, London.
Hofstede, G., 1991. *Cultures and organisations*. Harper Collins, London.
Jordan, P.W., 1997. Putting the pleasure into products. In *IEE Review*, November 1997, pp. 249-252.
Kandel, E.R., Schwartz, J.H. and Jessell, T.M., 1991. *The principles of neural science* (third edition). Elsevier, New York (illustrations of animal homonculi adapted from figures in Chapter 26).
Macdonald, A.S., 1999a. 'Aesthetic intelligence: a cultural tool'. In *Contemporary ergonomics 1999*. Hanson, M.A, Lovesey, E.J. and Robertson, S.A. (Eds), Taylor & Francis, London, pp. 95-99.
Macdonald, A.S., 1999b. 'Developing aesthetic intelligence as a cultural tool for engineering designers'. In *Proceedings of the International Conference on Engineering Design (ICED) Munich, 1999*. Lindeman, U. *et al.* (Eds), Technical University München, Munich, pp. 297-300.
Macdonald, A.S., 2000. Sensory encounter: the codification of 'soft' qualities. In *Contemporary Ergonomics 2000*. Ergonomics Society Annual Conference, Taylor & Francis Ltd, London.
Manzini, E., 1989. The Material of Invention: Materials and Design. *The Design Council*, **97**, 198, London.
Nagamachi, M., 1995. Kansei engineering: a new ergonomic consumer-orientated technology for consumer development. In *International Journal of Industrial Ergonomics*, **15**. Nagamachi, M. and Imada, A.S. (Eds), pp. 3-11.
Pirkl, J.J., 1994. *Transgenerational design; products for an aging population*. Van Nostrand Reinhold, New York, pp. 41-51.
Restak, R., 1995. *Brainscapes*. Hyperion, New York, pp. 22-23.
Sacks, O., 1995. *An anthropologist on mars*. Picador, London.
Seymour, R., 1996. Business masterclass. Glasgow International Festival of Design, Glasgow.
Smets, G.J.F., 1989. Perceptual meaning. In *Design Issues*, **5**, 2.
Toyota, 1999. Advert for the Toyota Yaris.
Wengenroth, U., 1999. From 'science versus art' to 'science and art' - reflexive modernisation in engineering'. In *Proceedings of the International Conference on Engineering Design (ICED) Munich, 1999*. Lindeman, U. *et al.* (Eds), Technical University München, Munich, pp. 1657-1664.

CHAPTER NINE

Emergence of Pleasure: Communities of Interest and New Luxury Products

Communities of Interest and New Luxury

Products: Emergence of Pleasure with Products

PATRICK REINMOELLER

Rotterdam School of Management, Department of Strategy and

Environment, Erasmus University, PO Box 1738,

3000 DR Rotterdam, The Netherlands

9.1 INTRODUCTION

'What is luxury?' is a question anyone can answer quickly, but few answers are entirely satisfactory. Literally, luxury means enjoyment of a rich, comfortable lifestyle and the indulgence of pleasure. As a business concept luxury has gathered a lot of attention in strategy and design. Recent mergers and acquisitions through companies such as LVMH, SMH, Gucci or Prada are testimony to the use of luxury as a strategic position (Nihon Keizai Ryutsu Shinbun 1999; Slywotzky and Morrison 1997). These companies have accumulated a set of different luxury brands in various product categories including apparel, watches, shoes and food.

The economic rationale behind this strategy is the superior profitability of luxury products (Table 9.1). Often the high margins are used to justify the acquisition of different luxury brands, or companies producing luxury products, despite the costs of sustaining different images and distribution channels. These developments, and the efforts to create new luxury experiences, show that luxury is becoming an important category in itself.

Table 9.1 Luxury Zone: Profitability in the Fashion Industry

	No. of Companies	**Sales (billion Lira)**	**Average Size, Sales**	**Profit %**
Individual companies	160	17.121	107	3.7
Company groups	30	30.423	1.014	6.1
Listed companies	12	14.726	1.227	8.2
Luxury companies	37	11.879	320	8.7
Top 10 luxury companies	10	9.212	921	11.0

Adapted from Il Sole 24 Ore 1999

Luxury is a complex and ambiguous concept. Alleres (1997) reports that Dior, Chanel and Yves Saint-Laurent are the most prestigious and well-known luxury brands, ranked according to share of awareness and share of attention by customers. Surprisingly, Benetton and Chevignon rank close to the top and Cartier, Louis Vuitton or Piaget rank much lower. This complexity and ambiguity does not only exist in the fashion and textile industry. In the car industry, for instance, different manufacturers, including BMW, Daimler Chrysler, Honda, Rolls Royce, Toyota and VW, target the market for luxury vehicles.

Luxury products provide unique value to customers and yet they are often criticised. The unique value includes prestige and, more important but often forgotten, pleasure (Kuethe and Reinmoeller 1999). Luxury products provide pleasure to such an extent that prices for luxury products can be very high. In the luxury industry the market constellation can be similar to a monopoly, where demand exceeds supply and manufacturers are able to sell the quantities they determine at prices they command. Such 'monopolies' are based on superior value creation.

Luxury products are often criticised as wasteful products. Owners pay premium prices and go to great lengths to enjoy luxury products and services. Luxury products are created by use of material, processes, packaging, distribution and promotion that exceeds the level of standard products to allow for pleasure and indulgence. The economist Frank (1999) interprets this effort beyond the average and reason as similar to Veblen's theory on conspicuous consumption, and reveals prestige seeking as an important reason for wasteful consumption. He describes social dynamics related to luxury products that limit free choice through peer pressure and misallocation of productive resources (Frank 1999).

These differences in understanding luxury products, as sources of pleasure and waste, show that luxury is a relative phenomenon. This paper explores the contingency of luxury as being a socially constructed pleasure and as a strategic opportunity to create value for customers. The main argument is that intellectual and physical experiences as well as social dynamics in communities of interest promote the emergence of luxury.

Several questions need to be answered. First, is luxury a public and socially shared, or a private and personal phenomenon? Second, is there a difference between luxury brands and luxury products? Third, are luxury brands and products limited to certain product categories?

The following section introduces the concept of luxury appreciation as a function of knowledge and context. In the third section communities of interest are introduced; complexity theory is used to understand the emergence of appreciation as a process of social construction. The fourth section presents four types of luxury; each is based on different knowledge processes in the context of communities of interest. The case study of the cellular phone category and brands illustrates the luxury types and conditions for emergence. The fifth part analyses the links between top-brands and luxury products. Implications for design strategy conclude the paper.

9.2 LUXURY APPRECIATION IS RELATIVE TO KNOWLEDGE AND CONTEXT

Historical, cultural, economic and individual contexts influence the appreciation of luxury. Appreciation of product experiences means to enjoy, understand and to judge products. Luxury products correspond to highest levels of appreciation; without appreciation there is no pleasure in luxury products. Such appreciation is a function of knowledge and contexts.

Knowledge influences luxury appreciation

Information affects knowledge by adding something to it or restructuring it (Machlup 1962). Because information is pervasive and accessible to everybody it is a commodity. Knowledge is context–specific, relational, dynamic, and is created through social interactions rooted in individuals' value systems and their commitment (Nonaka and Takeuchi 1995; Nonaka and Reinmoeller 2000).

There are two types of knowledge: explicit knowledge and tacit knowledge. Explicit knowledge can be expressed in formal and systematic language. It can be easily 'processed', transmitted and stored. On the other hand, tacit knowledge is highly personal and hard to formalise; it resides in the human mind and body. Subjective insights, intuitions, preferences and hunches are examples of such knowledge. Tacit knowledge is rooted in action, values and contexts. Therefore, it is difficult to communicate tacit knowledge to others. Both types of knowledge are complementary and the interactions between are key processes of knowledge creation (Nonaka and Takeuchi 1995).

Luxury appreciation involves both types of knowledge. Explicit knowledge on product specifications, for instance, does not justify high margins. The uniqueness of pleasure through luxury products lies in the tacit dimension. Tacit knowledge such as subjective pleasure and emergent meaning is based on direct experience and dialogue; it provides important value that can justify high margins (Reinmoeller 1996; 1997).

Contexts influence luxury appreciation

Luxury appreciation does also depend on context. Appreciation is an interaction process in context that involves more knowledge as it unfolds over time. Five kinds of contexts influence luxury appreciation: physical, temporal, natural, technological and social.

Physical proximity can determine experiences of luxury; being close to or moving towards something special can constitute luxury. The temporal perspective shows that luxury goods of the past, like tobacco, coffee or travel, are evaluated differently today. Technological change can accelerate changes of appreciation as, for instance, a comparison between the first cellular phones and current standards shows.

This paper focuses on the social context that is the key to appreciation of luxury. The dynamics of social consensus, shared value systems and beliefs go beyond prestige seeking; luxury appreciation emerges as socially constructed reality (Berger and Luckmann 1966). Social contexts determine the valuation of naturally and technologically unique products.

Luxury is socially constructed through repeated interaction between people sharing similar interests and knowledge. The creation of knowledge about a product and its valuation are continuing processes in social communities.

9.3 COMMUNITIES OF INTEREST AS SOCIAL BASE FOR LUXURY APPRECIATION

Communities of practice (Lave and Wenger 1991; Wenger and Snyder 2000) or communities of interest are groups of people informally bound together by distributed and shared expertise. The members share passion for products, qualities, ideas or other objects of interest. One prototype is the apprenticeship pattern describing how people acquire the skills and know how of their masters and peers. The interaction among members primarily supports mutual complementing of knowledge and continuous self-refinement (Lave and Wenger 1991). The members share interest in a domain. The purpose of such groups is learning and developing skills and knowledge. The members select themselves according to passion, commitment, and identification with the group's knowledge on the domain. Virtual communities meet regularly and interact directly with others that meet in chat rooms, networks or on the Internet (Rheingold 1993). Other communities are linked loosely by similar demographic data and indirect interaction, such as people who visit the same specialised markets.

Communities grow when interaction intensifies and widens; they decline when the special interest in the domain, the commitment of the members and interaction subsides.

Different communities have different cultures. While the underlying cultural differences between Asia, the US and Europe continue to be important, communities without boundaries such as professional communities of medical doctors, physicists, stock brokers, bankers or e-commerce entrepreneurs are becoming more important to individuals.

9.4 LUXURY AND EMERGENCE IN COMMUNITIES

Luxury is pleasure with products that emerges from communities. The communities construct knowledge and reality through social interaction (Wenger 1998). Such a process of co-creation involves a certain number of members who interact as in complex systems.

Complexity theory helps to understand self-organising properties and the inherent heterogeneity within such communities. The absence of central control is common to complex systems and communities of interest. Similar to complex systems, communities describe the emergence of meaning from the large number of members, their interactions and embeddedness in the environment (Granovetter 1985). Emergence is a product of coupled, context-dependent interactions. The interactions are non-linear; **the result impredictable; they cannot be infered by adding the behavioural effects of individual members.** But the complexity of emergence can be reduced by understanding the behaviour of single members taking into account the nonlinear interactions (Holland 1998). Thus we can understand highly complex emergent phenomena through the discovery of simple rules, the linear and non-linear interactions. Such emergence described in systems theory, theories of consciousness and complexity theory is based on competition and cooperation (Buetz 1997).

In a similar way public mental models emerge from activities and coupled interactions. The public mental models are based on large numbers of private, individual models that are iteratively altered through interaction with other people and the environment. A process of creation and adaptation helps shared mental models to emerge and evolve.

Shared mental models influence luxury appreciation. In communities the results of interactivity are a shared understanding and experiencing of products and emerging of appreciation of products. Shared understanding is important for an individual's appreciation of products.

9.5 LUXURY TYPES

Different knowledge processes (Nonaka and Takeuchi 1995) occur in communities of interest. The following luxury types are based on multiple-case study analysis and expert interviews in several industries including textile, consumer goods and entertainment. Two perspectives synthesise the results of in-depth case studies. The case of digital cellular phone systems in Japan and NTT Docomo 'I mode' illustrates both perspectives. The static perspective reveals four luxury types: emerging luxury, new luxury, luxury brands and luxury patterns (Figure 9.1). The dynamic perspective describes patterns of change between the four types (see section 9.6).

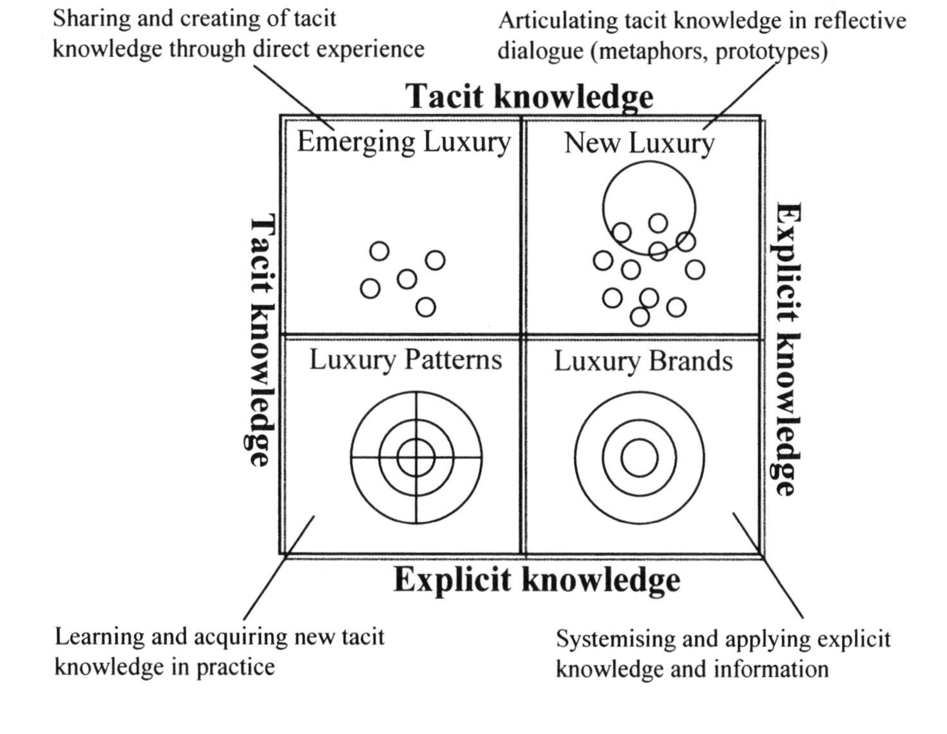

Figure 9.1 Four types of luxury: static perspective

Luxury patterns

Luxury patterns are based on the learning and internalising of explicit knowledge. In other words, in the developing of criteria by which products are appreciated as luxurious. Criteria for appreciation are often evaluation standards in the community. These luxury patterns include exclusivity, scarcity, high quality of materials, technology, price, aesthetics, manufacture, symbol and time (Kuethe and Reinmoeller 1999). These learned patterns of evaluation seem to be universal criteria; they are often linked to specific elements of design such as material, shapes, colours or ornaments according to the community.

Emerging luxury

Emerging luxury is often related to new experiences and new product categories. Sensing emerging luxury requires knowledge about the products, communities and contextual experience of use. Emerging luxury is based on sharing of tacit knowledge within small communities. The sharing of experiences of products in multiple contexts reveals new pleasures for the members of the community. This process of distributed accumulation of experiential knowledge is fundamental to emerging luxury; interation refines judgement and appreciation of these products.

New luxury

Articulation makes tacit knowledge of newly discovered pleasures and luxury products explicit within medium-size communities. By making tacit knowledge explicit, it can be shared with others more easily. Tacit knowledge is often articulated as metaphor, image, concept, diagram, narrative or prototype. Articulation facilitates appreciation of products and can create or establish new luxury products within communities. Further stimulation helps luxury to transcend the boundaries of communities.

The articulation of newly discovered luxury goods in community magazines, on websites or in chat rooms is becoming more important. It can expand the community. However, mouth-to-mouth communications on highly appreciated products in personal networks remain the most trusted source of knowledge.

Luxury brands

Brand image is built through a variety of factors, such as logos, product aesthetics, marketing campaigns, visuals and standards. Such sets help to communicate fast and efficiently within large communities. Brands and other forms of explicit knowledge are exchanged, reconfigured through symbols, documents or information networks. The communication of knowledge about brands is facilitated by the large amount of images and narrative available in print, broadcast and new media including the Internet.

Ironically, the more ubiquitous a brand becomes, the greater the danger that people will stop seeing it as a luxury brand. The luxury brand manufacturer must take care that the brand does not become seen as a commodity brand as sales increase.

9.6 CELLULAR PHONES IN JAPAN: 'I MODE'

When in 1979 NTT introduced mobile phones in Japan luxury patterns were used. These mobile phones were built into cars. This advanced technology was positioned as an expensive business instrument for executives and corporate use. The limited number of phones made the community of users an exclusive circle. The fundamental pattern was limited usage by the business elite. In 1987 NTT introduced the first mobile phones. Similarly pagers were positioned as communication tools for people in middle management. The use of the most advanced technology used for business purposes justified the indulgence. Later young teenagers and kids discovered their parents' pagers and began to use them for private communications with peers.

In 1989 DDI, a competitor of NTT, sensed opportunities in a growing market, entered the market for cellular phone services, and promoted the emerging luxury. In 1993 NTT began the digital phone service. Despite the elevated prices the growing community of users had begun to attract new companies and services. Before 1994 cellular phones were rented or leased to users, since then the phones are sold to the users. This has reduced the overall cost for users. It also created the possiblility to own phones and to enjoy the technology.

The cellular phone became a new luxury in the second half of the 1990s after the start of the PHS services, based on a different standard with much lower rates. The luxury of mobile communication was becoming more common. In 1996 the community of users had reached 10 million and cellular phones were established as a new luxury. The growing number of users created a lot of interest in the media and articulated the benefits and costs of mobile communication widely.

With the introduction of the second technological solution (PHS) for mobile communication, the number of different mobile services has grown. The category is established and expands market volume. In March 1999 the total number of cellular phone users reached 40 million, or one third of the Japanese population. The competition between technological standards and brands has intensified. Generic cellular phones and mobile communication have become common and the market leader NTT Docomo has a market share of 57.5 % and owns one of the best known luxury brands.

The establishment of competitive luxury brands is increasingly important. In digital services, for instance, the confrontation between two brands is notable. In 1998 the DDI group began to provide services in the European CDMA format. NTT Docomo then introduced the new 'I mode' service in February 1999. 'I mode' combines a cellular phone with Internet access. The development of 'I mode' Internet websites and exclusive contents has helped to establish 'I mode' as the luxury brand. In 14 months the number of 'I mode' users has reached 5 million. In order to strengthen the appreciation of 'I mode' services, NTT Docomo has developed new models for cellular phones in the future (Figure 9.2). They emphasise image processing through installed cameras and large colour displays for real-time video communication. NTT Docomo's challenge is to broaden the appeal of its product whilst maintaining a luxury image.

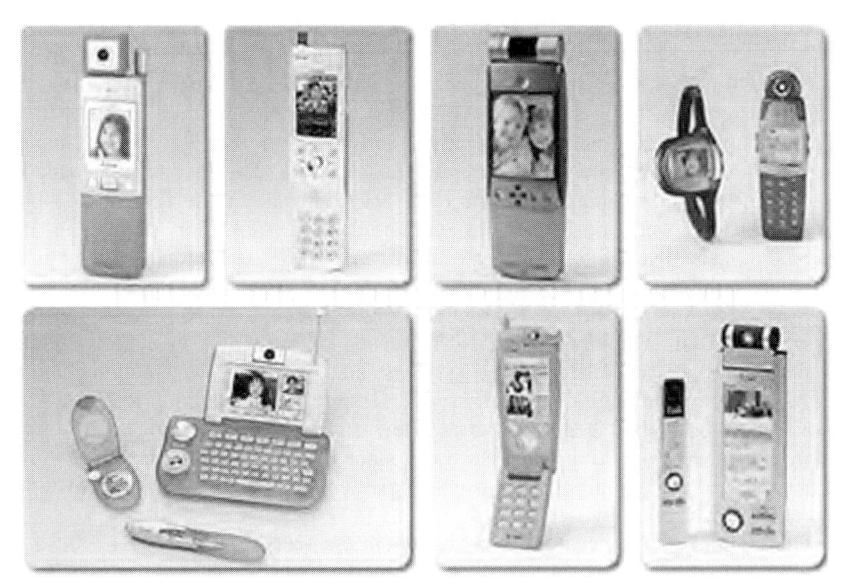

Figure 9.2 Prototypes for future NTT Docomo mobile communication (NTT docomo homepage)

9.7 PATTERNS, LUXURY AND BRANDS: THE DYNAMICS OF LUXURY

The dynamic perspective shows links between the different luxury patterns and evolutionary development over time. **These links point towards overtime**.

Luxury patterns are platforms for emergent luxury products. Adherence to established patterns first brings attention from core customers. If communities of interest begin to interact in relation to the product, luxury begins to emerge. Based on existing patterns new luxury can establish itself. Direct experience of the products by more members of the community, and complex interaction patterns among them, leads to appreciation of the products and recognition of them as luxury items. The case of NTT docomo shows that brands can maintain luxury appreciation by upgrading experiences.

The creation of luxury brands and differentiation often follows when new luxury categories are already established. If branding strategies and marketing techniques are successful the luxury brands become available easily. **Success in making luxury brands accessible, however, has to be balanced with avoiding that the products become mere standard or commodity**. The recent development of Starbucks in the US shows how replication changes the meaning of luxury brands. In the case of 'I mode' the position as leading value provider is not used to gain high margins but rather to increase diffusion.

Luxury brands can decline when they are perceived as mere pattern. Luxury brands employ specific ways of teaching markets about their value. Innovative ways of luxury branding, such as gifts for an elite or event sponsoring, lose their lustre when communities become familiar with such branding techniques. To avoid the impression of mediocrity, luxury brands need to innovate.

For small communities of interest luxury products are well known (luxury = brand), branding these products helps to go beyond the original community. Depending on several factors such as the size of the new, enlarged customer group the original community can be alienated (brand ≠ luxury). Luxury brands can strike a profitable balance between luxury for a limited community of interest and mass-market brand. The profitability of this strategy has attracted a lot of investment recently.

9.8 CONCLUSIONS AND IMPLICATIONS FOR DESIGN STRATEGY

Luxury products are the result of material and immaterial design. Luxury patterns employ traditional models, technologies and know how to claim superior evaluations. Luxury patterns are often established through the 'traditional' design work including the design of fonts, visuals, products, architecture, and so on.

Immaterial design involves nurturing social interaction processes that define the degree of appreciation by communities of interest.

The different types of luxury such as luxury patterns or brand were introduced. Four strategies can be identified. A generic focus on luxury patterns is indispensable. Specific strategies can focus on the importance of small communities for emergent luxury or on multiple or large communities (brands). The strategies can be integrated in a sequence including support of emergent new luxury products in small communities and then articulating luxury brands in large communities.

The success of these strategies depends on how communities of interest are facilitated and used. The task for design management and design strategies includes promoting self-organisation of communities as dynamic contexts for knowledge processes.

9.9 REFERENCES

Alleres, D., 1997. *Luxe...Strategies, Marketing*. 2nd Edition, Economica, Paris.

Berger, P.L. and T. Luckmann, 1966. *The Social Construction of Reality*. Doubleday, Garden City, N.Y.

Buetz, M.R., 1997. *Chaos and Complexity: Implications for Psychological Theory and Practice*. Taylor and Kenen, Washington, D.C.

Frank, R.H., 1999. *Luxury Fever: Why Money Fails to Satisfy in an Era of Excess*. Free Press, New York.

Granovetter, M.E., 1985. "Economic Action and Social Structure: The Problem of Embeddedness". In *American Journal of Sociology*, **91**, pp. 481-510.

Holland, J.M., 1998. *Emergence - From Chaos to Order*. Helix, Reading, Mass.

Il Sole 24 Ore, 1999. *Luxury Zone: Profitability in the Fashion Industry*. Italy.

Kuethe, E. and Reinmoeller, P., 1999. Fascination and Luxury: Design Strategies and Markets. In *Working Paper*, University of Cologne, Cologne.

Lave, J. and Wenger, E., 1991. *Situated Learning: Legitimate Peripheral Participation*. Cambridge University Press, New York.

Machlup, F., 1962. *The Production and Distribution of Knowledge in the United States*. Princeton University Press, Princeton, N.J.

Nihon Keizai Ryutsu Shinbun, 1999. *Sekai Seiha he "burando kantai"*, May 4, p. 1.

Nonaka, I. and Takeuchi, H., 1995. *The Knowledge Creating Company*. Oxford University Press, New York.

Nonaka, I. and Reinmoeller, P., 2000. "Dynamic Business System for Knowledge Creation and Utilization." In *Knowledge Horizons: The Present and the Promise of Knowledge Management*. Despres, C. and Chauvel, D. (Eds), (forthcoming).

Reinmoeller, P., 1996. *Marketing in Japan*. Metropolitan, Munich.

Reinmoeller, P., 1997. Towards a Synthesis in Knowledge. In *Design News*, JIDPO, **244/245**, pp. 82-85 (in Japanese).

Rheingold, H., 1993. *The Virtual Community: Homesteading on the Electronic Frontier.* Addison Wesley, Reading, MA.
Slywotzky, A.J. and Morrison, D.J., 1997. *The Profit Zone: How Strategic Business Design Will Lead you to Tomorrow's Profits.* Random House, New York.
Wenger, E.C., 1998. *Communities of Practice: Learning, Meaning and Identity.* Cambridge University Press, Cambridge, UK.
Wenger, E.C and Snyder, W.M., 2000. Communities of Practice: The Organizational Frontier. In *Harvard Business Review*, January - February 2000, pp. 139-145.

CHAPTER TEN

Carrying the Pleasure of Books into the Design of the Electronic Book

STAN RUECKER

Department of Art and Design, 3-98 Fine Arts Building,

University of Alberta, Edmonton, Alberta, Canada T6G 2C9

ABSTRACT

In an effort to design a better electronic book, a web-based survey and two focus groups were used to collect information on the opinions and practices common to frequent readers of conventional bound books. Although affect was included in the study as only one of five human factors, many participants used emotion as the preferred mode of discourse in addressing points in each of the factors. In other words, it was the many and varied pleasures of books and book culture that were key to the experience of frequent book readers; an electronic book design was subsequently outlined that attempts to add functionality without loss to either the existing functionality or this significant range of pleasures.

10.1 THE PLEASURE OF BOOKS

A variety of information was obtained from frequent book readers through two focus groups and a web survey. The focus group transcripts contained over 30,000 words, while the web survey contributed an additional 8,000 words. This information was intended to contribute to the design of an electronic book that would better suit the needs of frequent book readers.

In the course of analysing the results, it became clear that whatever the experience of frequent book readers might actually be, there seemed to be a tendency for them to characterise their experience in affective terms. Throughout the survey and focus group questions, which were organised into five human factors (physical, affective, cognitive, interpersonal and cultural) participants seemed to seize on whatever opportunity arose to deal with the subject in terms of either the pleasure or absence of pleasure afforded.

Respondents really liked books. They enjoyed reading books, lending and borrowing books, talking about books, buying books, and having books around them. One survey respondent suggested that books had saved her life.

Physically, they liked the fact that books are static and quiet. They liked the smell, feel and shape, and enjoyed holding books while reading. Some respondents enjoyed turning pages, the feel of paper, while others enjoyed marking their books.

In addition, people pointed out that books are portable, comparatively cheap and relatively permanent. Some mentioned liking pictures and others privileged scholarly apparatus. One respondent spoke of liking the variety, while many enjoyed collecting books.

10.2 PARTICIPANT PROFILE

The cohort consisted primarily of women, between the ages of 26 and 60, with at least one university degree. Half of them were currently graduate students. There was a total of 58 respondents, of whom 7 were also focus group participants. In general they preferred books to other media, although they tended to spend slightly more time reading from the monitor than they did reading from books. Most preferred to read in quiet places, often sitting down at home in the evening, although the majority of their reading was related to work or school. They used the university library far more frequently than they used the public library, and stayed longer on their visits. They bought nearly four books a month on average for themselves, and borrowed twice as many again.

The sample size used in this study gives approximately a 95% confidence level with a ±10% sampling error, given that the majority of participants tended to agree on key opinions (Salant and Dillman 1994; p. 55).

10.3 WHAT DO YOU LIKE ABOUT BOOKS?

The answers to this question had a powerful emotional charge, perhaps because it was the central affective question for the entire survey. Several of the respondents used words like 'love' and 'treasure', which to some extent reflects the self-selection that would naturally be part of responding to a survey, but is indicative nonetheless of the bounds of the discourse that seem appropriate to frequent readers in discussing their books:

'I can't think of anything I do not like – make that love – about books. I do love owning them and so, even though I administer an academic library, I don't like to borrow them. They are that important to me.'

Many of the people who answered this question felt a strong emotional attachment to books, and were able to articulate a wide variety of components that were constituents of the attachment. The gist of their comments can be summarised, however, in three words written by one respondent: 'the whole experience.' A more detailed list was created by another writer, as follows:

'The way you can take them anywhere. You don't need electricity. You don't need any hardware or software. They never crash. They smell good. You can flip back and forth with ease. You can read in a confined space. The way they're always there for you. The way they've saved my life more than once. The way they line up on a shelf so I can look at them whenever I want. The way they calm children. The way they enthrall children. The pictures of great artists are available to me even if I live in a remote location. And lots more. Books are beautiful.'

Other respondents added to the list of particulars, including the fact that books are static and quiet, that they can stimulate thinking and, in some cases, a sense of nostalgia; that they serve the dual purposes of escapism and education.

Several people mentioned physical factors such as the smell, feel and shape, and more than one person pointed out that there is something affecting in the fact that the reader holds the book while reading it. One respondent wrote of the usefulness of being able to annotate the pages of a book, while another saw the act of turning pages as a pleasure: 'I like to feel them and turn the pages.'

In addition, people pointed out that books are portable, comparatively cheap and relatively permanent. Some mentioned liking pictures, and others privileged scholarly apparatus. One respondent spoke of the universality of books, another of their variety, and still another mentioned that they were valuable for providing different perspectives.

Still others spoke of the reading process, suggesting that there was a pleasure in being able to identify various literary forms, to read a well-structured argument, or to see innovative uses of language. One respondent characterised reading as a meditative experience, while another mentioned the value of being able to control the pace so easily.

10.4 READING

Reading was characterised by the study participants as an activity that allows the reader more control than is common in other media. Reading was described as relaxing, private and self-directed. The reader can set the pace. Reading calls for self-reflection, mental engagement and imagination; it also calls more directly, in the opinion of one respondent, on emotional or moral involvement. The gratification is less immediate than in other media; reading also lends itself naturally to re-reading part or all of the book. Reading was also described as primarily verbal, although one respondent also explained that mental immersion was part of the experience: 'Somehow reading a book seems more tangible; somehow you can really feel yourself in the book; it seems more real, more permanent.'

More than one respondent felt that part of the virtue of reading lay in that it was a demanding activity: books required a time commitment; they took concentration; the experience was more in-depth. 'Books', in the words of one respondent 'kick start my critical faculties.' They also allowed more opportunity for exploring characters: 'I like books better because you can get behind the characters more, into their thoughts, feelings and motivations.'

10.5 CHOOSING BOOKS

In choosing books to read the respondents were looking for a variety of experiences, including relaxation, escape and pleasure. Some people emphasised their interest in genre or subject matter or the story. One respondent suggested that the search for a sense of belonging was part of the mechanism of choosing books, while another saw books as part of problem solving. One respondent felt that it was significant that in reading a book, the experience can be suspended and resumed.

Several respondents volunteered factors that indicated a complex terrain. One mentioned the attractiveness of some book store interiors; another that they tended to read more when they were sick. Another respondent pointed out that guilt was at issue in choosing to read some books: 'There are some books I should read.' Finally, the point was made that some people just like books, no matter what kind: 'It's a book. I'm drawn to all books.'

10.6 READING LOCATION

On more than one occasion participants discussed seeing people reading school work while at the gym, and expressed doubt as to their being able to learn in that environment. Several people talked about reading in the bathtub and indicated that there is a class of more or less disposable books that are selected for that purpose. Participants who do not read in the bathtub had different reasons for their aversion: one felt no books were disposable enough; another felt that reading was too work-related and therefore inimical to relaxing in the bath. More than one participant spoke of the pleasure of reading in bed.

10.7 WHAT DO THE BOOKS YOU LIKE HAVE IN COMMON?

To a certain degree the answers to this question could be logically divided into discussions of fiction and nonfiction, although some comments did apply to both kinds of books.

Fiction

In discussing fiction, people highlighted the importance of having a compelling story, and some of the respondents emphasised genre – mystery, fantasy. Several people wrote of the use of fiction as a form of escape or relaxation.

On a scale slightly expanded from the concerns of the individual title, one respondent suggested that continuity was important in the form of a novel series with on-going characters, and several mentioned having favourite authors. Others tended to read books that had been recommended to them by either public sources like Oprah Winfrey or private sources like family and friends, and one person said that the main thing their fiction choices had in common was that all the books had been received as gifts. Some respondents were able to be quite specific, with one suggesting that they looked for books with female characters, while another chose books that showed good defeating evil.

Nonfiction

Nonfiction titles are naturally associated with learning, and many respondents mentioned the value of learning something new on a topic of ongoing interest, or else learning something new on an entirely new topic. Some people mentioned a preference for certain non-fiction genres such as religion and self-help, while others pointed out that much of their non-fiction reading is required by work or school. A couple of people talked about preferring material that was up to date. One respondent mentioned the importance of the information being easy to navigate, while another brought up the fact that the books they chose to read tended to have a well-supported thesis.

Both fiction and nonfiction

Many of the responses could equally well apply to fiction or nonfiction. For example, people identified as common factors in the books they chose such elements as amusement, insight, challenge, inspiration and new perspectives. Others referred to competent writing, cohesiveness and the importance of being able to get caught up in a book. Special interests were mentioned in feminism and interculturalism. Finally, more than one respondent was unable to identify common features in the books they chose: 'They appeal to me on many levels and my choices are certainly eclectic so this is difficult to answer.'

Books as part of lifestyle

Participants generally considered books an important part of their lifestyle and, to a lesser degree, their social life. Books conveyed status, had lasting value, and were an indication of personality: 'I care more about the books I own than what car I drive.'

Participants borrowed books from and loaned books to their friends, and they all recommended books and had them recommended. They all discussed books with friends and colleagues. In addition, more than one participant spoke of the importance of reading books as a deciding factor in the establishment of personal relationships. It was felt that discussions of books helped both to establish characteristics of other people and to express one's own personality. One participant explained carefully that for her, at least, the choice of someone else's reading material was not as important an issue as the fact of the person's being a reader or not: 'I could not handle dating somebody who didn't read.'

One participant associated book buying with lifestyle:

'I have been collecting books for over 15 years now, it's one of my few pleasures in life and has introduced me to many interesting and friendly people whom I would otherwise not have had the pleasure of knowing. Scouring second-hand book shops is an integral part of my leisure activities.'

Personal libraries

Participants had fairly large personal libraries, with three-quarters of them owning more than 200 books at home; their work or office libraries were smaller – more than half had fewer than 200 books at the office.

Book collecting

Participants felt in general that there was a sense in which books are parts of collections that are an expression of identity, in much the same way that other collections express their collectors. They did not feel that one book could fill this niche. There was some discussion of book collections vs. CD music collections, and some agreement that the reading device could never be successfully disassociated from the titles for people who love books, although it might be a possible solution for a different cohort – perhaps those who were able to invest emotionally in a CD music collection, for example.

Participants also spoke of provenance, where books acquired value through having been owned by someone important in the past, whether a family member or whoever. Some books were also more valuable for representing in a physical artifact the effort that had gone into reading them, and books in general as signifying the owner's predilection for reading and learning.

In this respect one participant expressed doubt that any solution would be viable that involved erasure and over-writing. She felt that perhaps it was the relative permanence of a printed book that was a primary part of its value.

Replacing a conventional library with an e-book

The display of books on shelves was considered a very important feature of books.

And it's just such a great feeling, it's like the first time you get into a good library, and it's like, 'look at all these great books, I could read any book on the shelf.'

Participants who had books in storage through lack of apartment space felt the separation from their personal libraries as a hardship. One of the advantages of shelf display is that owners could spend time reviewing their collections to remind themselves what books they had and to note logical gaps where additional titles might be wanted.

In addressing what role an electronic book might play in relation to the personal library, it was generally felt that it would be a supplement only, which might provide additional tools or functions but which could never replace the collection.

Respect for books

One of the participants in the second focus group spoke of having respect for books, which necessitated a number of specific behaviours. She never folded corners of pages or kept books in the bathroom where they would be subject to damage from excessive moisture. She was also in the habit of making additional dust jackets out of brown paper to protect the covers – an activity which another participant claimed to have been part of her education in elementary school, where the children made similar covers to protect books owned by the institution.

Books and other media

Several respondents made intelligent distinctions between various media in terms of when each was desirable, and at least one gave a detailed description:

'On a Sunday morning I want my bagels, chai and at least one Sunday paper – usually two, the local rag and an international or larger city paper. On a Saturday morning I want to read comic books in bed until noon. On an exercise bike I want frothy pulp magazines that I don't have to concentrate on. I also subscribe to a few electronic newsletters, and read them as long as the sections are brief. I always read the CD covers and liner notes when I listen to music. Conde Nast Traveler magazine has a permanent home next to the toilet.'

Books compared to film

Participants agreed that the role of the reader in dealing with a book was quite a mentally active role, that texts necessitated engagement and interpretation. They did not agree as to the passivity involved in watching movies, with some participants using words like 'brainwashing' and 'helplessness' to describe certain movies, while at least one participant spoke of bringing critical tools to bear on the experience.

One participant spoke of the physical freedom of movement inherent in reading a book.

'I think because in many ways it's more active – it's a more active engagement, both intellectually, in the – if you're reading a fictional novel, you're creating the world in your own mind, and physically, you're turning the pages, you're sitting, you're – you can move around, I can lie down, I can sit up, I can walk up from my room. I can put it down for five seconds to brush my teeth and come back and pick it up again.'

Books compared to documentaries

Participants generally responded in favour of educational channels and documentary films, citing in some cases specific examples that they had enjoyed and learned from. One member of the second focus group differentiated between documentaries with and without an explicit narrative, preferring the former in that it provided a framework and some continuity that might otherwise have been more difficult to achieve in such a predominantly visual medium.

Another participant spoke of some of the problems inherent in documentary filmmaking. The first was the problem of the lack of scholarly depth, which she saw as fundamental to documentaries, in that the amount of content possible in a two-hour film might only translate to fifty pages of transcript* – which would be a fairly superficial treatment in writing. The second problem had to do with the provision of visual material which was both appropriate to the topic at hand and still interesting: 'Sometimes I find that the visuals trivialise the text.' There was no comparison, in her mind, between what could be learned from a book and what could be learned from a film. The film was laughable by comparison:

'It's pandering. It's interesting and fun to watch, and sometimes very valuable information ... but you can't compare it to ... what you could read in a book.'

Books as physical objects

In terms of books as objects, respondents felt strongly that books were personal in some way. Books could be stamped through use; 'it's easier to think of a book as a 'friend', or have a 'history' with it.' Books were commended as being portable, quiet, and providing an experience that was repeatable: 'Books are like old friends that you can come back to at any time.' Some people mentioned liking the way books feel. Others emphasised the importance of language, the charm of words, and the importance of the text being fixed rather than fluid.

People were variously enthusiastic about new books and old books, hard cover books and trade paper books and disposable paperback books. Details included the smell, the crispness of a new book, the comfortableness of a book that one has read repeatedly.

In direct physical terms it was remarked that books can be written on, and that their existence as physical objects makes subsequent reference to particular sections easier. Responses were in general quite enthusiastic about the virtues of books:

'Books are beautiful in their structure, educational, enjoyable, informative, insightful, and energising.'

10.8 THE DESIGN OF THE ELECTRONIC BOOK

'I would rather have a cheap book on paper than an expensive electronic thing. There's something 'right' about traditional books, I can't explain it. Why mess with a good thing? Unless an electronic book mimics and improves what I can get out of a real book, I don't think I'd want one.'

*In fact this is probably an underestimate, although the point is still valid enough. The standard script length for a 90 minute feature film is between 110 and 120 double spaced pages in 12 pt Courier.

Descriptions of current electronic books

In early 1998 three electronic books were announced, which were positioned not as hypertexts or digital stories, but rather as electronic reading devices. Earlier attempts at developing a commercially viable electronic book date at least to the late 1960s, with Alan Kay's Dynabook. Later entries include the Sony Bookman of the early 1990s, but no electronic book prior to 1998 could be seen as a commercial success.

The three electronic books currently available are the Rocket eBook, the SoftBook, and the EveryBook. These devices share the quality of being more like computers than they are like books. They have fixed display screens that give off light (the EveryBook differs from the others by having two screens, like facing pages); they hold multiple volumes in memory; the pages are turned or scrolled with a control.

The result is that although they provide some added functionality over the conventional bound book (primarily in making digital texts available in a portable reading device), they remove at least as many features as they add. It is these lost features that represent many of the pleasures of conventional bound books that were identified by participants in this study. In a somewhat cursory nod at this problem, all three of the existing e-books ship with leather covers or cases, in conscious imitation of traditional hardcover book bindings.

A summary of book-related pleasures

The following list attempts to summarise the pleasures of books that were identified by study participants. The pleasures have been divided into the four traditional human factors: physical, cognitive, interpersonal and cultural. Because several of the respondents used *Gestalt* terms such as 'the whole experience' or 'everything about books', it is probably best not to consider this reductive list as equating to the full experience of pleasure.

Physical pleasures

- holding and handling books as objects (smell, feel, shape)
- looking at books (inside and out)
- turning pages, feeling paper
- annotating

Cognitive pleasures

- reading (including following a story or line of argument)
- studying (accessing information)
- returning to passages previously read
- scanning spines on a shelf

Interpersonal pleasures

- borrowing and lending books
- recommending books
- discussing books

Cultural pleasures

- buying (both new and used books)
- collecting and displaying books
- reading book reviews

The process of carrying out these activities with the electronic books currently on the market is sufficiently different from the same process carried out for conventional bound books that there is virtually no similarity between the two means of delivering text.

10.9 DESIGN OF AN ELECTRONIC BOOK FOR FREQUENT BOOK READERS

The information collected from the web survey and focus groups was therefore used in the design of an electronic book specifically intended for this cohort. The assumption of the design was that an electronic book could be understood to exist between a pair of difference poles, with the laptop computer on one side and the conventional bound book on the other.

The design features books which are printed on demand using electronic paper that has been pre-assembled into signatures (see Figure 10.1). The printer is a device analogous to a conventional laser printer which downloads the digital text into the electronic paper signatures, then binds the signatures together into a book. The electronic paper is read by reflected light and statically retains its image in the absence of electricity. It can be erased or overwritten, and the signatures detached from each other in the printer so that the book is disassembled for reuse. It can be searched and annotated using a digital stylus. It can in its deluxe form store more books than it currently displays.

The cover and spine contain designs specific to each title, in much the same way as conventional bound books feature cover and spine designs. The titles are created by standard computer layout programmes and printed to a format that the book can display, so that fonts and graphics can be used in a manner similar to the conventional bound book. Its physical thickness represents the number of pages in the largest title currently stored, and readers can mark their spot with a slip of paper or can judge proportion remaining against total length at a glance or touch.

The digital or electronic paper that makes this design possible does not exist in a commercial form at this time, although there are at least four research companies working on tantalising prototypes.

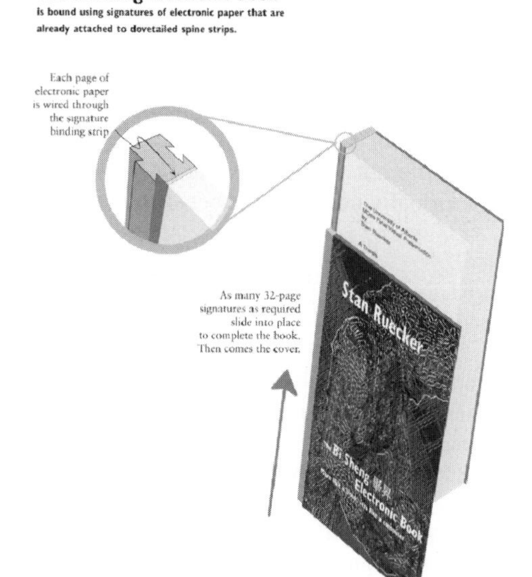

Figure 10.1 The Bi Sheng Electronic Book

10.10 REFERENCES

Asheim, L., 1955. The future of the book: Implications of the newer developments in communication. In *Papers presented before the Twentieth Annual Conference of the Graduate Library School of the University of Chicago*. Asheim, L. (Ed.), June 20-24, The University of Chicago, Chicago.

Berry, J.D., 1998. "Live text in your pocket." In *Upper and Lower Case: The International Journal of Graphic Design and Digital Media*, **25**, 1. International Typeface Corporation, pp. 12-14.

Bolter, D.J., 1992. *Writing space: The computer, hypertext and the history of writing*. Lawrence Erlbaum, Hillsdale, NJ.

Darnton, R., 1999. "The new age of the book." In *New York Review of Books*. http://www.nybooks.com/nyrev/ WWWfeatdisplay.cgi?19990318005F

Dorman, D., 1999. *"The e-book: Pipe dream or potential disaster?"* American Libraries, Feb. 1999, pp. 36-39.

Dyson, M.C. and Kipping, G.J., 1999. "The effects of line length and method of movement on patterns of reading from screen." In *Visible Language*, **32.2**, pp. 150-181.

EveryBook. "EveryBook opens onto the future." Accessed 99-04-23. Last Updated 05/12/98. Advanced Micro Devices, Inc. http://www.amd.com/advances/ columns_features/issue27/ everybook_ft.HTML.

Frascara, J., 1997. *User-centred graphic design: Mass communications and social change*. Taylor & Francis, London.

Gross, N., 1998. "Flatter, brighter – and easy to make? A new screen technology has the industry charged up." In *Business Week*, October 19.

Hsu, M., 1998. "The printed page is dead – again: rocket in your pocket. Still no substitute for the feel of a book." In *National Post*, November 12, , pp. B1-B2.

Lakoff, G., 1980. *Metaphors we live by*. The University of Chicago Press, Chicago.

Lücke, K., 1999. "Customized digital books on demand: Issues in the creation of a flexible document format." In *Visible Language*, **32.2**, pp. 128-149.

Mandel, C., 1998. "Taking a reading on paperless books." In *The Globe and Mail*. August 8, pp. C1, C10.

Mandel, C., 1998. "SoftBook a hard read." In *The Globe and Mail*. August 8, p. C10.

McMillan, S., 1993. "The future of publishing: the direction is digital." *In Communication Arts*, **35**, 6, November, pp. 220-226.

Miall, D.S., 1995. "Representing and interpreting literature by computer." In *Yearbook of English Studies*, **25**: Non-Standard Englishes and the New Media. Modern Humanities Research Association. Also available electronically at: http://www.ualberta.ca/~dmiall/complit.htm, pp. 199-212.

Nunberg, G., 1996. *The future of the book*. Nunberg, G. (Ed.), University of California Press, Berkeley.

Planar Systems Inc. *"Active matrix electroluminescence"*. http://www.planar.com/tech/tech.htm.

Rocket e-book, 1998. NuvoMedia. Site version 1.0.119, http://www.rocket-ebook.com/enter.HTML.

Salant, P. and Dillman, D.A., 1994. *How to conduct your own survey*. John Wiley & Sons Inc., Toronto.

Sebastian, L., 1992. "Ownership of software documents." In *Communication Arts*, **34**, 6, pp. 271, 274.

Silberman, S., 1998. "Ex Libris: the joys of curling up with a good digital reading device." In *Wired*.: Wired Magazine Group Inc., July, San Francisco, pp. 98-104.

SoftBook Press. *The intelligent reading system*. Menlo Park, CA. http://www.softbook.com/index.HTML.

Tschichold, J., 1991. *The form of the book: essays on the morality of good design*. Hadeler, H. (transl.) and Bringhurst, R. (Ed.), Hartley & Marks Inc., Vancouver.

Wiseman, R., Morgan, K. and Sless, D., 1996. "Testing, testing ... 1,2,3 ...". In *Communication News*. Communication Research Institute of Australia. 9:1, January/February, pp. 1-4.

Xerox Palo Alto Research Center. *"Electronic paper"*. http://www.parc.xerox.com/dhl/projects/epaper.

CHAPTER ELEVEN

Difficulties and Pleasure?

PIERRE-HENRI DEJEAN

CQP2 (Design and Quality of Products and Processes),

Technological University of Compiegne,

60203 Compiègne, France

11.1 INTRODUCTION

There is a general tendency to believe that difficulties prevent pleasure. For all work carried out on usability, marketing pleads implicitly for this thesis, without, however, defending it. In this paper we shall discuss a number of contexts of use in which usability and pleasure are in opposition to each other. The aim is to show the importance of pleasure-based as opposed to usability-based approaches to product design, in order to maximise the quality of user experience.

The starting point is based on some practical examples, where context or circumstances throw into question the link between usability and pleasure in use. Indeed, cases will be highlighted in which there is a clear link between usability and difficulty in use.

11.2 RELATIONSHIP BETWEEN DIFFICULTIES AND PLEASURE

Simplicity and simple pleasure...

Some of our research, carried out for the French Ministry of Industry, has shown that there is often a degree of stigma attached to the whole concept of simplicity. Indeed, this stigmatisation of simplicity is even inherent in the French language. The meaning of the words 'easy – comfortable' and 'easy-going' are associated with the concept 'without merit', and the expressions 'simple d'esprit' (simple minded) and 'plaisir facile' (simple pleasure) are not very flattering for the people they refer to.

Finally, and more concretely, there is genius in the ability to find simple solutions, on the condition that they do not devalue difficult or complex problems. It is likewise very valuable that a person is able to select either the most efficient product or the most simple and efficient method. On the other hand, the quality of the person who chooses this product or method will be disregarded. All the honour goes to the designer and not to the user: the first ambiguity of usability.

Difficulty and pleasure

Games are often appreciated according to their degree of difficulty. You might even talk about an increasing pleasure parallel to the level of difficulty. There are, however, some nuances to be taken into consideration. In the case of electronic games for instance, if the difficulty is sought after in the game itself – the service provided by the project in some way – simplicity and facility are appreciated in the access to the service; that is to say, the setting up of the game. Whereas in the instructions for starting up the game – the technical side of the operation – being as easy as possible is sought after. The interfaces, the directions for use, are developed for this purpose.

In the same way, in many activities, the increase in knowledge, skills or dexterity is preferred for the level of difficulty their acquisition represents. By analysis there are two kinds of pleasure: learning where only the result counts and acquisition where the process is more important than the result.

For the amateur cook, acquiring the knack might be preferred to the result. For the researcher the puzzle that the search for the applicable approach represents, or the equipment needed for setting up for his experiments, is equally if not more pleasurable than obtaining the result that he had already more or less anticipated.

Power and superiority

Difficulty is accepted in the case where the effort provides a power or a superiority compared to oneself or others. For instance if it is difficult to learn how to handle an ancient instrument like a sextant, a violin, ... once learnt it is a great pleasure to practice. The idea of the benefit of the effort made, the feeling of qualitative differentiation, must be linked to this fact.

The phenomena is already possible to observe in games or activities for amateurs, but takes on a very different dimension in professional life.

The monopoly on know-how or ability constitutes a power that creates difficulties for many people. On another level but in the same vein, cultural enrichment can be experienced as being a pleasure. The feeling of competence is valorising. In the studies of usability we have noted two attitudes that indicates this fact. The refusal or distrust in relation to systems which are too simple or too automatic, cause people to be afraid of losing their competence. The immediate consequence in the context of a country very affected by unemployment is that people fear losing their utility or uniqueness. A very simplified kind of job might be done by anyone. Indeed, was this not one of Taylor's fundamental thoughts ?

Reassuring difficulties

In certain circumstances, simplicity can create fear. This is the case when for instance actions involve high stakes and there is no time limit. The security of information, for example, applies to this attitude. Here an elaborated procedure is preferred to an approach which is too simple.

A comprehensible diagnosis, but one based on long explanations, becomes reassuring in relation to a troublesome worry. For some people medicine which is too pleasant will cause doubts to rise as to its efficiency.

The serious and rigorous aspect becomes a factor of credibility and hence of pleasure. In the layout of bank receptions the agents are very sensitive to aesthetics that will give the customers a serious impression; it might even go as far as causing an impression of austerity. With such a decoration, the banker will work with greater pleasure feeling more credible in relation to his/her customers thus more at ease, which will in turn make his/her work easier.

Fatigue and pleasure

Tired but happy! This is the expression corresponding to a satisfaction which has demanded quite a deal of physical or intellectual effort. Here as well, personal investment and pleasure are harmoniously united. To prevent pleasure being replaced by stress, it is, however, most important that the demands on availability of time and mind are strong. The feeling of satisfaction in relation to the result obtained might be related to the effort made to obtain this result. For many people sports give satisfaction by the physical effort more than the level of performance in itself; the important aspect is to participate. Marathons which have increased in number in cities all over the world is a good example of this.

Ethics and pleasure

We have placed religious as well as non-religious considerations in this category. These concern systems of value accepted by the social groups which have multiplied across the modern world.

We have observed that usability was badly regarded if it caused social problems such as increasing unemployment. The current ecological problems are showing tendencies in this direction and can even lead to active intervention. Cleaning and the salvage of waste products are not in themselves pleasurable activities even though the feeling of having respected nature can exist. Progress in ecological problems in ethical terms is both a basic tendency and an element of solution.

The pleasure of earning money on the stock market might be thwarted by the immoral aspect of the utilisation of the funds thus gained. This way of thinking has lead to the creation of ethical funds in the USA. Here the investor is assured that his savings are not placed in the support of projects that in his view are immoral. On the contrary, while gaining capital he will feel a pleasure in contributing to stocks that support sustainable development. Different religions have decreed rules forbidding certain pleasures like greed and lust, whereas technological progress has destroyed some of the disastrous consequences of these pleasures. The specialised 'low-calorie/health food' products contribute to the appreciation of good food without creating health problems.

Some people feel a certain pleasure in the punishments afflicted to them. More commonly, and in a rather lay sense a medication that leads to pleasure could be suspected of being inefficient.

11.3 MANAGING PLEASURE AND DIFFICULTY IN PRODUCT DESIGN

We have seen that pleasure in use is not always associated with usability. Indeed, there are occasions where difficulty in use can increase the pleasurability of interaction — for example by raising the users' sense of self-esteem.

It is evident that a cultural dimension ought to be expressed here, as well as in other components of the product (De Souza and Dejean 1999). Cultures face difficulties and pleasure differently. An exploration ought to be made of the cultural valuation of certain difficulties; the different cultural attitudes in relation to difficulties should be explored.

A possible dilemma for the designer is deciding how much of the product's mode of functioning should be revealed to the user. With 'black box' products operation may be very simple because the user is not required to understand anything of how the product actually works. However, if something goes wrong, then the user may have no idea how to correct this as he or she will not understand the process behind the way in which the product works. On the other hand 'glass box' products, where the user is more involved in the process may require more effort to learn. Once mastered, however, the user will be able to understand the workings of the product and may be able to come up with some innovative solutions should a problem occur. They have, in a sense, been empowered through having to make the extra effort to learn how to use the product in the first place.

In conclusion, the introduction of pleasure in the criteria of usability is essential. It is an element of renewal for a discipline which was for a long while oriented towards suffering ... and a benificial return of product ergonomics to the ergonomics of working conditions. The creation of a reference for pleasure is thus all the more necessary. Models and parameters might then allow us to approach this notion in a scientific way, thus approaching the very question whose complexity adds to the pleasure of its study!

11.4 REFERENCES

Cherfi, Z. and Dejean P.-H., 1999. *Quality and Ergonomics in Product Conception.* TQM and Human Factors, July 1999, Linjoping, Sweden.

Dejean, P.-H. and Soler, D., 1997. Ease of Use in the Next 20 Years. In *IEA 97*. July, Tampere, Finland.

Manzini, E., 1989. *La Matière de l'Invention.* Collection Inventaire, Centre Georges Pompidou, Paris.

Pereira, A., 2000. Design and Environment: Taking into Account the Interactions between the Social Actors. *Design Plus Research*, Politecnico di Milano, 18-20 May 2000, Milano, Italy.

Moles, A., 1972. *Théorie des Objets*. Editions Universitaires, Paris.

Souza, M. de, Bonhoure, P. and Dejean P.-H., 1998. L'Activité de Conception au Regard des Influences Culturelles. *IJODIR*, **1**, 2, INRIE, Paris, pp. 11-21.

Souza, M. de and Dejean, P.-H., 1999. Interculturality and Design: Is Culture a Block or an Encouragement to Innovation? (on line) In *Design Cultures 1999 - Third European Academy Of Design Conference*. March 30-April 1st, EAD: Sheffield Hallam University Sheffield, U.K.http://www.shu.ac.uk/schools/cs/ead

Weil, T., 1997. Comment les Enterprises Construisent des Reseaux dans la Silicon Valley pour Développer Leurs Compétences. {CD-ROM} In *2éme congrès International Franco-quebecois-Le génie Industriel dans un monde sans frontières*. 3-5 september, 1997. École des Mines d'Albi-Albi, France.

CHAPTER TWELVE

Envisioning Future Needs: From Pragmatics to Pleasure

JOHN V. H. BONNER

Department of Multi-Media and Information Systems,

School of Computing and Mathematics

Huddersfield University, Queensgate, Huddersfield, HD8 3DH, UK

and

J. MARK PORTER

Department of Design and Technology,

Loughborough University, Leicestershire, LE11 3TU, UK

12.1 INTRODUCTION

The success and failure of incrementally changing or radically new consumer products is ultimately decided in the market place. While there are many factors that influence the purchase and subsequent use of a product, a product that fulfils user needs should have a greater chance of survival. However, establishing future needs for future products that use a high level of interactive technology is fraught with problems.

Future needs have to be assessed against predictions of future behaviour and future developments in technology. Changes in behaviour and technology usually have reciprocating effects on each other, making it virtually impossible to draw any firm assumptions about what people will want and enjoy in the future. This process will always be more of an art than a science. But how can this 'art' be improved?

This chapter describes our attempts to address this issue by first examining contemporary user-centred methods and techniques that have enjoyed some degree of success or popularity in determining user needs in product design and other disciplines. We then go on to discuss how we adapted some of these methods so that we could specifically capture *future* user needs for new interactive products.

Two studies are reported which explore what type of user requirements methods are best suited for use by product interface designers and to capturing complex future user needs. Then, from this standpoint, some 'characteristics' of future user requirements methods and techniques are proposed.

12.2 CONTEMPORARY APPROACHES TO USER-CENTRED DESIGN

The inclusion of 'user-feedback' in the development of consumer products traditionally does not occur until the latter stages of the design process, when users' opinions on product variants are sought in 'hall tests' or consumer panels. However, with the increasing use of interactivity in consumer products, users now need to be involved in the interaction design process.

In the context of product design, different types of user-centred approaches have been reported, such as storytelling (Moggridge 1993), the use of modelmaking (Dolan and Wikland 1995), the use of focus groups (Caplan 1990), and more traditional usability testing methods (De Vries, Gelderen and Brigham 1994).

In comparison, user-centred design principles are more deeply rooted in the field of human-computer interaction (HCI). Apart from 'main stream' methods, which use observational or interview-based techniques, other approaches have emerged such as the use of scenarios to develop and refine user interface design proposals. An example of this is the use of cards, in discussion groups, to describe scenarios, which allow participants to specify and evaluate computer user interfaces. One method, known as CARD (Collaborative Analysis of Requirements and Design). This is coupled with a 'low level' approach known as PICTIVE (uses cards to facilitate the articulation of group task activity to establish system requirements to support this type of activity 'Plastic Interface for Collaborative Technology Initiatives through Video Exploration') which provides a close-up view of interface design proposals (Muller *et al.* 1995). The CARD approach has been modified by Lafrenière (1996). This modification, known as CUTA, enables a simple, user-derived task analysis to assist in interface design through the use of cards.

Both the CARD and CUTA methods use cards to depict elements of task activity such as task objects, for instance telephones and notepads, and process-based activities such as methods of working and situations; participants within the task activity are also depicted. Participants in both methods have identifiable and specific roles related to the task flows under discussion in the card sorting exercises. In both methods the cards are used as 'transitional objects' or points for discussion, in order to make explicit assumptions and interpretations about existing and future workflow methods. The methods require the participants to select task elements and place the cards in an agreed plan or sequence.

Many of these methods allow end-users greater possibilities to participate in and influence design decisions. Therefore, the user involvement in the design process moves from a consultative, or a merely user-centred role, to an active decision making or participatory role (Ehn 1988; Greenbaum and Kyng 1991). One of the problems with user participatory design, however, is how to provide adequate and appropriate support to users while they are engaged in design activity. Damodaran (1998) provides guidance by providing appropriate infrastructures within which users can operate. However, this guidance is aimed at organisations with management tiers with complex and interrelated decision making mechanisms. The challenge is to introduce effective participatory design at the product development level. Little has been reported on how true participatory design methods can be successfully implemented into product design organisations, but some attempts have been made to identify the hurdles in this process (Bonner and Porter 2000).

We therefore set about developing a user participatory design tool that could build on some of the methods described above, but which would also consider other factors that are important to the capture of future needs. These included:

- Encouraging users and designers to consider a wide range of speculative, innovative or novel aspects of interface design. This was important because consumer product interfaces vary in design characteristics more than computer user interfaces. Design factors such as display size demand less conventional interaction styles. Furthermore, the introduction of convergence-based technologies (telecommunications, computers and consumer products) suggests there will be many new commercial applications in the future.

- Allowing the consideration of pragmatic user requirements such as usability, and less tangible requirements such as relevance to lifestyle and pleasure. All these factors are important to the product's success.
- Any proposed design tool has to be quick and easy to implement, without involving a high level of training either for the designers or the users. Also, most importantly, the design tools need to be developed in such a way that designers take ownership of them and continue to use them without further intervention or support. This is a critical and major departure from other methods described above, where ownership and the motivation to use participatory methods belongs to the facilitators. Our approach was to train designers to become implementers of the design tools; to convert them to use such methods. We had to design the tools so that the benefits and rewards of such a process quickly become obvious.

One of the main problems with user participation is that users find it difficult to communicate ideas or concepts which are beyond their own experiences. Criticisms often levelled at introducing user participation methods are that they do not know what they want or cannot articulate their needs. If they can, they do not know what is possible or impossible. They find it difficult to be perceptive about how they might use a product or conversely they respond superficially.

Therefore, the use of focus groups, questionnaires and product clinics often depends upon users being able to anticipate scenarios of product usage with which they are familiar, or being able to reliably describe or interpret their own usage behaviour with existing products. While these methods can reveal very useful information if performed correctly, designers can often be left with a huge amount of data that provides little insightful information which can actively assist in the specification and design of a product interface. Data gathering methods may require considerable analysis before any meaningful results can be revealed.

A 'card sorting' tool appeared to offer an approach that could overcome many of these problems. We could see four main advantages over more traditional user requirements capturing methods:

- The cards provide a discussion mechanism, or act as 'transitional objects' allowing more critical contextual thinking to occur.
- The process would allow cards to describe novel interaction styles without having to design the interface to support such a concept. This also allows participants to interpret or define concepts on their own terms.
- Cards can suggest concepts divorced from defined or existing technology, therefore product functionality not currently possible can be discussed.
- Card sorting exercises can provide a physical schema or representation of tasks which may assist in interface design.

We devised a set of card-sorting exercises that we thought would be appropriate for designers. The objective of the two studies was to establish if this kind of approach was acceptable to designers and whether it was effective. If so, we wanted to know how the design tool should be structured to ensure that it is used successfully and produces valid and meaningful results.

The first study involved practising designers working on the development of advanced interfaces for a cooking appliance. The second study involved post-graduate design students in the initial stages of an individual, interface design project.

12.3 STUDY 1

Three designers from a large manufacturing organisation of domestic appliances volunteered to participate in the study. The designers agreed to use the proposed workshops to develop some proposals for a current design project related to advanced cooker interfaces.

Procedure

Four workshops were conducted over a two-day period, and all three designers were involved in the majority of the workshops. They were provided with verbal and written instructions on how to implement the card sorting tool, and were briefed on how to intervene and gather user requirements. They were instructed to take notes on the perceived effectiveness of the card sorting design tool and also on any design ideas that emerged from the exercises. The workshops and each subsequent exercise were introduced and explained by the researcher. During the exercises the researcher took little active part but took notes throughout the four group sessions.

The participants in the four workshops consisted of volunteer factory and office workers from the one of the manufacturing plants. Each group contained five or six male and female participants who were randomly assigned to each group. All workshops were video recorded for further analysis. Each group was given an introductory explanation of the purpose of the exercises.

All groups were given the four exercises to do, which were presented in the same order to each group. They were told that they had to prepare and cook a meal and explain the process by placing a set of cards on a table. It was explained that they had to decide how and where they placed the cards. The only rule they were asked to comply with was that they must discuss the process as a group and arrive at consensus agreement if any differences in opinion were found. They were also encouraged to talk to each other about the exercises. A time limit of 15 minutes was set on each exercise.

The first exercise required users to place cards depicting sub-task activity on a table (depiction of the sub-task activity had been devised by designers prior to the exercise) in such a way that the placement and selection of the cards indicates the users' representation of the overall task. In this exercise we asked participants to prepare and cook a meal using a series of cards that described cooking sub-tasks such as 'Check carrots to see if they are ready'. Sub-tasks were described in high level or general terms, with the addition of many other 'peripheral' activities to encourage alternative ways of representing the overall task. The intention was to allow the participants to discuss the whole cooking process and allow open discussions about the steps required in cooking.

In the second exercise participants inserted cards depicting descriptions of possible cooker functions, which can or could be found on a cooker control panel. These cards were divided into three groups: low, medium and high technology functions. These groupings were colour-coded. Low level functions included heater controls and selector switches, medium-level functions included auto timers and temperature probes, while high-level functions included cooking and menu planners.

The third exercise was devised to enable the participants to think about a week in the life of a cooker, and to place cards depicting typical cooking activities under cards labelled with the days of the week. The purpose of this exercise was to discover if participants would be able to identify aspects of cooking that they enjoyed or disliked and to make inferences about how their cooking habits could result in changes to the cooker interface.

The last exercise required the placing of cooker function cards under different character profiles. Participants were asked to match cooker functions to different profiles that reflected different levels of interest and enjoyment in cooking and the use of cooking technology. The intention for this task was to find out if users could make 'third party' design decisions on behalf of fictional characters.

In the final exercise one of the participants was asked to read out the character profiles from the cards provided and then to discuss what type of cooker features would be most suitable for each character profile. At the end of the exercises, participants were asked by the designers to make comments and to reflect on the card sorting process. After the participants had left, the researcher asked the designers for their thoughts and where improvements could be made.

Results

The design exercises definitely provoked discussion about cooking methods and strategies. However, this was not controlled or steered by the designers and therefore discussions were at times irrelevant to the objectives of the exercises, with many discussions resorting to personal experiences of cookers at home. Participants were observed adding function cards with little thought for their consequences to other related function cards, or the implications in terms of interface design. There was also a tendency to select functions based on normal cooking habits rather than on the cooking task presented.

The designers expressed initial concern about the card sorting exercise and were hesitant about a process over which they had no direct control. They were not sure how participants would react to many of the vague or unclear proposals depicted on the cards. During the exercises they appeared to be unclear of their role in the process, although they were more prepared to get involved in discussions at the end of the sessions.

The designers were surprised to observe participants demonstrating often conflicting user needs. For example, some of the participants expressed a strong suspicion of technology but were happy to consider some quite radical and advanced proposals during the exercises. When the designers were asked why they had not taken notes for use later on, they said they did not feel this was necessary. They felt that from the card sorting exercises they had a clear understanding of the direction they could take with future cooker interface proposals. They reported that the process had provided them with many new ideas and they found the exercises extremely illuminating and worthwhile. One designer said 'in the five years I've been here I have never been able to gather as much useful information from users as I've been able to do here'. They were encouraged and, indeed, surprised at the way the participants dealt with the situation, and thought that some of their ideas were very useful.

12.4 STUDY 2

In the second study, we decided to continue the development of the design tool, using post-graduate design students studying MA Digital Design at Teesside University. The key difference to the first study was the introduction of a further set of exercises once the card sorting exercises were complete. These were introduced as 'scenario design' based activities, where more refined design proposals from the card sorting were tested using real-world scenarios.

Procedure

Six students volunteered to undertake the optional 15-week module in 'Interaction Design'. They were aware that some of their studies would involve untried methodologies and that therefore clear guidance would not always be possible due to the use of design tools currently under experimental development. Five of the students were graduates in industrial design and one in fine art. All students worked on their own interface design proposals.

Students were given tuition on the card sorting tools with a practical demonstration. They were instructed on how to conduct their own card sorting exercises, with at least 3-4 potential users, for their own proposed product. Some of the problems in the previous study were presented to the students and they were advised to consider these problems in the design of their own card sorting exercises. On completion of this work, students were asked to provide a methodological description, their findings and a critique of the exercises.

After this, the scenario design tool was introduced using one of the student's proposal as a case study – an interactive tourist map and navigational unit. The student was asked to prepare a block model of the device and be able to describe in high level terms the functionality of the device and how a user might interact with it. The students then developed a scenario in which a tourist must find their way from a railway station to a Tourist Information Centre. We then conducted a role playing session, where one of the students acted the role of 'tourist' at the local city railway station using the 'prototype' device, while the 'designer' then talked through the features and functions as they were requested by the 'tourist'. This exercise provided all the students with an opportunity to understand the objectives of the scenario design tool, and also to consider how this approach may be adapted for their own needs.

After this workshop, the students decided to conduct some scenario design work between themselves before working independently. The students then carried out their own scenario design activity and written reports were provided on their experiences.

Results

In general, there was a positive response to the use of the cards, with many responding that having done it once they would have a much better understanding of how to conduct exercises in the future. Again, some reported surprise at the level of insight their participants demonstrated. One of the design students had used the card sorting tool with children and reported that this method was readily accepted by the children, although he did find a heavy degree of peer pressure in the conclusions the children arrived at. All provided constructive criticism on how card sorting could be improved.

The design students expressed uncertainty about how to introduce, control and conclude design issues during scenario design activity, but were surprised at the quality and inventiveness of suggestions made by the participants used in the scenarios. Most of the student designers reported problems in recording and noting participants' comments but felt that this would improve with practice. All the students enjoyed working with the participants although they did find problems with the quality of the feedback.

12.5 DISCUSSION

Overall, in the first study the card sorting tool was regarded as successful by both the participants and the designers. The designers found it a useful device for gleaning user requirements and testing some speculative interface design proposals. Our concern was that despite the designers having read the design tool instructions and having our support throughout the planning and execution of the card sorting exercises, they were still unclear about their role, how they should control the exercises and also what they should be concluding from the outcomes.

We felt that to overcome some of these problems we had to improve the design tool so that the involvement and role of the designer was more explicit (which occurred in the second study). The objectives of the card sorting exercise needed to be clearly defined and made more explicit both to the designers and the participants. Although using high level or broad description cards generally worked well, particularly in promoting active discussion amongst the participants, cards needed to be composed more carefully and more accurately to reflect potential user needs. The designers also needed to be actively involved in the process and be able to effectively glean information from the exercises.

The second study provided an opportunity to test some of the conclusions we had drawn from the first study, in particular the notion of participatory design involving more real world scenarios and making the process more explicit and transparent. Real scenarios were introduced to see if many of the learning problems experienced by both the designers and participants could be improved, and also to reduce dependence on 'contrived' tasks. This was indeed effective, but the game playing rules still appeared to be unclear. While further attempts were made to clearly describe the steps involved in conducting the card sorting exercises, the only effective mechanism was through experiential learning. Students frequently stated this in their reports. They progressed through 'doing' rather than through instruction.

There was no doubt that this method of gaining user requirements had an affect on some of the students. One student reported that his whole perception of design had changed in light of using the design tools. Despite this a dilemma still remains. On the one hand, both the designers and their participants appeared to enjoy the experience and thought it worthwhile. On the other hand, there was still too little evidence that meaningful outcomes were being gained from the process.

It should also be noted that, in the second study, design students were being used. As students they are motivated by assessment which will inevitably distort their design behaviour and attitudes towards the design tools. Furthermore, design students have more time, are more willing to experiment and can take more risks than practising designers. The results from the second study must therefore be treated with a degree of caution, as demonstrable acceptance with student designers is not necessarily an indicator of success in a commercial environment.

12.6 CONCLUSIONS

Clearly the studies demonstrate a paradox. Both the designers and the users or participants expressed enjoyment in using the design tools, but there is little tangible evidence that valid and reliable outcomes are being obtained. The design tools, however, have demonstrated that designers can influence the way that users articulate their perceptions and intuitions about future products and their associated needs. It is encouraging that many of the participants could articulate feelings beyond usability, and could begin to consider strong and deep subjective feelings relating to their future expectations of and preferences towards very speculative design proposals. Designers in both studies remarked on the insightful comments that participants were capable of making. In spite of this both studies suggested that much of the design data gathered by the designers were a result of serendipity rather than of systematic and planned investigation.

In summary, if this type of approach to capturing future, highly subjective user requirements is to prove successful, subsequent generations of these methods will need to consider the following factors:

1. Bringing designers and participants together through scenario game playing appears to be an effective and enjoyable experience and provides ample opportunity to discuss speculative design proposals, but the objectives of any exercise need to remain clear and explicit throughout the process
2. For designers to adopt and implement such methods there must be a very quick acceptance of their effectiveness. Our studies suggest that this can only be achieved through experiential learning.
3. Process mechanisms still need to be developed that help to control and guide discussion within the design space or problem. These mechanisms need to help both the designers and the participants to understand their roles and the scope of the design problem in order to ensure that viable outcomes are obtained. Our studies suggest that prescriptive approaches do not work.
4. Improvements still have to be made to how design suggestions or proposals are prioritised and consolidated, again to achieve measurable and viable outcomes.

12.7 REFERENCES

Bonner, J.V.H. and Porter, J.M., 2000. Introducing user participative methods to industrial designers. In *Proceedings of the International Ergonomics Association and Human Factors and Ergonomics Society 2000*, San Diego.

Caplan, S., 1990. Using focus group methodology for focus group design. In *Applied Ergonomics*, **33**, 5, pp. 527-533.

Damodaran, L., 1998. Development of a user-centred IT strategy: A case study. In *Behaviour and Information Technology*, **17**, 3, pp. 127-134.

Dolan, W.R., and Wikland, M.E., 1995. Participatory design shapes of future telephone handsets, Design for the global village. In *Proceedings of the Human Factors and Ergonomics Society, 38th Annual Meeting*, San Diego.

Ehn, P., 1988. *Work-oriented design of computer artifacts*, Arbetslivcentrum.

Greenbaum, J. and Kyng, M., 1991, *Design at work: Co-operative design of computer systems.* Greenbaum, J. and Kyng, M. (Eds), Lawrence Associates, Hillsdale, NJ.

Lafrenière, D., 1996. CUTA: A simple, practical and low cost approach to task Analysis. In *Interactions*, **3**, 5, pp. 35-39.

Moggridge, B., 1993. Design by storytelling. In *Applied Ergonomics*, **24**, 1, pp. 15-18.

Muller, M.J., Tudor, L.G., Wildman, D.M., White, E.A., Root, R.W., Dayton, T., Carr, R., Diekmann, B. and Dykstra-Erickson, E., 1995. Bifocal tools for scenarios and representations in participatory activities with users. In *Scenario-Based Design: Envision work and technology in systems development*, Carroll, J.M. (Ed.), John Wiley and Sons, New York.

De Vries, G., Gelderen, T. van and Brigham, F., 1994. Usability laboratories at Philips: Supporting research, development and design for consumer product and professional products. In *Behaviour and Information Technology*, **13**, 1 and 2, pp. 119-127.

Section 2
Design Techniques

CHAPTER THIRTEEN

Designing Experience: Whether to Measure Pleasure or Just Tune In?

JANE FULTON SURI

IDEO, Pier 28 Annex, The Embarcadero, San Francisco, CA 94105, USA

13.1 INTRODUCTION

What are the characteristics that make one product more enjoyable or attractive than another? Why do people find some interactions exciting and affirming but others depressing or unnerving? How much are our experiences the result of wired-in characteristics of the human nervous system? How much do social and cultural values influence our perceptions? And what about context?

As practitioners in the design business, we confront these questions daily. Our colleagues and clients look to us to provide facts, direction and wisdom concerning the kinds of experience that people will have with the things we design.

Both the design and business communities now talk about the design of experience – 'user experience' or 'customer experience' – as a conscious element of their offerings. There seems to be mainstream acknowledgement, at long last, that the things we design do not exist in a vacuum. Rather, it is understood that they are in dynamic relationship with people, places and other things; that they carry personal, social and cultural connotations; and that the quality of people's experience changes over time as it is influenced by multiple contextual factors. Given that the human factors profession has long claimed interest in the relationships that develop between people, products and systems, it's rather ironic that we are not better equipped to provide answers to the questions above. There are vast areas of human experience that we have barely begun to explore – particularly those to do with people's emotional responses to things.

The chapter begins by briefly exploring the complexity of people's experience of the designed world and what the human sciences have to say about it. It concludes that there is little explicit knowledge to help designers make predictions about people's emotional responses, although there are some useful frameworks to help us think about the issues. The final section uses examples from design practice to illustrate an approach that helps us, as designers, become more sensitive to the emotional quality of our own and other people's experience. Rather than develop an understanding of people's affective responses through analysis, the goal is to 'tune in' to people's perceptions by engaging or witnessing them directly and thus guide and inspire design.

13.2 THE COMPLEXITY OF EXPERIENCE

Experience is a dynamic, complex and subjective phenomenon. It depends upon reactions to multiple attributes of a design – for example, its behaviour, logic, sound, mass and texture, look and smell – that are interpreted through filters relating to personal, social and cultural significance (Macdonald 1998). To take a simple example: What is the experience of a run on a snowboard? Its quality is influenced by the weight and material qualities of the board, the fit of your bindings and your boots, the snow conditions, the weather, the terrain, the temperature of air through your hair, your level of experience, your current state of mind and even the mood and expression of your companions. The experience of even simple artefacts is influenced by a dense interplay of contextual factors. And increasingly designers are concerned with the experience of much more complex artefacts in which hardware, software, environments and services converge.

Products as elements of service

Products and environments these days are rarely experienced independently of some kind of service offering. The overall quality of interaction with a telephone or a hotel, for example, has as much to do with characteristics of the service you encounter as with the design of the physical elements you interact with. Brand and reputation largely depend upon factors beyond the artefacts themselves. In fact, it seems that as Internet and e-commerce applications develop, it will be *services* that drive the development of many kinds of products, rather than the other way round. For example, products such as personal MP3 players, Web-enabled phones and electronic books exist as a means of providing people with access to services – without the service they have no purpose. Inevitably, as people's perceptions of a company's offerings broaden to include services in addition to environments and products, the scope of human factors design must expand also to consider how these elements work together to contribute to the total 'user experience.'

13.3 CONTRIBUTIONS FROM THE HUMAN SCIENCES

In the human sciences, the areas of psychophysics, cognition, emotion, interpersonal behaviour and human culture have traditionally been studied separately. In applying human science to design and human experience, we face the reality that these aspects are inextricably linked. And the human factors discipline, the science concerned directly with application to design, has traditionally integrated only those aspects of the science which relate to functional aspects of products and systems. Open any of the classical texts – for example, Wickens (1992), Sanders and McCormick (1993), Salvendy (1997) – and you will find an emphasis upon performance-related issues of safety, usability and physical and cognitive fit. There is no mention of enjoyment. Until recently, perceptual-motor abilities, perception, memory and cognition have dominated our understanding of people's interactions with products and systems.

But the human mind is much more than an information processor and people's reactions are rarely an outcome of cognitive processing alone. Think about the many consumer products that have been very successful despite the fact that their design often makes them difficult to use – Swatch watches, VCRs and computer keyboards, for example. Especially in designing the kinds of products which people have a choice of *whether* to buy and/or use and of *how* to interact with them, we need to consider emotional issues of appeal, fun, aesthetics, taste, ritual, image, lifestyle – the entire range of personal, social and cultural fit.

Some limited explorations of these issues have occurred within psychology, sociology and anthropology.

Aesthetics

In his autobiography, Raymond Loewy (1951) – regarded in the USA as the father of industrial design – has a chapter entitled 'Design and Psychology.' He writes:

'Little if anything has been written about psychology applied to design... The sensory aspects of the normal human being should be taken into consideration in all forms of design. Let's take the Coca-Cola bottle, for instance. Even when wet and cold, its twin-sphered body offers a delightful valley for the friendly fold of one's hand, a feel that is cozy and luscious.'

His interest was in what he called the 'psycho-physiological science' which he believed could offer an understanding of the ways in which visible and tactile quality, colour and form affect people's feelings and perceptions of products. Even today there is little emerging from the science of psychology that can provide the kind of direction that Loewy hoped for.

Ray Crozier (1994) gives a good overview of work formally attempting to understand human aesthetic responses to properties of places and objects in the world. He refers to explorations by Gustav Fechner in the 1870s, and more recent researchers, of the role of the 'golden section' in judgements of aesthetics. He discusses the *Gestalt* school's explorations of the significance of 'pattern' and principles of unity, symmetry, regularity and harmony in human perception. Crozier also reviews work influenced by Berlyne's theories (1971; 1974) linking aesthetic preferences with optimal levels of physiological arousal, whether by sensory properties of the object itself or characteristics of the person's experience of it, such as novelty or emotional association.

But none of these theories can explain or predict the way people will react to designed objects or spaces in practice, not least because aesthetic preferences represent only one part of people's experience with things.

Social and cultural factors

Last time you bought a bag or briefcase you probably thought about aesthetics and such practical issues as the capacity and features you would need. You probably also considered the image it conveys about *you* – does it say 'serious', 'professional', 'arty', 'hip'? Your choice may have been influenced by other social or cultural considerations – is it like one used by someone you admire? Is it of a design or a brand your peers respect?

Social science has recently begun to acknowledge what consumers, designers and anthropologists have long known; Dittmar (1992) writes that:

'... material possessions have a profound symbolic significance for their owners, as well as for other people......they influence the ways in which we think about ourselves and about others. Moreover... self-conceptions, identity, values, group membership, stereotypes and perception of other people are central concerns of social psychology. One would therefore expect to find that the theoretical perspectives within the discipline which deal with these topics have something to say about the meaning of material possessions. But they virtually ignore the material substratum of our existence.'

Miller (1997) writes of the myriad ways in which 'objects' have significance socially, such as supporting non-verbal communication (Goffman 1971) and establishing meaning about our lives and ourselves (Csikszentmihalyi 1991). He suggests that we need to develop a much more conscious 'social psychology of objects' which he says 'might lead to a 'social ergonomics' to parallel the 'cognitive ergonomics' which has been promoted by works like Don Norman's 'The Psychology of Everyday Things'.'

Macdonald (1998) discusses the role of culture in people's aesthetic judgements and presents a framework by Hofstede (1991) who describes human perceptions relating to meaning at different levels: those that are uniquely personal, having significance for an individual because of his or her own associations; those that have significance to a specific social or cultural group through shared meaning; and those that are universal, related to human nature at an innate or fundamental level.

Limitations of measurement

Work in the human and social sciences has tended to focus in an analytical way on the relationship between people's responses and the designed world. This approach, necessarily limited to exploring a few issues at a time, has so far offered little that is applicable to the real-world context of design practice. Qualitative and comparative types of measurement certainly enable us to learn some things about particular issues relating to perceptions – for example, how specific shapes, textures and colours for a toothbrush handle will affect their appeal to children. But in many circumstances such methods are not helpful. Measurement, by its nature, forces us to ignore all but a few selected variables. Hence, measurement is useful when we are confident about which variables are relevant. Frequently designers are concerned with developing new products and services which are not yet in existence or for use in new contexts or by populations who have had little or no prior exposure to them. In these cases, part of the design effort itself is to explore just which design attributes and contextual issues will be important, and how such variables might interact to create a specific kind of experience for people. Here traditional measurement and evaluative techniques are not appropriate.

Frameworks

There may be limitations in directly applicable knowledge, but we do have access to some useful frameworks that represent the complexity involved in people's affective experience of things. Macdonald's (1998) adaptation of Hofstede's (1991) framework of personal, social, cultural and universal layers has been mentioned already. Other useful frameworks help us to structure thoughts around physical, cognitive, social and cultural human factors at various levels; Maslow's (1970) hierarchy of human needs from the most basic survival to self-actualisation adapts readily to people's affective experience of things; Jordan (1998) refers to similar ideas through Tiger's (1992) framework of physiological, psychological, social and ideological levels of pleasure. Simply having structures to think about the variability of human experience provides a starting point for understanding. Even though the frameworks themselves do not provide answers, given any design problem we can use them as a kind of checklist to prompt questions about the implications of design decisions as we consider them.

Tuning in to experience

In practice of course we must add detail to these frameworks. The work of design involves three main activities related to designing appropriate emotional quality and experience:

- *Understanding existing experiences*, to discover those characteristics that will be critical considerations in achieving a desirable design outcome.
- *Exploring and evaluating design ideas*, to optimise the application of particular design details to enhance people's experience of a new system or product.
- *Communicating proposed design solutions,* to give other people a convincing sense of what it will be like to interact with a new product or system before it is built.

Each of these activities prompts questions, either from designers or clients, about how specific design attributes will relate to people's experiences at multiple levels – physical, psychological, social and cultural. One approach to answering such questions, instead of seeking and presenting evidence in the traditional manner, is to create situations in which people can personally appreciate experiences by directly observing or, better still, participating in them. This is what I mean by 'tuning in'. It involves not only the opportunity to make personal discoveries but also to reflect upon and discuss them. Both project teams and audiences need to articulate and capture their discoveries to reach a common point of view about what is important for the population that will be eventually served by the design.

The remainder of this chapter illustrates this approach as it is being developed at IDEO.

Sharing personal discoveries

Project teams are typically made up of individuals representing different types of expertise – in research, design and engineering relating to human, product, environment, business and technology issues – who work together to accomplish project goals. We have learned that team-based, as opposed to purely specialist-based, exploration and sharing of discoveries helps better integrate contributions from traditionally distinct disciplines. When designers and clients are encouraged to reflect upon their own and other people's experiences they quickly reach a common vision about important human issues, including those relevant to the emotional quality of interactions with systems and products. Such an approach greatly diminishes the demand for facts and figures in support of particular design ideas. With respect to experiential quality, shared insights speak for themselves.

This emphasis on team discovery, in turn, changes the role of human factors in design. Rather than just acting as an external provider of information and expertise, the human factors professional's role becomes one of devising and championing ways to facilitate personal discovery about what it would be like for people to engage with a product or system we are designing. It also requires that 'human factors' entertain a broader, more holistic view of the relationship between people and the designed world than the discipline has historically.

The methods that have proved to be most useful are those which respond to the intrinsically subjective nature of experience. These are of two types: 'empathic research and design' which helps participants identify with the thoughts and feelings of people other than themselves; and 'experience prototyping' which represents design ideas in more experiential ways than the more traditional forms that represent appearance or mechanism. Both are related to the idea of 'tuning in' because they provoke participants to become more sensitive to the factors which affect experience – either their own, or by observation or extension, someone else's. Let's look at some examples as they relate to each of the three main design activities.

Understanding existing experiences

The goal of activities here is to discover what qualities and contextual elements currently contribute to the positive aspects of people's experience. Such qualities represent opportunities for design to preserve or enhance them. Additionally, it is important to uncover negative aspects of the existing experience, both in terms of what is missing and elements to avoid in the future.

There are many ways to learn about existing experiences. Contextual observation of various kinds and subjective reports gathered through interviews and other means are widely used. Typically though, these exercises involve real potential users in real contexts and are carried out by a trained researcher who then analyses results and communicates them to a design team. The examples below show some adaptations of these methods to develop exercises which enable shared discovery directly by design team members.

Personal narratives about the experience of making images

This example involves an essay-writing assignment given to team members as a way of developing sensitivity to qualitative aspects of their own experiences in a familiar domain:

In a project related to the development of digital cameras, a first step was to understand as much as possible about the sources of enjoyment and frustration in taking pictures. The design team, including representatives from marketing, strategy, technology research, engineering and design were all asked to write a personal account of a recent experience in which they had taken or used images in some way. Their instructions were to:

'think of a specific situation – a day out, a report or presentation you were preparing – when you were engaged in using a camera or dealing with images. Write a couple of paragraphs (more if you like!) describing your experience. Tell what you did, what happened and try to recall your thoughts, feelings and intentions.'

The idea was to prepare team members to think 'subjectively', rather than 'professionally' about imaging issues. It was a warming-up exercise prior to making observations of other people's image-making activities and aimed to develop sensitivity to interpreting people's words and actions in terms of the thoughts and feelings behind them. The anecdotes covered a wide range of events including a wedding, inspecting the plumbing, putting a slide show together, sending digital images home for the first time, taking a cruise and making a birthday card. Sharing just a few of these stories helped the team start to develop an intimate perspective on some of the rituals around picture taking, what makes it fun, the value and downside of instant pictures, what is valued in favourite cameras and what really matters about image quality. The team was able to start developing a user-centred framework that captured ideas about the pleasure of creation, joy in improvisation, emotional value of pictures, the personal relationship that develops with a camera, the socially invasive nature of large cameras and 'the technological potting shed syndrome' as one designer described that aspect of recording events with video-cameras that effectively takes people (often men!) out of the social action. The main value of this exercise was that it expanded the team's point of view about what was important to people. Later discussions about design and technology options were informed by vivid understanding of the social and personal values attached to cameras and the communicative and creative purposes of picture-making.

A simulation of patients' experiences

This next example builds upon people's own imaginations and the use of proxy devices to recreate the essential elements of a personal experience that would not otherwise be available:

The project was to design product and service related elements for a cardiac telemetry system. The system would involve both face-to-face and remote doctor – patient interactions as well as automated supervision for patients with chest-implanted automatic defibrillators. Before embarking upon design solutions, the team wanted to know what system characteristics would be needed to ensure as positive an experience for patients as possible? What would it be like to be a defibrillating pacemaker patient? The design team set up circumstances to produce a similar experience. The aim was to provoke insights into important functional and emotional issues and inspire thoughts about how to deal with them. One of the designers distributed pagers and a recording kit to the other members of the development team (Figure 13.1*)*. The pager signal was to represent a defibrillating shock that would be of sufficient impact to knock a person off their feet. Participants were paged at random times during the week and asked to note their circumstances for each occasion – where they were, with whom, what they were doing and what they thought and felt knowing that this represented a shock. After this, team discussion about personal experiences ranged from anxiety around everyday activities like holding an infant son or working with power tools, to social issues about how to make sure that onlookers knew what was happening and how to get proper medical help. The participants, including electrical and mechanical engineers, bio-technologists and representatives from marketing and product planning, quickly translated their own experiences into patients' emotional need for reassurance. They were aware of the need for explicit information display and feedback in their next generation of the system and products.

Figure 13.1 Each recording kit contained a single-use camera, notebook and pen for participants to record surroundings and impressions when the pager signalled that they had received a defibrillating shock

Exploring and evaluating design ideas

The central task in exploring and evaluating design ideas is to discover which design attributes will give rise to the desired emotional qualities – whether delight, reassurance, excitement – and, ultimately to design an experience that embodies them. The activity involves the creation and refinement of ideas through an iterative process of representation and evaluation. Design ideas are expressed as sketches and models of various kinds that take on increasingly high fidelity as the design progresses. Representations are evaluated, formally or informally, by designers as they work, by people who represent the eventual users and customers, as well as by client or sponsor. Typically design prototyping activities emphasise aspects of the artefact itself – its look and feel, how a mechanism might work – rather than the experience of use in context. Notable exceptions are user scenarios and screen-based dynamic simulations that specifically explore experience (for example as described in Houde and Hill 1997; Fulton Suri and Marsh 2000). The examples below illustrate some other ways of getting in touch with experiences and the people having them, at different levels of design resolution.

Role playing an improvised train journey

This example is focused on a very early stage of design exploration and uses role-playing, 'body-storming', and improvisation techniques to explore and generate design ideas around different people's experiences of interactions in a range of simulated contexts:
In embarking on the design of elements of a new rail service, a group of designers decided to use improvisation methods to help come up with design ideas to provide positive experiences for different types of travellers. They devised roles – a commuter, parent with a small child, an elderly woman – and various situations during the different stages of the train journey, including entering the station, ticketing, waiting, riding the train and connecting to other means of transportation. On this occasion a professional actor, familiar with improvisational theatre, served as the process moderator. Each situation was introduced by distributing a card containing the rules of that 'scene', explaining the goal and defining the basic the roles of the participants. For example, in one scene the moderator gave one designer the role of 'parent', another 'toddler', and gave the parent the instruction to buy a return ticket for yourself and your child. Another designer played the role of a ticket-vending machine. The moderator then instructed the designers to take on particular characteristics, e.g. 'Now you have gloves on', or to make assumptions about environmental conditions 'It's dark and windy' or characteristics of the designed objects 'It will only take coins, no notes', or 'The ticket machine is very helpful and friendly.'

This improvisation was an opportunity to experience potential interactions but also to generate initial design ideas through 'body-storming.' Body-storming (which is physical brainstorming in context) occurred either during or between scenes in response to problems that were uncovered. Some of the ideas were verbally expressed but many were expressed physically and came about spontaneously through interaction with proposed design elements, or quickly improvised stand-ins. For example, in this improvisation of the ticket machine interaction, someone picked up a piece of foam-core to provide a shelf for the toddler to rest his colouring book on. The 'ticket machine', when told that it was out of order, turned away from the intending purchaser to indicate its status through a radical change of appearance. The dynamic physical situation itself stimulated those interesting design responses in situ. There were also breaks after each scene to allow for group reflection, discussion and capture of what was learned through participating or witnessing each scene.

Experiencing feedback from a channel changer

This example uses off-the-shelf components to improvise the essential physical and behavioural qualities of a discrete interaction so that designers and others can experience it directly:

The intent here was to explore some new ideas about ways of providing control over channel selection for televisions. The designer wanted to understand and demonstrate what it would be like to change channels through different kinds of interaction rather than button pushing. His goal was to explore and evaluate the experiential quality of this detailed interaction by abstracting its essence and, for the moment, ignoring other aspects of functionality or look and feel. He created what he termed 'behavioural sketches' which were simple electronic circuits containing a few lines of code (Basic Stamps™), encased in translucent off-the-shelf soap dishes. The two experience prototypes were controlled by a tilting gesture, which had the effect of switching channels up or down. The first provided visible feedback in the form of moving light bands whose apparent speed increased with tilt. The second offered tactile feedback using vibrations that varied by rate and intensity. By tinkering with the simple software program, he was able to efficiently develop and test many subtle iterations of the product's behaviour and examine the resulting experience of using it. The improvised nature of these prototypes proved to be vital for exploring this conceptual approach with other designers and the client. The prototypes were expressive enough for everyone to have a sensual and compelling experience, but abstract enough to avoid constraining people's imagination about ways of enhancing the experience, or thinking about other applications where such gestural and multi-sensory control methods might be appropriate.

Using experience prototypes of a picture communicator

This example involves a trial of robust working prototypes by users in a real-world context which allows them to enjoy a more realistic experience as they begin to integrate the device into their everyday lives and to try out new behaviours:

The EC-funded 'Maypole' project involved exploration of family communications. In partnership with Nokia, the Netherlands Design Institute and Universities of Helsinki and Vienna we developed several design concepts for communication devices, one of which was a 'picture communicator' with a touch pad display on its rear side. A working prototype of that interface was built to check that it was intuitive and workable. But what would it be really be like to communicate pictures with such a device? Nokia built a set of fully communicating prototypes that were used in field trials in Helsinki and Vienna. The prototypes were robust enough to distribute to children who could take them away and play with them unsupervised for days at a time (Figure 13.2). The communicators were a great success. The experience of being able to take pictures and send and receive proved so compelling that the children almost forgot about the inconvenience of the batteries and transceiver unit that they had to carry around in a backpack to make the prototypes work. As an observer of user evaluations, one knows very quickly if the designed experience is a good one, as in this case when people get so involved in the experience that they forget about the limitations of the prototype.

Figure 13.2a Children used fully working prototypes of the picture communicator. Their pleasure was evident ...

Figure 13.2b As they sent and received pictures from their friends ...

Figure 13.2c Despite the heavy backpacks containing batteries and drivers for the system ...

13.4 COMMUNICATING DESIGN IDEAS

Communicating the design intent behind concepts is a key activity in design. First it is important that a sponsor or client is helped to understand whether a specific design proposal will really lead to a desirable experience for the people who will eventually interact with it. Second, communication must be compelling enough to create a shared vision about the critical but subtle details that contribute to the overall experience. Without this vision there is danger that such details will be overlooked or reinterpreted such that the final implementation engenders a very different kind of experience from the one that was intended. Appearance models, drawings and specification documents may be important for reference but they do not support the creation of that vision in a compelling way. The examples below show ways to give other people, clients or developers for example, a powerful sense of what it will feel like to interact with a new design.

An experience prototype to specify interaction architecture for a digital camera

This example demonstrates the communication and motivational value of providing a portable and tangible prototype in getting design ideas across and keeping them alive. It involves a prototype that lets people experience the design ideas directly and make that experience available to others.

In the early phases of a project on digital photography, we used traditional communication techniques such as scenarios and interactive screen-based simulations to explain our vision of the interaction architecture. But these techniques did not help the client to fully grasp the intended user experience and camera behaviour. The breakthrough came when the designers built an integrated hardware and software prototype that represented only the intended 'look and feel.' The prototype bore little resemblance to a desirable product in shape, form, size, configuration, or weight. There was a sizeable cable running from the camera to a desktop computer where all the processing occurred (Figure 13.3a).

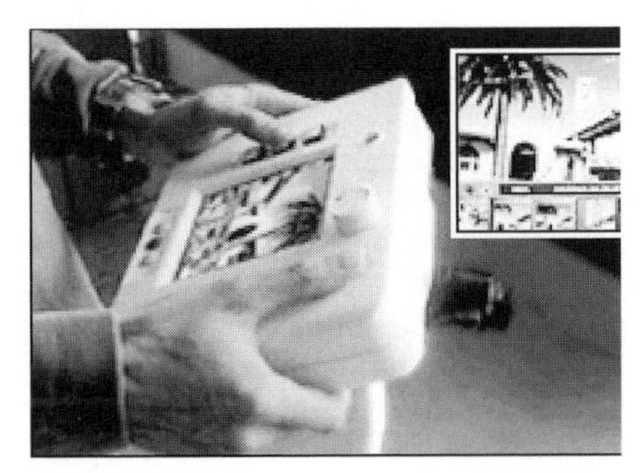

Figure 13.3a The 'user experience prototype' communicated dynamic aspects of the control system screen behaviour for a digital camera interaction architecture using processing power cabled from a personal computer

The prototype itself contained a small video camera attached to an LCD panel, encased in a box. The size of the LCD panel was determined by the desired resolution, a key aspect of the proposed user experience, rather than by the desired physical size. The prototype had a live video-feed and could capture still photos with audio-annotations in real time. Since one of the critical variables of the user experience was the system response time, the prototype was set up so that it could be fine-tuned through the desktop computer. It was easy then for the design team and the client to appreciate the impact of delays in response time on the quality of the user experience.

Eventually the client's developers asked for multiple copies of this prototype to serve as a 'living specification' to maintain the vision and verify new design concepts within their design organisation. We were told that there have been many pressures to change the resolution and the speed of response, but that the prototype enabled the developers to see, feel and resist the negative impact of such changes (Figure 13.3b).

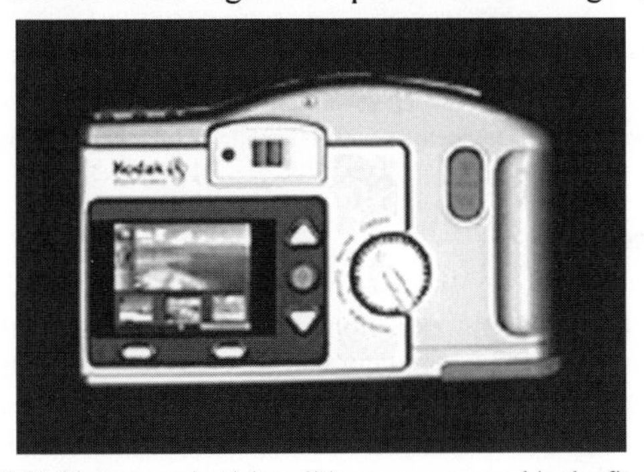

Figure 13.3b These experiential qualities were preserved in the final implementation

Using video to set the mood for Kiss Communicator prototypes

This example illustrates how traditional communication methods and experience prototypes can work hand in hand to establish a suitable emotional tone and context to successfully communicate a proposed experience.

The 'Kiss Communicator' was a concept prototype built to explore ways of using technology to communicate with another person in a subtle, sensual way. 'We wanted to keep the nature of the physical object as simple as possible, so the interaction was more about the experiential message' the designer said.

Designed to facilitate the exchange of emotional content between people separated by physical distance, the 'Kiss Communicator' uses wireless technology to transmit the digital equivalent of a personal gesture, like a wave, wink or a kiss. Each communicator is able to connect only with a specific corresponding module, resulting in a secure and intimate one-on-one exchange. To let a partner know that you are thinking of her or him, you squeeze the communicator gently. It responds with a slight glow to invite you to blow into it and create your 'message' in the form of an animated light sequence as the device responds to your breath. The 'message' shows while you blow and if you are happy with it, you simply relax your grip and it is sent to the corresponding communicator. On the other end, the partner Kiss Communicator indicates that there is a message but waits until its owner squeezes it to play back the light sequence.

There are some important conditions necessary to really appreciate the experiencing of using this prototype: An intimate relationship, two distant people, sending a gesture, etc. Most of the time when we show this concept we are talking to clients in their business suits in a conference room somewhere. To help convey the experience in this alien context we produced a short video sequence which shows a pair of communicators in use by a dreamy couple who are separated by distance. Using conventional devices like soft focus and a romantic soundtrack the video creates, at least temporarily, an atmosphere that is a helpful precursor to the hands-on experience of the physical prototypes themselves (Figure 13.4).

Figure 13.4 A video of people using the Kiss Communicator helps to set an appropriate context for people to appreciate the intimate and subtle behaviour of the prototypes themselves

13.5 CONCLUSIONS

Design practice involves making decisions which affect the quality of experience that people will have with products, environments, services and systems. There are limitations to our detailed understanding of how specific design variables contribute to people's affective experience in practice. Other than in fairly circumscribed conditions, where it is clear which variables matter, we cannot make generally applicable or convincing predictions about people's emotional responses to collections of design attributes. Though some useful frameworks exist, the complex interaction of multiple variables and the subjective nature of experience often makes it very difficult to measure effectively in ways that are useful to designers.

On the other hand, the use of empathic methods and experience prototyping can help designers, in three key ways:

- In *understanding* essential factors of existing experiences.
- In *exploration and evaluation* of ideas by providing inspiration, confirmation or rejection of ideas based upon the quality of experience they engender.

- In *communicating* the experiential quality of design proposals and providing convincing support for the critical contributing elements.

They do this principally by allowing people to have their own experiences and, through reflection and discussion, make discoveries about what is important. Of course, there are drawbacks inherent in such a subjective approach. One person's experience cannot be assumed to be similar to, or predictive of, another's. This is why reflection and discussion is essential. Even empathic methods that emphasise careful observations of other people or role-playing, to be successful, must be based upon good choices about what characteristics of people and context are relevant to explore. Experience prototyping also inevitably makes assumptions about what features and behaviour of a design to model. Particularly when exploring and evaluating design ideas, it is important to think of prototyping as involving multiple iterations, rather than an isolated event. Using this approach does not need to exclude the use of more objective, human factors methods.

But, as the foregoing project examples suggest, in many design applications it is more useful to focus on enhancing designers' and clients' sensitivity to experiential issues in design, than to focus solely on measurement. Such experiential methods have the advantage that they preserve the intrinsic richness and subjectivity of experience and emotional phenomena. They also set up processes that lead to shared perspectives among members of design teams and their client or sponsoring organisations and a sense of commitment to the design of desirable experiences.

13.6 ACKNOWLEDGEMENTS

Thanks to colleagues and clients for sharing examples of their work and especially for stories and inventions provided by IDEO San Francisco and by: Leon Segal; Marion Buchenau; Alison Black, Thomas Stegmann and Alexander Grunsteidl; Duncan Kerr, Heather Martin and Mat Hunter.

13.7 REFERENCES

Berlyne, D.E., 1971. *Aesthetics and Psychobiology*. Appleton-Century-Crofts, New York.

Berlyne, D.E., 1974. *Studies in the New Experimental Aesthetics*. Hemisphere, Washington D.C.

Crozier, R., 1994. *Manufactured Pleasures: Psychological Responses to Design*. Manchester University Press, Manchester, UK.

Csikszentmihalyi, M., 1991. Design & Order in Everyday Life. In *Design Issues*, **8**, pp. 26-34.

Csikszentmihalyi, M. and Rochberg-Halton, E., 1981. *The Meaning of Things: Domestic Symbols and the Self*. Cambridge University Press, Cambridge.

Dittmar, H., 1992. *The Social Psychology of Material Possessions: To Have is To Be*. Harvester-Wheatsheaf, Hemel Hempstead.

Fulton Suri, J. and Marsh, M., 2000. Scenario Building as an Ergonomics Method in Consumer Product Design. In *Applied Ergonomics*, **31**, 2, pp. 151-157.

Goffman, E., 1971. *The Presentation of Self in Everyday Life*. Harmondsworth Pelican: originally published Anchor Books 1959.

Hofstede, G., 1991. *Cultures and Organisations*. McGraw-Hill International, Maidenhead.

Houde S., Hill C., 1997. What do Prototypes Prototype? In *Handbook of Human-Computer Interaction* (2nd edition). Helander, M., Landauer, T. and Prabhu, P. (Eds),

Elsevier Science B.V., Amsterdam.
Jordan, P.W., 1998. The Many Faces of Pleasure: Design and Product Appeal. In *Innovation*, Summer 1998, pp. 30-33.
Loewy, R., 1951. *Never Leave Well Enough Alone*. Simon and Schuster, New York.
Miller, H., 1997. *The Social Psychology of Objects*. Paper presented at Understanding the Social World Conference, The Nottingham Trent University; UK.
Macdonald, A., 1998. Developing a Qualitiative Sense. In *Human Factors in Consumer Products*. Stanton, N. (Ed.), Taylor & Francis Ltd., UK.
Maslow, A.H., 1970. *Motivation and Personality* (2nd edition). Harper and Row, New York.
Salvendy, G., 1997. *Handbook of Human Factors and Ergonomics* (2nd edition). Salvendy, G. (Ed.), Wiley Interscience, New York.
Sanders, M.S. and McCormick, E.J., 1993. *Human Factors in Engineering and Design* (7th edition). McGraw-Hill, NY.
Tiger, L., 1992. *The Pursuit of Pleasure*. Little Brown and Company, Boston, USA, pp. 52-60.
Wickens, C.D., 1992. *Engineering Psychology and Human Performance*. Harper Collins Publishers Inc., New York, NY.

CHAPTER FOURTEEN

Using Video Ethnography to Inform and Inspire User-Centred Design

KATE TAYLOR

Serco Usability Services, 22 Hand Court

Holborn, London WC1 6FJ, UK

MARTIN BONTOFT

IDEO Europe, White Bear Yard 144a Clerkenwell Road,

London EC1R 5DF, UK

and

PROF. MARGARET GALER FLYTE

Ergonomics and Design Group, Department of Human Sciences,

Loughborough LE11 3TU, UK

14.1 INTRODUCTION

This chapter is as an attempt to document the values and pitfalls of applying the ethnographic technique to the formative stages of design. Based upon research conducted with IDEO Product Development, this paper also intends to highlight the problems encountered, and the compromises that must be made to ensure a more integrated fit between ethnography and design.

An aim of the chapter is to raise awareness of the value of introducing the method of video ethnography at the formative stages of the design process. While ethnography has long been a central methodology for anthropologists and social scientists, the present surge of interest in ethnography (related to the fields of market research, advertising and design), can be seen as a relatively recent phenomenon. To keep pace with this trend, many design consultancies are now looking to explore the value of utilising the ethnographic method. However, the transition from applying the technique within an academic context, to applying it within a design context, needs careful consideration and management. The human factors specialist (with an awareness of the ethnographic methodology), may be best placed to understand the dynamics of the design process and to work alongside anthropologists in order to aid the integration of ethnography into its new context. With proper management, it is felt that video ethnography has a beneficial part to play in encouraging the uptake of user-centred information. The method of ethnography looks set to evolve into a highly valuable tool that will enable designers to gain a more holistic picture of potential end users and their product requirements.

14.2 WHAT IS ETHNOGRAPHY?

Ethnography derives from the discipline of anthropology and its concern with observing, documenting and understanding distant cultures. Put in more simplistic terms, the purpose of ethnography is to report the way of life of people who lie beyond the experience of our own way of life. Clearly this method is in sympathy with the objectives of the human factors specialist, by ensuring that the design process remains strongly user-centred. Based upon first hand observations, this technique employs qualitative methodologies in order to convey the 'inner life and texture' of a particular social group. Johnson (1994) describes ethnography as being 'an intensive contextual and holistic approach which aims for depth rather than coverage'. By aiming for depth, it is possible for this technique to generate incredibly rich, descriptive data. To quote Lareau and Schultz (1996) 'Rather than being interested in how frequent a behaviour is, they [ethnographers] seek generally, to understand the character of day-to-day life of the people in the study'.

This method differs from other qualitative research methods in four main ways:

- The first is the fact that the researcher becomes involved with directly observing the participant within the study. Unlike an interview, the researcher is witness to actual events taking place.
- The second major difference is the fact that participation on the part of the researcher is actively encouraged by this technique. When conducted with sensitivity, it is believed that this form of 'embodied understanding', encouraged by the experiential methodology of ethnography, will enable the researcher to understand the concerns of the observed more fully.
- The third main distinguishing feature of ethnography involves the comparatively lengthy time that the researcher traditionally spends in the field (usually around one year).
- The fourth main difference concerns the fact that research is conducted within people's naturalistic settings. The goal is to foster a relaxed and natural setting for the research to take place. The places that the participants select to conduct the research can be very telling indicators of their personal identity, and help to add to the rich descriptive data that is generated.

14.3 THE ETHNOGRAPHIC TEXT

Ethnographic knowledge reaches its audience via an 'ethnographic text'. Although most commonly the format of an ethnographic text is that of a written account, the communication of ethnographic knowledge is not solely restricted to this medium. Whilst field diaries, written accounts and photographs have traditionally been used to communicate ethnographic knowledge, other ethnographers have explored the potential of using moving image cameras to record and communicate events in the field. As video has become cheaper and more accessible, this method of communicating ethnographic knowledge has become increasingly popular. However, the medium of video has found itself at the centre of a recurrent debate concerning its 'claim upon the real' or mimetic qualities (refer to Henley 1998 for a discussion of the use of film in ethnography and the resulting 'documentation versus documentary debate').

The research conducted at IDEO and the opinions expressed in this chapter are concerned with video ethnography, rather than other forms of capturing and communicating ethnographic knowledge. The medium of video was chosen for the following reasons:

- Design is a visually led discipline, therefore video is an appropriate visual communication medium.
- Video is potentially one of the most direct forms of communication, as it does not require a mediator.
- Understanding video requires designers to be actively engaged in the communication process.
- Video is a richly descriptive form of communication, providing extra layers of information that may inspire design.
- Video has power because it is associated with the notion that 'seeing is believing'.

These five points are expanded in section 15.3 (Sensory Aspects) to justify the role of video as an appropriate communication medium for the design context.

Recently ethnography has been the subject of a backlash against the opinion that an ethnographic text could provide a valid account of the 'native's perspective'. Although a discussion of 'The Crisis of Representation' (Clifford and Marcus 1986) is beyond the scope of this chapter, it is worth pointing out that ethnographies are never wholly accurate accounts, and it is inevitable that the resulting data is subjective to some degree. However, if approached sensitively and if the researcher's influence is acknowledged, then the ethnographic technique can still yield some highly valuable contextual insights into people's opinions and behaviour.

14.4 WHY THE NEED FOR ETHNOGRAPHY IN THE DESIGN PROCESS?

During the 20-year history of user-centred design, human factors specialists have been responsible for advocating numerous approaches intent on improving the usability of products. Whilst improved usability is a desirable concept, many of the more traditional approaches to user-centred design have come under criticism for the fact that they focus solely on the ability of a person to interact with a product on a physical and cognitive level. This approach denies the fact that a person's relationship with a product can extend beyond the physical and cognitive, and can incorporate social and emotional aspects of people's identities. Macdonald (1998) supports this view when claiming that 'Ergonomics is usually about avoiding discomfort or injury rather than proposing a qualitative threshold – one which could enhance the users' experience'.

In response to this debate, a recent movement has emerged that aims to promote a more holistic, person-centred version of design, whereby the potential end user is consulted in order to understand people's social and emotional responses to products, alongside the more traditional usability criteria. Jane Fulton Suri (1995), head of human factors at IDEO, one of the world's largest design consultancies, supports the need for a more holistic understanding of people's responses to products, and claims that the traditional approach of human factors underplays its role with design. 'The new challenge is to make products and environments that do better than keep us from getting sick, damaged or irritated. [...] we need to explore new ways of learning about people in dynamic interaction with products, and new ways of implementing what we find out.'

The method of ethnography attempts to meet this challenge. The intention is not to 'do away with' more traditional usability methodologies, nor is it to rely solely upon ethnography as the only user-centred input to the design process. Instead, the aim is to utilise ethnography in order to access subjective values and latent user needs, and to combine this information with more traditional usability techniques to inform product concepts. As highlighted by Macdonald (1998), 'The objective nature of [the human factors specialists] traditional methods has not given them a language to explore and express such ideas.' It is hoped that ethnography may help to alleviate this problem and to provide the human factors specialist with a tool that will enable them to explore and express the more subjective and emotional experiences of product use.

14.5 THE ETHNOGRAPHIC PROCESS AND THE DESIGN CONTEXT

The ethnographic method is not simply a one-way activity. Instead it takes a connected and iterative route through the methodological stages. In this sense data gathering, analysis, editing and communicating do not exist as discrete events but are strongly related, and exert a profound level of influence upon each other. 'Whereas in most research, analysis follows data collection, in ethnographic research analysis and data collection begin simultaneously. An ethnographer is a human instrument, and must discriminate among different types of data and analyse relative worth of one path over another at every turn in the fieldwork, well before any formalised analysis takes place' (Fetterman 1998). Research, sampling, analysis, and communication go hand-in-hand.

Generating user requirements

One of the values of using ethnography to generate user requirements lies in its ability to uncover the (frequently sub-conscious) 'say/do' dichotomy. As Miller (1998) states 'we are constantly faced with everyday discrepancies between what people say matters to them, and what they actually give their attention to'. Ethnography is able to alleviate this problem because it is possible to obtain a knowledge of what people claim they do (via discussion), and then to actively observe their behaviour – what they really do. Therefore ethnography allows a much more in-depth understanding of people's complex responses towards products and services. There is no substitute for actually 'being there' as an event unfurls.

The other major value of ethnography when communicating with potential users is its ability to access latent needs. Finding methodologies that enable users to articulate their needs, especially with regard to a novel product concept, has been a problematic area for human factors specialists. Initially, within the field of user-centred design, it was commonly believed that user requirements could be 'captured'. This concept implies that people were fully conscious of all of their product requirements and were simply waiting for the opportunity to tell designers what they needed from a product. Obviously this is not the case, as it is impossible for the potential end user to anticipate all of their needs in relation to an abstract product concept. Requirement **generation** therefore became one of the main aims of user-centred design methodologies. The human factors specialist aimed to introduce stakeholders to a new product concept by creating scenarios and role-plays in order to actively generate user requirements. Ethnography is able to generate user requirements because it provides a method that is able to access users' latent needs through interaction with an analogous product, or by encouraging people to think about their sub-conscious behaviour as they perform their everyday activities in the presence of the researcher.

More to the point, not only may a potential user be unable to fully envision a future product, but designers and human factors specialists are also likely to overlook some vital research criteria because they are unable to fully appreciate how the future product would operate in reality. The ethnographic method allows the researcher to witness and uncover useful design insights that may have been completely overlooked if other research techniques were utilised. This point is supported by Whyte (1955), who wrote 'As I sat and listened, I learned the answer to questions I would not have had the sense to ask if I had been getting my information solely on an interviewing basis.'

Analysing and editing the video data

Although a considerable amount of literature exists concerning the value of utilising the approach of video ethnography, the actual mechanics of **how** to decide which parts of the footage should be left in the video and which edited out, remains relatively undocumented. Referred to in the industry as a 'Black Art', the editing process certainly seems to be an elusive subject area.

Ethnographic research is highly contextual, meaning that its results remain tied to the site of production. In order to make the research findings applicable to design, it is necessary to look beyond the participant's viewpoint and to understand what is happening from a broader perspective. Having gained an in-depth understanding of the particular participant, the researcher must take a step back to see how the findings relate to the 'bigger picture'. 'The ethnographer's task is not only to collect information from the emic or insider's perspective, but also to make sense of all data from an etic or external social science perspective' (Fetterman 1998).

Two ethnographic techniques are available to aid the researcher in this task:

- The first is that of triangulation. 'Triangulation is a basic in ethnographic research. It is at the heart of ethnographic validity – testing one source of information against another to strip away alternative explanations and to prove a hypothesis' (Fetterman 1998). Without applying this technique, the findings generated from ethnographic research are little more than stories. While they may prove interesting and engaging as stand-alone entities, they bear little relevance as a human factors research tool. By using triangulation to gauge the main similarities and differences between each set of data, it is possible to map the differing perspectives along a continuum, and to build up a portrait of different potential product users, their values and product requirements.
- The second technique used to aid the analysis and editing phase is known as 'crystallisation' and refers to the distilling of the main issues. Video-based ethnographic research generates massive amounts of video footage. Clearly this information must be condensed in order to make it both manageable and useful. When analysing and editing the ethnographic data, care must be taken to ensure that generalisations and stereotypes do not result from the distillation process. While some anthropologists would argue that the researcher must stand back from the data to some degree and let the culture 'speak for itself', the design process necessitates a different approach due to time limitations and design practice. The task of crystallisation within a design context is to find video clips that succinctly identify design insights and generate opportunities for design. Simply displaying the main themes of the research is not enough. These themes must be interwoven and connected to create a meaningful story that relates each particular participant of the research to the other participants, in terms of their needs, values and beliefs, in relation to the potential product concept. Only then does the ethnographic method become a tool that is useful to the human factors specialist.

Communicating with designers

Macdonald and Jordan (1998) have highlighted the problems caused by the 'communications gap' resulting from the dichotomy that exists between the visually orientated artistic profession of design, and the scientific discipline of ergonomics. As human factors specialists, our task is to generate and understand user requirements and to communicate these to the designers. The need to achieve effective communication is supported by Eason (1995) when he claims that 'the 'media' by which the human factors message is conveyed are very important. The analysis of the design process shows that even the best contribution could be wasted if it is directed at the wrong people, in the wrong form, or serves only to threaten the job or work of existing designers'.

Method of delivery

The design process is known for being a highly constrained activity, where obstacles (such as time, resources, relevance, and 'fit with practice') affect the up-take of user-centred design (Eason and Harker 1991). These constraints affect the role that the human factors specialist must adopt when delivering user-centred findings to designers. As user-centred design has developed to embrace more participatory forms of design, the human factors specialist has needed to adapt his/her role in the communication process. Rather than simply passing information to designers, in the form of technical reports or guidelines, the human factors specialist has needed to take a more active role in the communication process, thereby adopting the role of a 'facilitator' (Eason 1995).

Finding an appropriate communication medium

Design has always been a visually led discipline, whereas human factors have tended to be dominated by technical writing and data generated from physical measurements. Clearly there is a conflict between these two communication media, as highlighted by Macdonald and Jordan (1998). 'Whilst designers may prefer visual communication aids, many human factors specialists may favour communicating by technical reports. This sort of reporting can be seen by designers as dull, produced in a difficult-to-use format, and more suited to a university laboratory than a commercial design studio. The language that ergonomics uses has limited the extent to which it can communicate its own field effectively to others. On the other hand, visual tools can help orient human factors material to suit designers' strengths as well as assisting all those in the product development teams facing difficult qualitative value judgements'. Video ethnography gives the facilitator a visual medium capable of communicating directly with design teams to show them the requirements, opinions and latent needs of the potential product consumers (Figure 14.1).

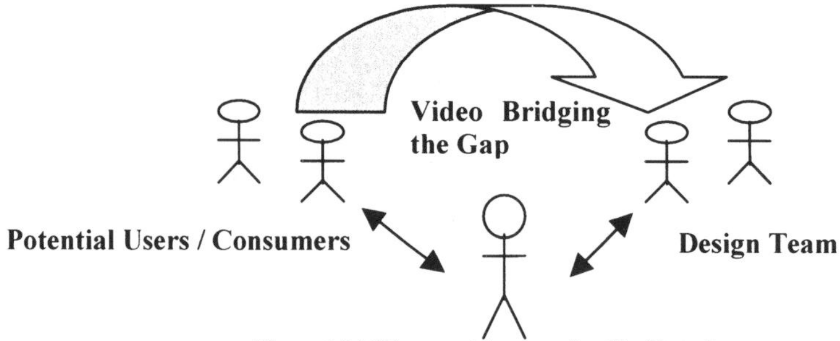

Figure 14.1 Ethnographic researcher 'facilitator'

During recent years, designers have increasingly relied on the use of video within the design process. Most commonly this usage has been restricted to operating the video as something that can 'document' the various stages of design development. It has also been used by human factors specialists to aid the process of measuring user performance. The use of video in conjunction with ethnography proposes a new role for the medium of video within the design process, one where video can make an inspirational contribution at the formative stages of design. The use of video as a communication tool is clearly of value to the design process, given its visual nature. However, video also has four other characteristics that make it incredibly beneficial when articulating user-centred accounts:

- **Video is direct.** Video is potentially one of the most 'direct' forms of communication, as it does not require a mediator. Although the researcher is responsible for analysing and editing the footage, video still enables participants to 'speak for themselves'. This enables designers to witness potential end-users in a more first-hand manner than other media allow.
- **Video is believable.** By utilising video to communicate ethnographic accounts to designers, it is possible to communicate experiences that lie beyond the designer's individual range, and to develop an empathic closeness between the user and the designer. As Suchman and Trigg (1991) claim, 'What we see consistently is that the closeness of designers to those who use an artefact [...] directly determines the artefact's appropriateness to its situation of use'. By using the medium of video, it is possible to really foster this sense of closeness. Video is an appropriate visual medium to communicate to designers, and has added power because video is accompanied by a sense that 'seeing is believing'. It seems to be the case that if a person witnesses an event on video, they are more likely to accept it as a truth than if someone just told them of the same event.
- **Video is richly descriptive.** Video is also such a rich media that it is able to communicate the main themes of user research, while still enabling designers to draw significance from 'background events'. Video communicates the message of the user-centred research, while also delivering extra layers of information. This extra information gives further scope to spark design ideas of relevance to the potential user population.
- **Video involving active engagement.** Whilst editing enables the video to communicate the researcher's findings, the richness of video means that the media can be interpreted in multiple ways. Video is, as McLulah has pointed out, 'a highly participatory media, where participation and empathic engagement has to be invested to make sense of the material'. The medium of video therefore enables designers to become more actively involved in co-defining users needs and requirements; it provides a media for on-going negotiation and reflection (Binder and Buur 1999).

14.6 CASE STUDY – APPLYING THE ETHNOGRAPHIC METHOD TO IDEO'S DESIGN PROCESS

IDEO were responsible for proposing an MSc. thesis research topic that aimed to explore the value of using video ethnography as a way of communicating user-centred issues to designers, at the formative stages of the design process. IDEO provided the equipment and facilities necessary to conduct the research and Loughborough University provided academic support and guidance. In order to conduct this research, a hypothetical design brief was proposed that aimed to discover people's attitudes towards developing the mobile phone as a 'fashionable' artefact. The ethnographic method was applied to the design brief, with the intention that an edited video would be produced to communicate the main 'insights' to the designers. Whilst conducting the research, notes were taken to document the process, and to highlight the positive and negative aspects associated with using the technique. The following paragraphs contain a summary of the main issues that arose.

Data gathering

Traditionally ethnography has been characterised by the lengthy amount of time that the researcher spends immersed in the culture of an 'other'. With such a highly constrained activity as the design process, often the time available to spend with a participant is reduced to only a matter of hours. This results in a dilemma for the design-based ethnographer. Time spent in the field is a valued characteristic of ethnography because it enables rapport to be built between the researcher and participant, thereby establishing an honest understanding of the participants' everyday way of life. If the time available to conduct research is limited to a matter of hours, then it becomes difficult to gain an 'insider's perspective' when conducting research with a complete stranger.

However, several other strategies are available that enable the researcher to rapidly establish links with people. The following four strategies suggest ways in which to achieve these rapid connections:

- **Enlist acquaintances as participants.** Involving an acquaintance can be a useful strategy to adopt when research time is short. However, this approach is not always appropriate. The use of an acquaintance can only really help to quickly establish a rapport between participant and researcher, when the researcher's identity is similar to that of the user population being researched. It would be wrong to assume that just because the researcher is vaguely acquainted with the participant this automatically leads to a deep understanding of the participant's lifestyle, values and needs, during a short period of research. If the experiences of the user group extend markedly beyond the researcher's own experiences, then other strategies are needed to help to rapidly foster a deeper empathic understanding.
- **Research as a team rather than an individual.** Obviously there is a question of balance to be addressed here, but provided the team is small (e.g. 2-3 people), then it can be useful to conduct research in small teams. Traditional research involves long-term immersion that enables the researcher to break down their outsider's view, and to begin to understand from within. When conducting rapid ethnography it is tempting for the researcher to feel they have an understanding of the participant's view point, but to then impose their own perspective onto this understanding. By researching in teams and de-briefing each other after fieldwork sessions, it is possible for the team to begin to gain a broader perspective of events and their meanings, because a multiple-viewpoint is adopted.
- **Use an opening questioning technique.** Another strategy that helps to link the researcher to the participant in a rapid way involves using a very open questioning style. By adopting a 'researcher as apprentice' stance (observing behaviour while asking relatively naïve questions) as advocated by Contextual Inquiry, it is possible to encourage participants to begin to reflect upon their tacit knowledge (Beyer and Holtzblatt 1995). Questioning participants openly and sympathetically makes it possible to get a deeper understanding of their experiences.
- **Reveal information about yourself.** Whilst traditional ethnography would not advocate this strategy because it may bias the data, this approach can be a useful way of quickly establishing links with people. The most appropriate way to achieve

effective results using this strategy is to listen to the participant first and gauge their perspective. A sensitive researcher can then pitch themselves at the right level and tell stories that encourage the participant to respond. Rather than simply asking questions (which can seem rather accusative), it is possible to tell the participant 'about the time when.....'. This approach encourages the participant to open up to the researcher and usually results in the participant communicating an informative tale back to the researcher. Obviously this strategy can potentially bias the data but a sensitive researcher may still find this technique of use in rapidly generating information when time is short.

Whilst some anthropologists may claim that such a rapid approach to ethnography invalidates the resulting data, it is important to remember that we are attempting to adapt the ethnographic methodology to fit the design process. Even when conducting a rapid form of ethnography, the methodology is still capable of generating valuable user-centred insights that extend beyond the realm of most other research techniques. Some of these beneficial aspects include the following:

- The resulting information is more realistic than other interview techniques, due to the fact that it is contextual and therefore places the participant, product and researcher in the real life setting where the potential product would most probably be used.
- The research is conducted in the participant's 'naturalistic setting', rather than in a usability laboratory or design studio. Not only does this aid the participant in feeling more relaxed with the research, it also provides further contextual information that carries significance in conveying people's identity and everyday behaviour.
- A third value of the technique is the fact that emotional and social responses are also conveyed to the designers. This results in the designers establishing an empathic awareness of experiences that extend beyond their own.

Analysis and editing

One drawback of this technique, and the use of video in particular, is the fact that it is incredibly time consuming (both to conduct and analyse). The mobile phone research produced 400 minutes of video footage. This was eventually condensed into a 7 minute video that aimed to communicate the main design insights. Even when taking a rapid approach to ethnography, the tasks of transcribing, coding, analysing and editing are still incredibly time-consuming.

Efficiency of the ethnographic method

The other major drawback of using the ethnographic technique in the way outlined by this case study is the fact that it generates masses of data that extends beyond the research topic in question. If this data is not carefully managed (i.e. transcribed, logged and archived), then this information becomes waste. Unfortunately, this is the current state of much commercial ethnography. By approaching ethnography from a project-by-project basis, the method becomes highly inefficient and difficult to justify.

14.7 RECOMMENDATIONS THAT AIM TO EVOLVE THE INTEGRATION OF ETHNOGRAPHY AND DESIGN

Communication issues

The following recommendations resulted from a debrief session at IDEO. Members of the design team were invited to attend a presentation to watch the video produced in response to the 'mobile phone as fashion' design brief. The following points summarise some of the suggestions made by members of the design team, in response to the communication of the rapid ethnographic research:

- The designers responded well to the video. Careful, succinct editing can really help to highlight the diverse needs and values across the range of potential end users. Clips played one after another, in quick succession, can communicate the dramatic differences between people and their subsequent product requirements (both articulated and latent).
- Video was seen to be a transitory medium, which resulted in the designers feeling that they wanted something physical to take away from the viewing. It was felt that subsequent viewings of the video were unlikely to occur in reality, and therefore it might be more realistic to consider producing a paper-based summary of the video content for future reference. This summary could contain digitised images and accompanying quotations that would operate as 'triggers' to what had been seen in the video.
- The length of the video is vitally important. It was considered that between 5 and 10 minutes would be long enough to absorb the main insights relating to the range of users. Any longer would just result in over-burdening the designers with information.
- The suggestion that seemed to hold the most merit in attempting to evolve the use of ethnography within the design process, concerned producing a number of different video edits to serve different purposes. Whilst the 5-10 minute video was considered adequate to communicate the main findings to the design team, it was suggested that the designers should interact with the video far sooner in the process. This approach would involve the researcher producing a very raw edit that communicated the main themes that were emerging from the research, while the research phase was still ongoing. In this way, the process of analysis becomes much more interactive as designers' opinions are gauged very early in the process. Not only would this be more likely to result in producing information that was more in sympathy with designers work, but it would also help to promote the ethnographic method and to encourage designers' to develop a sense of ownership of the resulting user-centred knowledge. This would hopefully increase the acceptance of the ethnographic method, and would enable the video to contain knowledge that designer's would find inspiring. Pursuing this type of approach may be particularly worthwhile if the researcher has little previous design experience, and especially if the researcher comes from a purely anthropological academic background.

Research approach

The research conducted at IDEO, and summarised in the preceding section, involved a rapid form of ethnography applied to a specific design brief. However, approaching the use of ethnography in the design context in this rapid manner may not be the only way that ethnography can be applied to the context of design. Rather than solely researching on a product-by-product basis, it may be more efficient to build up a searchable 'database' of ethnographic knowledge. This would enable a slightly more traditional form of ethnography to be pursued, due to the fact that a longer fieldwork period would be available to the researcher. The researcher would be able to immerse themselves in a household previously unknown to them, and to spend several days, or maybe a couple of weeks, attempting to gain a holistic understanding of the participants. A longitudinal database approach to research would also enable participants to become more accustomed to the presence of the researcher and video camera. Obviously this technique would be unable to research in relation to specific product briefs, but it would be appropriate for accessing broader themes and general trends that may lead to the identification of design opportunities. It is over time that the value of the ethnographic technique reveals itself – the rapport builds, the insights are generated and the data becomes richer and more descriptive.

The following table (Table 14.1) highlights some of the main advantages and disadvantages associated with both the rapid and more long-term ethnographic approaches. This table simply aims to highlight the fact that ethnography can be employed within the design process in different ways. There may even be a case for exploring the establishment of a searchable database (for identification of current trends and future design opportunities), alongside the use of rapid research (conducted on a product-by-product basis). It is suggested that further research be conducted to evolve the method of rapid ethnography, and also to assess the feasibility and value of pursuing a database approach to ethnography.

Table 14.1 Highlights of the main advantages and disadvantages associated with both the rapid and more long-term ethnographic approaches

Rapid Approach	**Searchable Database Approach**
Advantages: • This approach can quickly generate information relating to a range of user requirements. • The research is conducted in direct relation to a specific product brief. This ensures that the resulting information is valuable to the designers. • The resulting user-centred information can be delivered directly to the designers at the point of need (rather than necessitating that the designers must go and search a database to see whether the information exists). • It is possible to charge the research to the client because it has been conducted specifically for the purpose of meeting the client's design brief. • The rapid approach encourages the	**Advantages:** • This approach allows the true value of ethnography to emerge because it is conducted over time. This results in a more in-depth understanding of the participants' experiences, and an opportunity to discover people's latent needs. • The use of a searchable database means that the information generated is used more efficiently. The information is well archived, resulting in less waste. • This approach is particularly of value when wanting to research current and future trends towards broad subject areas. By looking at a broad topic, such as 'technology within the home' it is possible to identify current and future design opportunities. • It may be possible that a keyword search could lead to a serendipitous

discipline of ethnography to evolve the method into a usable tool that is applicable to the design context. **Disadvantages:** • The time available for fieldwork may not be long enough to gain a real sense of 'closeness' with those whose experiences lie beyond the realm of the researcher's. • The data can be less richly descriptive and less honest because the participant is not comfortable enough to move beyond the 'stock answers' and to express their real opinions and beliefs. • This approach can be very inefficient because it may generate data that could be useful to future projects, but without careful (and time consuming) archiving this information may be wasted.	discovery that may prompt further design concepts. **Disadvantages:** • This kind of database would need to be searchable. Such a scheme has massive implications for hardware and software requirements and their associated costs. • Although longer-term research is likely to generate more insightful information, it is pointless unless it is well publicised and made available and easily accessible to designers. • The information held by the database may not be detailed enough to meet designers needs for certain projects. (However, it could provide a useful overview and help to direct a further period of research). • It is possible that a searchable database may result in information being taken out of context. It becomes possible to jump to conclusions about a specific video clip without appreciating its associated contextual meaning. • The management of so much data would be an important consideration, and may necessitate a full-time team of people dedicated to its cause. • How will this approach work financially? With a generic database it may become more complex to charge clients for the research. However, perhaps it would be possible to charge clients a subscription fee for use of the database or to charge a flat rate for use of the research.

14.8 CONCLUSION

The ethnographic methodology appears to be an extremely valuable tool to add to those already at the disposal of the human factors specialist when communicating user-centred knowledge to designers. Not only can it provide a more holistic understanding of the complex relationship between people and products, it can also identify latent needs, and provide design insights to spark innovation, whilst simultaneously ensuring that these ideas remain grounded in the context of everyday usage. However, the context of the design process differs markedly from ethnography's usual field of use, and a significant degree of adaptation is needed to ensure it is able to fit into the design context. As a result of the joint research with IDEO, some suggestions have been put forward that may help to evolve the ethnographic method and to aid its integration within the design process. The use of ethnography within design is still at the early stages of development. This chapter has highlighted its applicability to fill such a role, but further research and application of the methodology in a design environment will be needed, in order to develop ethnography into a practical and fully integrated user-centred design methodology.

14.9 REFERENCES

Beyer, H. and Holtzblatt, K., 1995. 'Apprenticing with the customer: A collaborative approach to requirements definition'. In *Communication of the ACM,* May.

Binder, T. and Buur, J., 1999. *Video as design material in collaborative design.* School of Art and Communication, Malmö.

Clifford, J. and Marcus, G.E., 1986. *Writing culture: The poetics and politics of ethnography.* Clifford, J. and Marcus, G.E. (Eds), University of California Press, Berkeley, CA.

Eason, K., 1995. 'User-centred design: for users or by users?' In *Ergonomics 1995,* **38**, 8, pp. 1667-1673.

Eason, K. and Harker, S., 1991. 'Human factors contribution to the design process'. In *Human factors for informatics usability.* Shackel, B. and Richardson, S. (Eds), Cambridge University Press, Cambridge.

Fetterman, D., 1998. *Ethnography: Step-by-step* (2nd edition). Sage Press.

Fulton Suri, J., 1995. *Ergonomics in designing for the future.* IDEO Product Development, San Francisco.

Henley, P., 1998. 'Film-making and ethnographic research'. In *Image-based research: A sourcebook for qualitative researchers.* Prosser, J. (Ed.), Falmer Press, London.

Johnson, R., 1994. *The dictionary of human geography,* (3rd edition). Johnson, R. (Ed.), Blackwell Publishers Ltd., Oxford.

Lareau, A. and Schultz, J., 1996. *Journeys through ethnography.* Westview Press.

McLuhan, M., 1964. *Understanding media: The extensions of man.* MIT Press.

Macdonald, A., 1998. 'Developing a qualitative sense'. In *Human factors in consumer products.* Stanton, N. (Ed.), Taylor & Francis, London.

Macdonald, A. and Jordan, P., 1998. 'Human factors and design: Bridging the communication gap'. In *Contemporary ergonomics 1998.* Hanson, M.A. (Ed.), Taylor & Francis, London.

Miller, D., 1998. *Material cultures: Why some things matter.* Miller, D. (Ed.), UCL Press, London.

Suchman, L. and Trigg, R., 1991. 'Understanding practice: Video as a medium for reflection and design'. In *Design at work: Co-operative design of computer systems.* Greenbaum, J. and M. Kyng (Eds), Lawrence Erlbaum Associates Inc., New Jersey.

Whyte, W., 1955. *Street corner society,* (14th edition). University of Chicago Press, Chicago.

CHAPTER FIFTEEN

Linking Product Properties to Pleasure: The Sensorial Quality Assessment Method — SEQUAM

LINA BONAPACE

Ergosolutions, Consultants in Design and Ergonomics,

Via Slataper 21, Milan, 20125, Italy

15.1 INTRODUCTION

The search for pleasure has always been at the centre of human endeavour. Jordan discusses this aspect in the paper 'Human Factors for Pleasure Seekers': 'Since the beginning of time humans have sought pleasure ... A source of pleasure comes from the artefacts with which we are surrounded. For centuries human beings have tried to create decorative and functional artefacts, artefacts that have raised the quality of life and brought pleasure. These objects were rough-hewn from rock or made out of bronze or iron by inexperienced hands that only intended to make something for personal use. Then, when trading and bartering developed, these people became skilled craftsmen who created artefacts for the use of others in the community.'

The buying public today is very much aware of the qualities of 'user friendliness' and 'ease of use' in a product but even more, it tends to take for granted that the product will be usable and he/she will be disappointed if it is not. The concept of 'user friendliness' is flexible and in order to create positive benefits that users will really notice, the human factors field is presently expanding beyond usability, to more holistic 'pleasure-based' approaches.

Many manufacturers of consumer products, such as household appliances, electronic devices and motor vehicles, are working on quality and on the promise of supplying higher levels of comfort, ease of use and pleasure experienced by customers in owning and using their products. Product manufacturers are involving the marketing, industrial design and human factors field more and more in order to gather input on customers' needs, desires and buying intentions. Many have realised that there is a competitive advantage to be gained by designing quality products with 'good design' for people.

The relationship between the designer and the human factors specialist is central to the creation of products which fulfil user requirements of safety, functionality, usability and pleasure. The 'design team' works well when there is a common knowledge among designers, engineers and technicians of ergonomic methods and user trials and when the human factors specialist is knowledgeable of the phases, constraints and request of the design process.

There is a need to understand industrial design 'whose main aim is to add value to a functional device ...' and that '... to attempt to understand emotive responses to products is to attempt to understand the very basis of industrial design itself.' The human factors discipline can participate in the design process by providing assistance to designers, in facilitating their practice '... the ultimate aim of the research need not be to arrive at prescriptive design methods, but to help develop assistive methods or techniques which help design teams do more effectively what they already do.'

It is important to understand the design process and how data will be assimilated and used by inter-disciplinary design teams usually composed of:

- Designers who make use of strategies more than methods; they work in the future tense; they need information in a concise and readily assimilated format.
- Production engineers and technicians who are used to quantifying issues. They usually want numerical specifications and clear requirements that leave little to interpret.
- The product manufacturer who is interested in the various qualities and aspects of the product that can contribute to its success on the market. He is also concerned with development costs and time constraints which usually characterise the production process.
- Human factors specialists who are the *user's advocate:* they are concerned with user needs and desires. They are also *facilitators for the designers:* by informing and inspiring the members of the design team while delivering reliable data and suggestions for re-design or new concept design.

The overlap in ergonomic features and design qualities of a product is often present according to Macdonald (1998); 'Often an ergonomic feature has a tangible aesthetic quality. Designers have experience in 'fine-tuning' not only a product's aesthetic but also its ergonomic qualities, and here they rely on important information from the ergonomist.'

In the discussion and conclusion of the book 'Human Factors in Product Design' Green and Jordan (1999) state that 'The discipline of ergonomics/human factors is at an important point in its development. It is a discipline that has always been linked with design, whether of work practices or products or systems, but developments in the recent past have placed a new emphasis on the nature of those links... The electronic revolution has precipitated an unprecedented invasion of hitherto esoteric problems into the common domain, and together with the greying population of that domain, has created new demands on ergonomists and the design profession.'

There is an increasing interest in understanding and investigating those aspects in products that give people pleasure. Recently, human factors, as a discipline, has begun to ponder ways of investigating and gauging user sensations of pleasure and how they can be introduced into new products. The aim is also to communicate the collected data in ways that will inspire the members of the design team. An example of this is SEQUAM (Bonapace 1999), (Bandini Buti, Bonapace and Tarzia 1995) which involves analysing the link between physical properties of a product and users' responses to the tactile, prehensile, thermal, functional and acoustic contact with the product. Another example is Kansei Engineering (Nagamachi 1986, 1997) which investigates the link of the broader design characteristics of a product and user responses to that product.

New investigative techniques have also been developed. Jordan (1998) and Macdonald (1998) discuss case studies on the analysis of products and suggest the links between particular elements of their design and particular pleasure of use. Jordan and Servaes (1995) carried out semi-structured interviews and recorded users reported reactions to pleasurable products and the product properties associated with these reactions. Other techniques have been developed and applied by Fulton-Suri of IDEO such as body-storming with designers, personal narrative with all members of the design team including product manufacturers management and empathy through role playing. Video ethnography, the creation of personaes, story collecting of user experiences, scenario creations are some of the new techniques that facilitate the communication of user requirements, needs and realities to the design team.

15.2 USER INVOLVEMENT IN THE DESIGN PROCESS

The user-centred design focus calls for user involvement in the product development process (Figure 15.1).

When referring to user involvement in the implementation of functionality and usability aspects in product design, many authors report positive results. According to McClelland (1995), 'User trials can provide information which is at least useful, and at times, essential if products are to successfully accommodate users' requirements.' User trials are the most valuable source of information about a product's performance and can provide the best quality of data on which to make a decision to change a design or to make a new product.

In principle user trials can be employed at all stages of the development process. The type of user trial carried out, its specific objectives and the criteria used, all have to be discussed as project-specific issues. Product testing can directly contribute to the cycle of development by focusing attention on user issues which are critical to the effectiveness of the product. Each test produces fresh information and the experience from one test will help to refine the design and also define the needs for the next development stage.

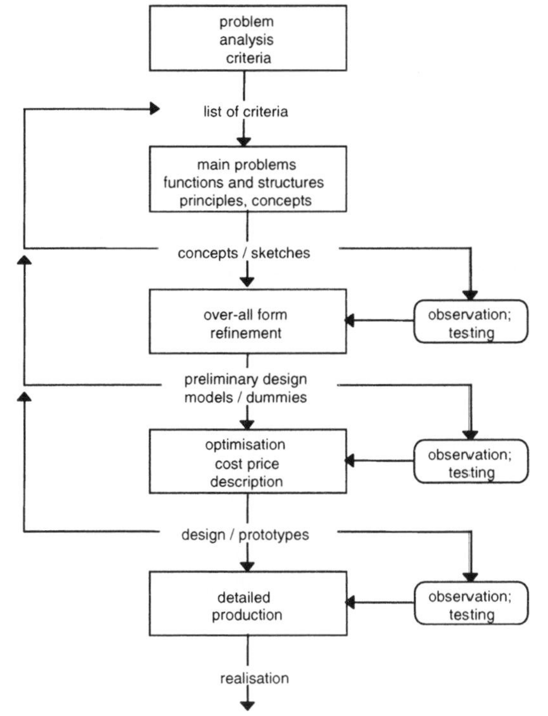

Figure 15.1 The process of user-centred design (from Den Buurman 1997)

This does not simply mean considering user requirements while designing and asking users their opinions, but also assessing the performance of users when using products. This takes place during various moments of the design process: Analysis, creation and evaluation phases (Jordan, Thomas and Taylor 1998) (Figure 15.2).

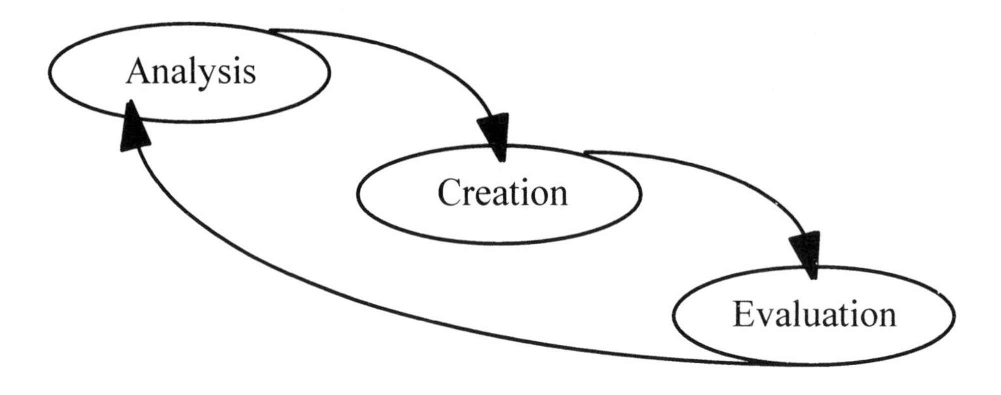

Figure 15.2 Generic design phases (based on Jordan, Thomas and Taylor 1998)

The various moments of the user-centred design process can also be defined according to the activities and aims of user research through the involvement of users in the various phases (Figure 15.3):

- **Explore**: new product concepts, the effectiveness of preliminary design proposals, what people do, what they want, need and like, using creative techniques, focus groups, on-site observations, etc.
- **Assess**: design proposals regarding the usability of operations, various aspect of the product (tactile, acustic, aesthetic, size, shape,...), pleasure in use, etc.
- **Verify**: design choices according to standards, benchmarks or specifications and refine the final design solutions.
- **Monitor**: product use, ease of use, pleasure in use, new and unplanned uses, from sales through duration of ownership to the purchase of another product.

ERGONOMICS	**DESIGN & DEVELOPMENT**
the final user	the product
PRIOR USER RESEARCH	*INTENTION*
Existing information Existing methods Existing technology	Re-design New product development
EXPLORE	*CONCEPT DESIGN*
Explore what users do Explore what users need & desire Explore new product concepts Explore effectiveness of concepts Explore other existing products	Product brief Concept development Sketches and illustrations
ASSESS	*DESIGN DEVELOPMENT*
Assess functionality aspects Assess usability aspects Assess pleasure aspects Assess other existing products	Design & detail development Mock-ups and prototypes
VERIFY	*FINAL DESIGN*
Verify the design according to norms Verify the design according to benchmarks	Final development of design Production prototypes Manufacture
MONITOR	*USE*
Monitor product during promotion & sales Monitor product during various uses Monitor product when being discarded Monitor other existing products	Product release Use & maintenance Discard

Figure 15.3 User involvement in the design process

15.3 SENSORY ASPECTS

People perceive the world through their senses, mainly by sight, hearing, touch, and our awareness of our body in space (kinaesthetic) and then by odour and taste. The world that surrounds us can be considered as a supplier of stimuli that reach people through sensory channels which triggers the stimulus and response mechanisms: input (sensations), perception, process (cognition), action and output. Input is the stimuli that the environment sends us and that can be quantified, measured and reproduced, perception is how much the individual actually takes in after the signal has passed through the person's psycho-sensorial filters and the meaning the person attributes to it. Process (cognition) is related to elaboration and decision making, action is the result of the decision and output is the result(s) of the action.

The way people perceive the sensations from the outside world are at the centre of user-product interaction (Figure 15.4). In other words we are talking about users sensorial experiences with artefacts. Macdonald (1998) states that 'a user's subjective response to a product is a complex phenomenon, but it is important to understand some of the subjective factors influencing product choice and acceptability.' When examining a product for the first time, people unconsciously sum up the sensations that the object communicates and that are perceived as being pleasurable, indifferent or of repulsion. Over a period of usage, people receive other stimuli from the objects and the judgement becomes more precise. Nevertheless, before being able to judge the object for its real features, there has already been made a judgement of 'pleasurability' that heavily conditions the opinion of the product.

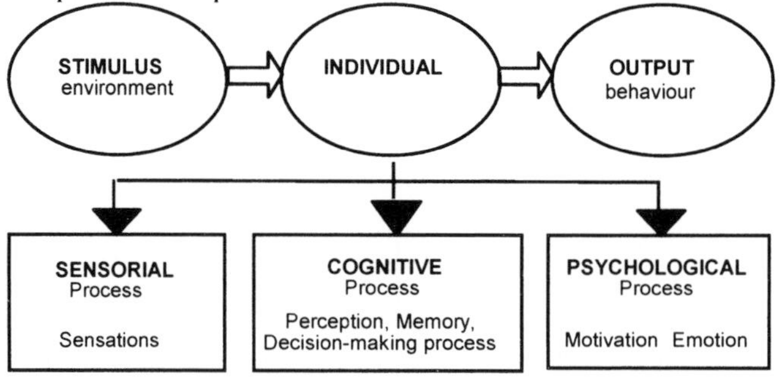

Figure 15.4 User - Product Interaction (based on Darley)

Nothing in the outside world is the private property of any one sense. The senses operate together, and the environmental information that gives people the picture of our world ultimately belongs to the brain, to do with it what it will. We can do more than one thing at once, but the brain is selective about the information it takes in, about the information we pay attention to. The brain filters, integrates, decides what is important, and lets us know. A great deal of the brain is devoted to the reception and analysis of sensory information. The homunculus or 'little man' is used in various ways to graphically represent the relative dedication of the human brain to the senses (input) and to movement of body parts (output). A large portion of the brain is devoted to hands, face, and genitalia. The hand which comes into contact with most man-made products is given a large amount of motor and sensory space in the brain and we are greatly dependent on those for sensory information.

The sensory pleasure of objects related to the qualities perceived through a person's senses are beyond any absolute measurement that would make them valid for everyone for all time. Subjective pleasure is strongly linked to the shifting aspects of individual and cultural variables, and may thus be dependent on the culture of the ethnic and social group the users belong to, the sensory modalities being governed by the different individual perceptive characteristics, in turn dependent on the elements of the 'tastes' of the society in general and of the individual.

The *culture* variable constituted by the social or ethnic group to which the user belongs would mean that, for example, an Asian surely has a historical and cultural background very different to that of a European, not unlike the differences within the same grouping between someone who lives in contact with nature and a city-dweller.

The *sensory aspects* also stand as a dividing element because each individual has his or her own characteristics that derive from functional reductions that are acquired or congenital (an example would be age), or deriving from the type of work they carry out (there are surely strong differences of tactile perception between a user, who uses his hands for heavy work, such as a bricklayer and another user who uses his or her hands in a lighter way, such as an office worker).

Questions of *time* concern the change in time of taste: for example certain colour combinations that were judged by past generations as being extremely vulgar or strange, are now seen as 'recherché' and pleasing; the same goes for forms, shapes, sizes, materials and finishes. Changes in taste today are not only very frequent and very different according to age groups, they also vary very quickly.

Hofstede (1991) proposes a useful three-level model concerned with cultural 'filters': *Personality:* Is specific to individual, is inherited and learned; *Culture:* Is specific to group or category, is learned; *Human nature:* Is universal, is inherited. He discusses how cultural differences manifest themselves through symbols at a superficial level through heros, rituals and finally values at a deeper level.

15.4 HUMAN FACTORS AND PLEASURE

During user trials people express themselves on the experience of utility, functionality and ease of use of the objects but many also freely put forward opinions and choices that are more attributable to its sensorial aspects. The person tends to express further reasons of pleasurability for the objects depending on the sensations transmitted to them by way of the products sensorial qualities. In fact perception can be activated by stimuli that can be pleasant or unpleasant to the user. This leads one to believe that alongside safety, well-being, comfort, and ease of use, the pleasurability transmitted by the object to the individual is an important aspect in the search for quality.

A constructive basis from which human factors as a profession can contribute to the creation of pleasurable products has been set.

Hierarchy of needs

Jordan (1999) has taken the idea of a hierarchy of needs from Maslow (1970) and applied it to human factors proposing a hierarchy of user needs as follows:

Level 1 – functionality, Level 2 – usability, Level 3 – pleasure.

A similar hierarchy was proposed by the members of the Ergonomics and Design work group of the S.I.E. Italian Ergonomics Society (see Bandini Buti 1998) regarding product requirements for user needs:

Level 1 – respect of the environment, Level 2 – health & safety, Level 3 – usability, Level 4 – pleasure.

In conclusion the hierarchy of user needs when interacting with products is proposed in Figure 15.5:

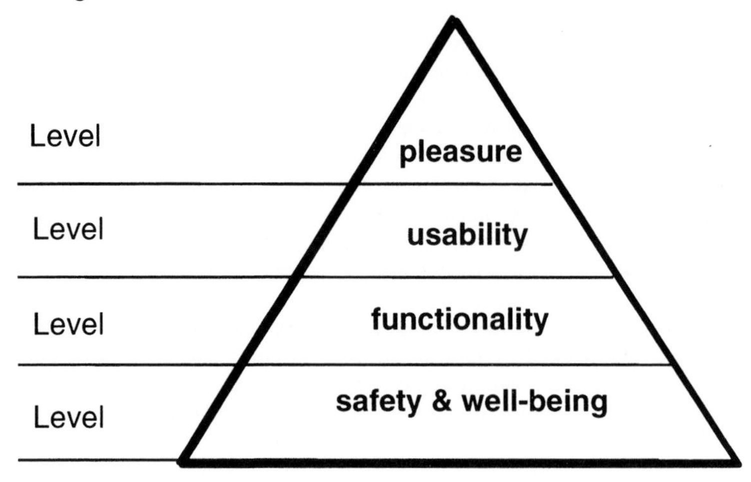

Figure 15.5 Hierarchy of user needs when interacting with products

The four-pleasures framework

The structure of pleasure proposed by Jordan (1999) is divided into four parts:

- **Socio-pleasure**: is the enjoyment derived from the company of others. For example, having a conversation or being part of a crowd at a public event. Products can facilitate social interaction in a number of ways such as coffee makers which provides a service which can act as a focal point for a small gathering.
- **Psycho-pleasure**: it is the type of pleasure that traditional usability approaches are perhaps best suited to. In the context of products, it relates to the extent to which a product can help in accomplishing a task and make the accomplishment of that task a satisfying and pleasurable experience. For example, it might be expected that a word processor which facilitated quick and easy accomplishment of, say, formatting tasks would provide a higher level of phycho-pleasure than one with which the user was likely to make many errors.
- **Ideo-pleasure**: refers to the pleasures derived from 'theoretical' entities such as books, music and art. In the context of products it would relate to, for example, the aesthetics of a product and the values that a product embodies. For example, the use of bio-degradable materials in a product embody the value of environmental responsibility in those persons particularly concerned about environmental issues.
- **Physio-pleasure**: are the pleasures that have to do with the body and the senses. They include pleasures connected with touch, taste and smell as well as feelings of sensual pleasure. In the context of products physio-pleasure would cover, for example, tactile and olfactory properties. Tactile pleasures concern holding and touching a product during interaction such as in the context of a telephone handset or a remote control. The surface qualities are of paramount importance. The acoustic properties are also relevant with products that are activated, give acoustical feedback or simply the sound that are created by touching the product.

If the aim were to gauge the quality of a handle, aspects concerning its use (how it is grasped, ease with which its functional direction can be assertained, etc.) and concerning its safety (if there are sharp edges, corners, etc.) would have to be assessed. However, if the handle were only to be gauged by these aspects it might be assumed that many are indistinguishable from others because they produce identical effects on the people that use them. In reality though, no one would say that the sensations from grasping a wooden handle are the same as those from grasping one in brass or ceramics. For example, in the instance of case handles an evaluation should rate their use characteristics – they have to be big enough and well shaped so as not to cut into the hand – but to gain a fuller understanding of the users' experience it is also necessary to gauge the sensation the material transmits during the contact with the fingers. The leather handle will give a sensation different to that of a metal product even if the shape is the same. But even coarse leather will give a sensation of touch that is quite different to that of a smooth leather.

The concept does not change if applied to the sensations associated with complex products such as motor vehicles. In trying out a car a user opens the car door, sits inside, touches the steering wheel and the gear change knob. Before even starting the engine the user has formulated a judgement on the vehicle, a judgement that is mainly related to tactile sensations. These sensations, then, are strongly present in the relation between user and objects.

Further developments and contributions

Jordan (1999) states that having developed a hierarchy of user needs, a definition of pleasure with products and a framework for different types of pleasure is a constructive basis from which human factors as a profession can contribute to the creation of pleasurable products. But in order to achieve concrete results it is necessary to tackle the following three issues:

- Understanding users and their requirements;
- Linking product properties to emotional responses in order to fulfil these requirements (pleasure benefits);
- Developing methods for the investigation and quantification of pleasure.

Traditionally the human factors field has mainly developed studies concerning the physical and psychological well-being of users related to posture, strain, physical and mental strain. Fulton (1993) states that 'the human factors profession was actually developed specifically to respond to problems and deficiencies – to errors, discomfort, injuries, delays and low productivity. The interest in understanding human physiological capabilities and limitations has been largely aimed at eradicating such problems.' She also states that 'the new challenge is to make products and environments which do better than keep us from getting sick, damaged or irritated. Products and environments can enhance the quality of learning about people in dynamic interaction with products and new ways of implementing what we find out.'

The quantification and metrics of pleasure are essential for human factors as a field if it wishes to contribute to the creation of pleasurable products. In the past the human factors field has contributed a great deal in terms of enhancing the functionality, safety, adaptability and usability of products. This was made possible through an understanding of user needs and requirements, ways of quantifying and measuring the interactions and methods for analysing, assessing and designing 'ergonomically sound products'.

Jordan states that 'being able to quantify usability has enabled human factors specialists to set usability specifications which have then been included as part of the overall product specification ... Talking in terms of 'usability' or 'quality of use' can seem vague and 'woolly', particularly to technical colleagues, who are used to numerical specification. Quantifying the issues gives an unequivocal signal as to what is required and enables an equally firm judgement as to whether or not the criteria has been met. It follows that if human factors as a profession is to take the lead in ensuring product pleasurability, then the development of measures and tools for quantifying pleasure will be beneficial if not essential ...'

Questions of *subjectivity as a measure*: at first sight the question of subjectivity may seem a problem that is insurmountable. It is possible to examine size, weight, strain, structure – they can be measured with tools and qualified numerically with values that hold true both in time and in space; for example in considering a way of measuring weight it is possible to be certain that, within statistical limits, in another place and in another occasion the same results will be obtained (objective data), whereas in examining sensations it is not possible to obtain data with the same repeatability, because the aspects under investigation are principally subjective. Nevertheless subjective judgements are presently the most effective way of investigating the overall quality of products.

The necessity to gauge the sensations felt by the individual when interacting with a product is therefore ever more important. According to Jordan (1999) these should examine not only effective evaluation methods but method for capturing user requirements and methods for early concept evaluation. Human factors has methods that are effective in evaluating pleasure level such as interviews, questionnaires with ranking and rating scales, focus groups and has methods that can be adapted to the investigation of pleasurability. Chapanis (1959) and Pepermans and Corlett (1983) discuss the uses of phycho-physical methods in ergonomics research in which observers make judgements about the sensations they experience (e.g. comfort). Pirkl and Babic (1988) developed a model which translates quantitative data for the senses into guidelines in the area of transgenerational design. Jordan (1999) suggests attitude scales for measuring pleasure; potential behaviour correlates to pleasure and displeasure and the frequency of product use.

The introduction of higher quality levels in all that is perceived by the senses, as well as the undoubted sensitivity of the designer, also requires devising and applying new and refined methods and tools. Not only should one be in a position to identify pleasurable sensations for the users, but it should also be possible to rate them rigorously, propose them in projects, describe them in the related documentation, reproduce them in the objects and perfect them by comparing the proposals with reality.

In the same way that it is not sensible to assume that just because the designer finds a product usable anyone can use it, the same applies to sensory aspects. It is necessary then, to find a measure of sensorial quality that is independent from and complimentary to the judgement of the person designing the product.

So the challenge today is to move beyond the stage of functionality and usability towards a more complete pleasure-based approach to product design.

15.5 SEQUAM – SENSORIAL QUALITY ASSESSMENT METHOD

Product design today calls for special attention to the quality of the sensations perceived by the end users while using consumer products. SEQUAM was developed as a means of exploring and analysing user-product interaction with the objective of generating data for design purposes.

SEQUAM was developed in SEA – Società di Ergonomia Applicata by Bonapace and Bandini Buti in 1992 in response to Fiat Auto's needs to increase perceived pleasure of product image and use. This regarded interior components with which users come into contact for short and long periods of time such as the steering wheel or the door handle which leaves a clear impression of the car 'feel' when he/she touches or uses it. There was a clear need to create transmittable data concerning the physical pleasure aspects through the creation of pleasurability guidelines to be used by all the components of the interdisciplinary design team (designers, technicians, producers, human factors specialists, marketing people, etc.). It was also necessary to clearly end effectively communicate with the designers and assist them during the creation of new products with the objective of creating a high level of pleasure in use.

SEQUAM has been applied on the following car components since 1993: steering wheels, automatic and manual gear shifts, internal and external door handles, column mounted lever systems, push buttons and turn knobs for heaters and air conditioning systems, internal door panels and car information systems.

SEQUAM was also applied to other fields such as domestic armchairs and washing machine control panels for the Whirlpool Corporation.

SEQUAM is a quick and effective technique which involves analysing the link between physical properties of a product and the users' responses to the tactile, prehensile, thermal, functional and acoustic contact with the product. It purports to be a formal and systematic approach to linking products' physical properties with user responses relying on statistical analysis to make these links.

The results obtained give indications of the degrees of pleasure that can be used in design. The objective is to gauge and reproduce these aspects of pleasurability and to turn them into specifications that can be introduced into the manufacture or supply of products in the same way that the technological performances required from the materials and manufactured products are specified. This means going from the sphere of speculation to that of rigour. This is a must in order to communicate with technicians, engineers and management who are used to numerical specifications and quantified issues. In other words the study of the sensory quality is approached in a manner similar to usability in which clear usability specifications are laid out, so as to be able to obtain the performance promised by a product without penalising aspects of well-being or comfort and achieving pleasure in its use.

15.6 SEQUAM FRAMEWORK

The framework of the method is based on two main aspects:

- The link between objective parameters (physical product properties) and subjective sensations;
- User involvement in various phases of the investigation: exploring, assessing and verifying pleasurability.

The link between objective parameters and subjective sensations

All objects used by man can be characterised by the following two types of parameters (Figure 15.6):

- **Objective parameters** (quantified and/or measured), that is the elements measurable using what has been elaborated by the various branches of scientific knowledge;
- **Subjective sensations** that are closely linked to characteristics and reactions that can be gauged by techniques devised through cognitive psychology.

OBJECTIVE Physical product properties	+	SUBJECTIVE Sensations perceived by users	=	Pleasurability Parameters

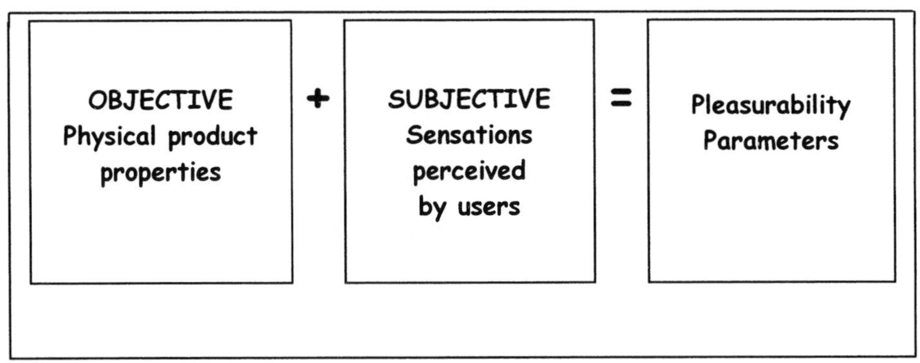

Figure 15.6 Links between objective parameters and subjective sensations

INVESTIGATION TECHNIQUES

Focus groups
Interviews
Questionnaires with rating and ranking scales
Observations of subjective investigation (3 phases, visual, tactile, use)

Figure 15.7 Examples of product quantification and metrics

The subjective sensations felt by a person (user sensations) in contact with artefacts are separated into sensorial categories in order to better analyse and understand them. These in turn refer to product specific physical properties. This will also help in effectively communicating the necessary input and advice on how to create a product which will elicit particular responses to designers (Figure 15.7).

It is important to remember though, that the design of a product is not the result of the sum of its parts, but is seen as a whole, from the *Gestalt* point-of-view. Designing is a creative act put forward by the designers in which choices and compromises constantly occur.

USER SENSATIONS	PRODUCT PHYSICAL PROPERTIES
Tactile sensations	surface quality, softness and grip
Prehensile sensations	object shape, size and dimensions
Functional sensations	ways of using, activating and manipulating the object
Thermal sensations	related to the conductivity and thermal capacity of the materials
Acoustic sensations	sonority of materials, of object when being activated, of feedback
Visual sensations	surfaces and body colours, surface finishes and softness, forms and shapes

Figure 15.8 User sensations and corresponding product physical properties

There are also the sensations of taste and smell, that are secondary for gauging the sensory characteristics of the objects (Figure 15.8). It is, however, true that a 'newness' smell can be considered more or less pleasurable according to the circumstances and can have a considerable influence on the appreciation of a product. On the other hand our approach to sensorial features would be totally different if we were considering food; right from 2000 BC the Chinese attributed equal importance to three elements for the pleasure of eating: taste, smell and sight. In fact the ingredients of any dish, for example 'canton rice' are also chosen on the basis of their colour mix.

The surface qualities of products are of paramount importance in the perception of pleasure with products due to the great extent given to the tactile sensations of the hand. These sensations are also be perceived through other senses such as sight. Taylor, Roberts and Hall (1999) observe that according to designers 'users' visual perceptions of the quality of a product are based entirely on surface information'.

User involvement in various phases of the investigation: exploring, assessing and verifying pleasurability

The subjective investigation is carried out in three different phases that fairly correspond to those of the design process. The names given to the phases refer to the objectives of the investigations carried out – exploring, assessing and verifying pleasurability. The investigation can be carried out in all of the phases or in single phases depending on the types of products under study and the time available. The outputs of phase 1 and 2 are mostly design guidelines regarding pleasurability whereas phase 3 is concerned with assisting the design team in the design of products which are characterised by a high level of pleasurability.

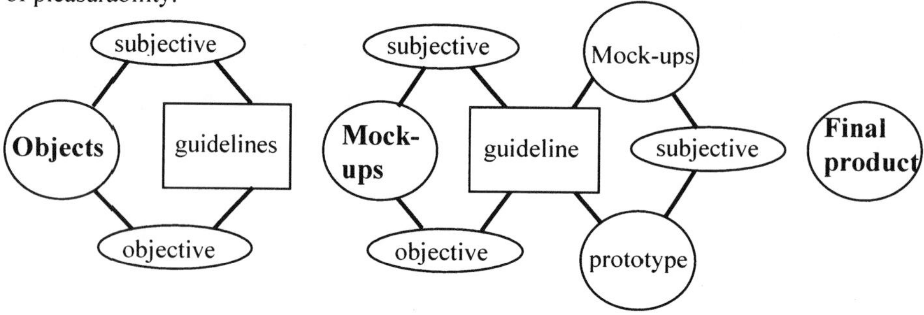

Phase 1 - EXPLORE **Phase 2 - ASSESS** **Phase 3 - VERIFY**

Figure 15.9 The SEQUAM process

*Phase 1. **Explore** the pleasure aspects with existing products*

The first phase constitutes a study of a selection of objects currently on the market deemed particularly interesting for their various aspects of pleasurability. This is in order to give the broadest possible range of product properties. In fact the commercial output is the expression of the proposals and the creative verve of numerous designers and technicians. The purpose of this phase is to quantitatively rate the appreciation of the product properties and to direct the design of the experiment study models needed for a systematic study of the pleasurability of product properties.

Steps:

- Overview of products on the market;
- Qualitative interviews with 5-6 users in order to identify the sensory parameters that characterise the objects under examination (e.g. touching, rotating, opening, closing, etc.) and identification of the objective qualities – product properties that can be measured *(e.g. size, weight, surface finish, etc.);*
- Selection of the objects-products (max 10) for the user trial that possess the qualities being studied in various degrees. Each object is to be defined by its measurable (size, hardness, weight, sound, etc.) or qualitative features (family of shapes, sameness of shapes, surface characteristics, colours, features of action, noise features, etc.);

- Quantification and measurement of product parameters;
- Preparation of the subjective investigation (test plan, ethical guidelines and permission sheet, questionnaires and materials, test-site or set-ups with equipment, etc.);
- Programming of subjective investigation (define user group and the characteristics of the sample of users for the trial, recruit persons for various test sessions, etc.);
- Carry out subjective investigation with briefing of the investigation objectives, observation techniques, questionnaires, interviews and debriefing;
- Correlation of objective and subjective data;
- Data elaboration; and
- Creation of design guidelines containing the results.

The advantages related to this phase are the following: it is fairly quick and easy to get hold of samples for testing and almost all the variables (product properties) can be tested because the object is finite and functioning.

Some disadvantages are that the object may conjure up stereotypes that are difficult to eliminate (well-known or famous objects that the person knows or uses habitually). When working with existing products it is possible that only usual properties will be found; therefore innovative ones will have to be tested in phase 2.

*Phase 2. **Assess** the pleasure aspects of design proposals with mock-ups*

The study of what is on the market gives indications as to the pleasurability of what is currently being produced and is thus meaningful for the here and now; however it does not tell us anything about innovative and future trends. Above all, it should be noted that what is offered on the market today has been conceived a fair time back. It is hence necessary to introduce a study phase where objects purposefully designed for the research are analysed. This must enable:

- The research of product properties that in the first phase required more in-depth study;
- The analysis of objects that completely respond to the preferences that emerged in the preceding phase;
- The analysis of the trends that emerge from the study but were not followed through due to a lack of suitable objects; and
- The study of innovative trends and design proposals.

This phase consists of planning, creating and experimenting with a homogeneous series of study models and in reading the responses to the varying of the individual product properties in a situation with very few variables. The phase is needed to check the variability limits of the properties studied by submitting mock-ups that may even have exaggerated features for judgement (for example grips that are exceptionally large or exceptionally small).

It is important to note that the projects devised during the study should not be seen as true and proper design projects in as much as they have been conceived for study purposes (hence for checking and very often for exceeding the limits of pleasurability) and not as part of an organic design context. The results of the study are to be seen as a stimulus and as guidelines for the overall design project and as a means for rigorously gauging the aspects of pleasurability.

Steps:

- Create groups of objects in mock-up forms to be tested according to the qualities, preferences, innovative trends or design proposals identified in phase 1;
- Develop mock-ups adapted to the type of trial and the features being investigated;
- Quantification and measurement of product parameters;
- Preparation of the subjective investigation;
- Subjective investigation preferably done with the same users as in phase 1;
- Correlation of objective and subjective data;
- Data elaboration; and
- Product guidelines.

An advantage associated with this phase is that with the creation of specially devised models the single product property can be analysed to great effect by comparing the models of objects that are identical in all their parts apart from the property under study (for example the cross-section of the grip if the object is to be carried or transported). Another advantage concerns the time required for developing this phase which is not excessively long even if slightly greater than that of the preceding phase in that the projects need to be developed and mock-ups in various materials need to be constructed. Functioning models are more effective and at times are necessary for the correct testing of the pleasurability aspects but can also take more time and be more expensive to construct.

*Phase 3. **Verify** the pleasure aspects of final designs with prototypes*

For the final check of the pleasurability aspects of the products, one has to be able to carry out studies on functioning prototypes that have all the features of the finished products and that are inserted in a coherent surrounding from both a functional and a formal point-of-view, and used actual conditions by a cross-section of potential users.

When the development of the projects allows the production of prototypes that are fairly close in terms of finish and function to the finished product, the results of the first two phases can be checked in real use conditions by a large cross-section of users.

During the development of the product, which should include the ergonomists as part of the design group, the various stages of the product can be checked, combining the tests with the data collected in the two phases.

Steps:

- Creation of the work group (designers, human factors specialist, producers, etc.);
- Development of the product based on the product design brief and on the pleasurability guidelines (from phase 1 and 2);
- Creation of mock-ups for testing on smaller groups (10 persons) in static situations;
- Testing, data elaboration, reporting results and suggested changes;
- Development of prototypes for testing on small groups (10 persons) in dynamic situations or in a similar final use context;
- Testing, data elaboration, reporting results and suggested changes;
- Final prototype for testing with a large group (30-40 persons) in dynamic situations or in the final context (e.g. car prototype); and
- Testing, data elaboration, reporting results and last suggested changes.

15.7 SEQUAM COMPONENTS OF THE USER INVESTIGATION

Users

User trials require a sample of users which in some way reflect the product user population as a whole. This means selecting a group of users who do not just have the same characteristics as the user population but who reflect the extent to which these characteristics vary. Ways of determining the user profile, the process of selection and user numbers are widely discussed in literature (McClelland 1995).

The following are basic considerations regarding user groups:

- The users should be grouped per age and sex according to the characteristics of the potential user population.
- The users should not be specialists with respect to the product being studied, such as testers, service personnel, employees of the producer company, because they could be conditioned by familiarity with the object or by private interests.
- They should possess a certain sense of criticism and the capacity to imagine real situations starting off from more or less realistic simulations (that go from the isolated object in the laboratory or workshop to the complete functioning objects and presented in simulated conditions of use).
- The size of the user sample varies and should be carried out with larger groups than those suggested for usability testing. We suggest 20 to 40 persons when investigating products, according to the phase in which the user trial occurs (explore, assess, validate) and 5 to 10 when singling out the preliminary information during inteviews.

The objects of the investigation

The evolutionary nature of the product development is quite evident when referring to the product development process scheme. The forms change depending upon the development stage, the product type and the aspect which requires investigation.

The object of the investigation can be a complete product or collection of objects for a specific function (e.g. steering wheel, ambient thermostat, CD player, etc.). Product physical properties are context-dependent; therefore choosing objects homogenous only in terms of type of functioning (levers and switches not attached to a specific function) should be avoided. Studies on objects for the same function though activated or used in different ways (e.g. manual and automatic gear change, driver and front passenger seats, etc.) are likewise to be avoided.

The group of objects can also be mock-ups, full and part prototypes or simulations. The groups must be homogeneous, such as only prototypes or only similar looking mock-ups, especially if the objects are to be compared with one another. For an in-depth look at effective testing with prototypes see Rooden (1999).

Types of trials

Three different types of trials can be carried out:

- Laboratory tests with objects isolated from their use-context and assembled on structures specifically devised for carrying out the studies. Carrying out trials studies on isolated objects in laboratories or workshops enables the testing of a relatively high number of samples, it speeds up the testing and enables optimum administration of trials times.

- Testing objects in a mock 'natural' context (household appliances in mock kitchens, parts of cars assembled on supports). These are preferable conditions inasmuch as they are closest to real use, although bringing the users to the trials area already creates a certain degree of artificiality (see Kanis 1999).
- The condition closest to reality consists in the observation of user behaviour in their habitual place of activity (in context). This type of trial is very effective when examining the person in his or her own environment using manufactured products or functioning prototypes. It is not as effective when investigating with a large quantity of non-functioning objects.

Methods and techniques

- Direct observation and filming of users during three distinct moments of interaction with the objects: visual, tactile and contact use (Figure 15.10);
- Face-to-face interviews;
- Thinking aloud;
- Questionnaires with rating and ranking scales, factual statements and open-ended questions;
- Briefing and debriefing sessions.

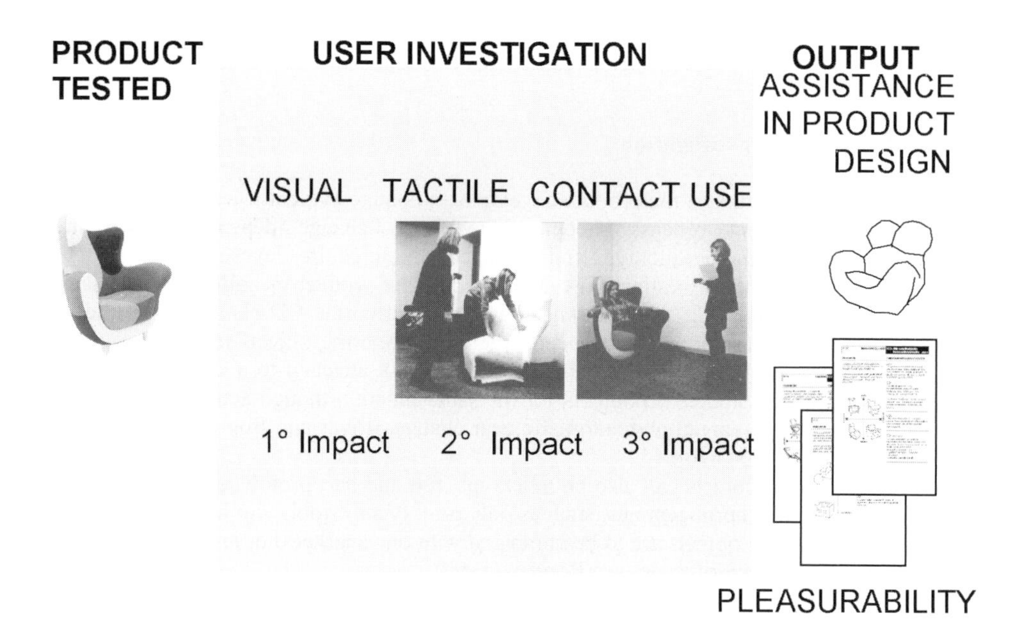

Figure 15.10 The three successive moments of the subjective investigation

Data analysis

The gathered data is statistically analysed in order to make the link between the product physical properties (objective data) and user responses (subjective data).The subjective data is also statistically analysed. Correlations are done between various types of subjective data and between the subjective and objective data.

The measures and quantifications of the product physical properties (objective data) previously carried out are also used in data elaboration.

15.8 OUTPUTS

Guidelines are an effective and systematic means of transmitting findings and communicating results to the interdisciplinary design team. The pleasurability guideline is proposed as a tool for introducing pleasurability in the design project through indications that will not limit the creative freedom of the designers. The pleasurability guidelines can also be considered as a means of supporting dialogue among the design team on the sensory qualities of objects based on objective and subjective data.

Three main characteristics of the guidelines are:

- Effective transmission: the data contained in the guidelines is presented in a concise manner which can be read by users from different backgrounds (suppliers, technicians, designers, management, etc.).
- Rapid and readily assimilated format for consultation: the results are illustrated with diagrams, charts and schemes for clear and rapid consultation.
- Reading on a number of levels: with reference to the background study documentation and study data that justify the findings.

Pleasurability guidelines contain the data and findings in various forms:

- Each product physical property is presented in the form of a diagram with the pleasurability ratings placed vertically and the various objects tested placed horizontally in order of preference. Scales highlight the range of satisfaction for each property thus leading to the formulation of indications.
- The diagrams are always confirmed by and compared with the impressions and suggestions gathered from the interviewee comments.
- *Subjective data is presented in quantities such as the percentage ratings of responses and qualitatively with the analysis of free responses.*
- Findings will be given as optimum maximum and/or minimum figures, trends and indications that refer to specific product – or user categories (e.g. sports-cars, older users).

The data on pleasurability evolves in time and in space, hence the investigations will have to be repeated in time. Pleasurability guidelines will have evolving information. As mentioned by Taylor, Roberts and Hall (1999): ‘Design is always context-dependant both environmental and human. Perceptions and attitudes to products change over time, influenced by personal factors as well as by new products entering the market place. The design context is therefore dynamic and we are aiming at moving goals.’

Assistance in product design

The findings contained in the pleasurability guidelines are the starting point from which the human factors specialist and the other members of the design team proceeds in developing the design project. The human factors specialist assists the design team by:

- Adapting trials and techniques to the severe time constraints;
- Proposing the necessary trials and evaluations according to the specific objectives;
- Making the results accessible in a format that is usable for design purposes (for designers);
- Supporting design decisions with data on the origin of the decisions and the benefits obtained (for management or technicians);
- Highlighting the advantages and disadvantages of the various proposals for the end-users; and
- Participating in all the phases of the development process.

15.9 CASE STUDIES

The following are examples of the application of SEQUAM to various products. They also differ according to the desired outputs – the creation of guidelines or assistance in product design.

Internal car door handle levers

The creation of pleasurability guidelines for Fiat Autos design and ergonomics groups and for their components suppliers (Figure 15.11 to 15.13)

Phase 1: Singling-out of product physical properties with 10 users during interviews, subjective investigation with 18 users on 14 cars gathered in 2 test phases with 7 cars each. The product properties investigated were: shape and size of lever and of lever recess, lever position, effort required to deploy, sound of deployment, grasp angles, surface finishes.

Phase 2: Subjective investigation with 18 users on 13 functioning mock-ups of complete handle lever proposals.

Figure 15.11 The 11 internal door handles tested in phase 1

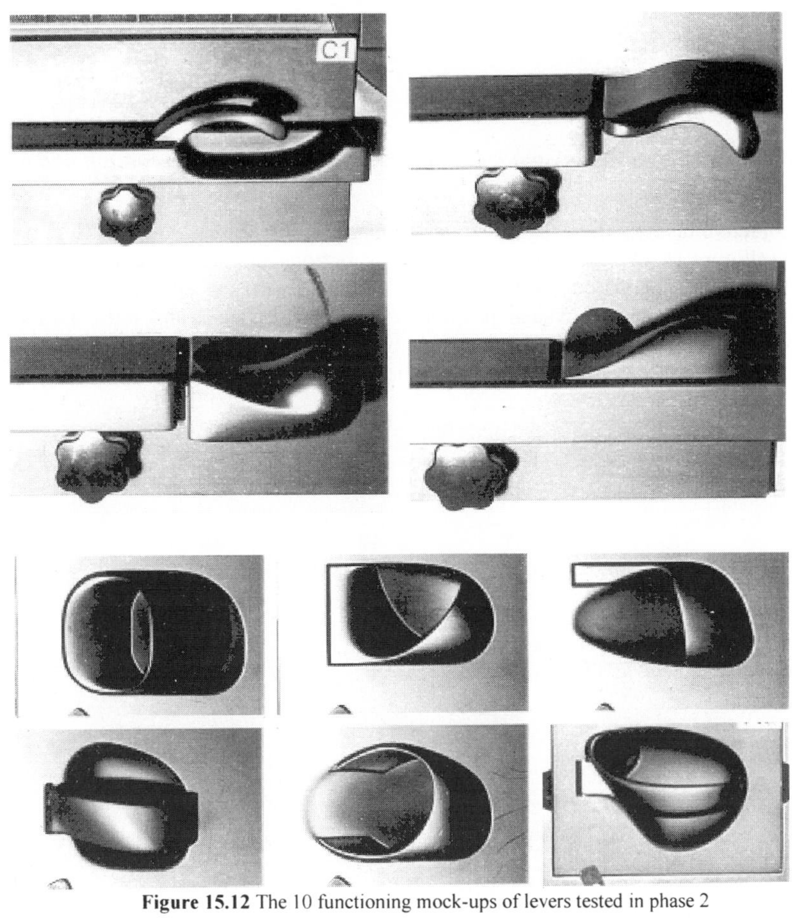

Figure 15.12 The 10 functioning mock-ups of levers tested in phase 2

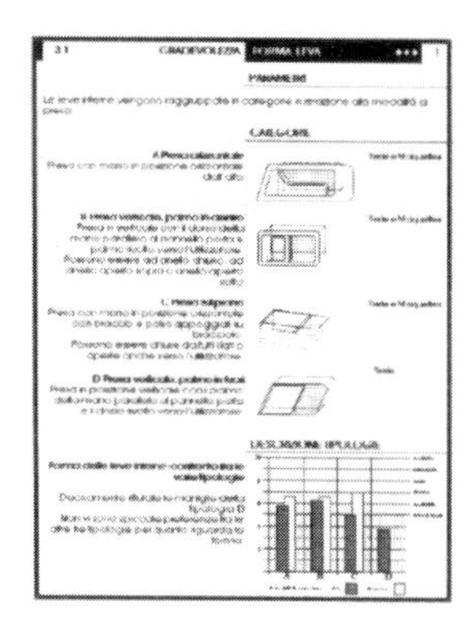

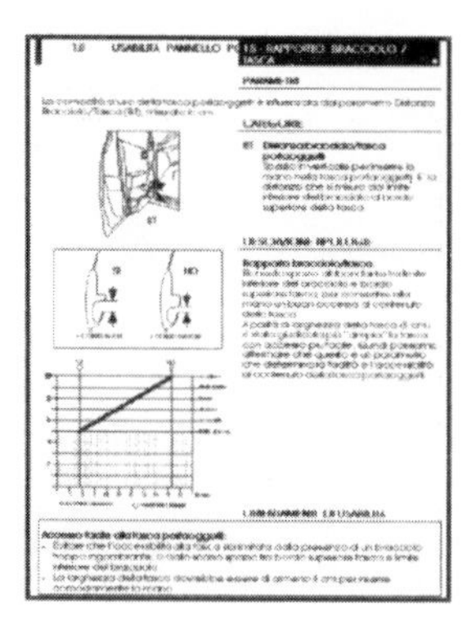

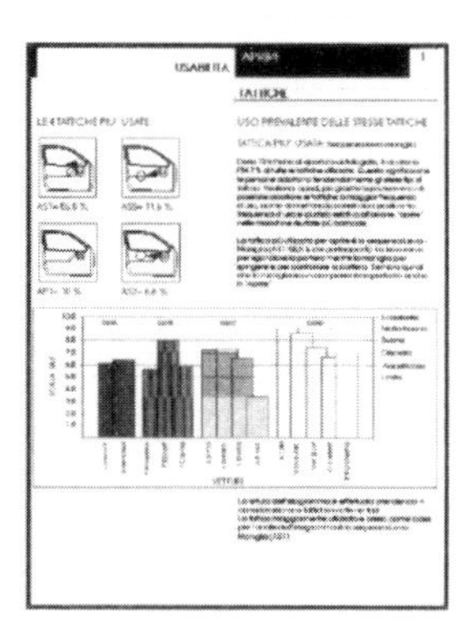

Figure 15.13 Examples of pleasurability guidelines for internal door handle levers

Clothes washers

The creation of pleasurability guidelines for Whirlpool Corporations design and usability teams (Figure 15.16)

The subjective investigation regarded various parts of the clothes washer – the washer as a whole (Figure 15.14), the control panel, knobs, buttons and the drawer (Figure 15.15).

The users evaluated both product features such as attractiveness and user friendliness and product physical properties such as the forces and noise factor of the mechanisms used (push buttons, turn knobs, levers) and drawer movement, surface finishes, sizes and shapes of buttons, knobs, lever and handles, colours.

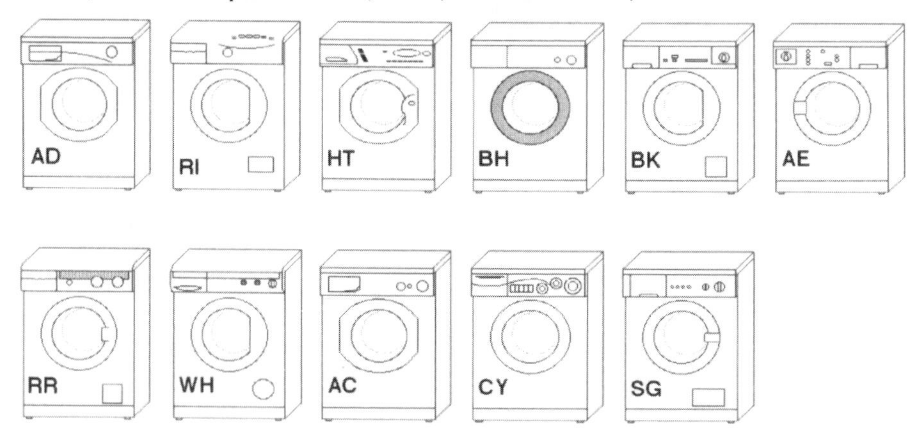

Figure 15.14 The 11 clothes washers of the user investigation

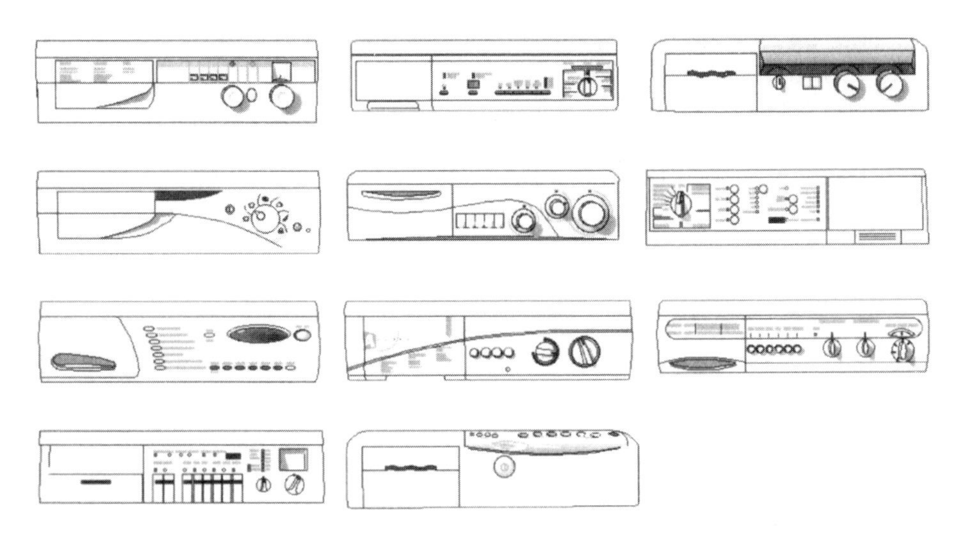

Figure 15.15 The control panels of the clothes washers

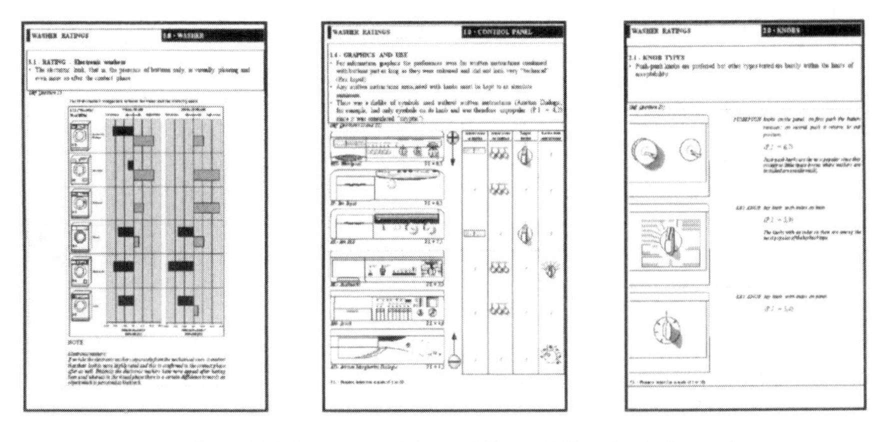

Figure 15.16 Examples of pleasurability guidelines for clothes washers

Armchairs and car seats

The creation of pleasurability guidelines for a multi-client group of furniture manufacturers to be used by their R&D and design sectors

This project was also developed with Fiat Auto in order to gather inspiration and information from domestic seating to be transferred to the automotive sector (Figure

15.17 and 15.18).

Product physical properties were singled-out with 10 users during interviews observing photographs of domestic and car seats.

The subjective investigation with 30 users observing, touching and using 8 domestic armchairs from the current production. The individual tests and questionnaires were divided into three parts – visual, tactile and use, in order to compare the varying level of perception between each phase.

The product properties investigated were: shapes and sizes, softness and density, colours, finishes, materials, textures, drawings and patterns. Product features such as comfort, safety and adaptability were also investigated.

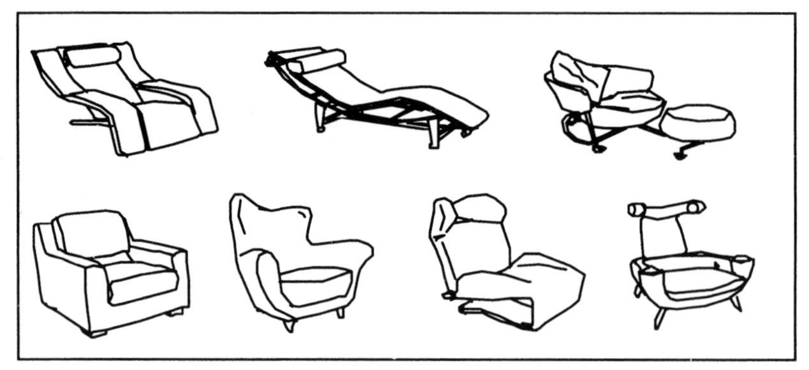

Figure 15.17 The armchairs used in the subjective investigation

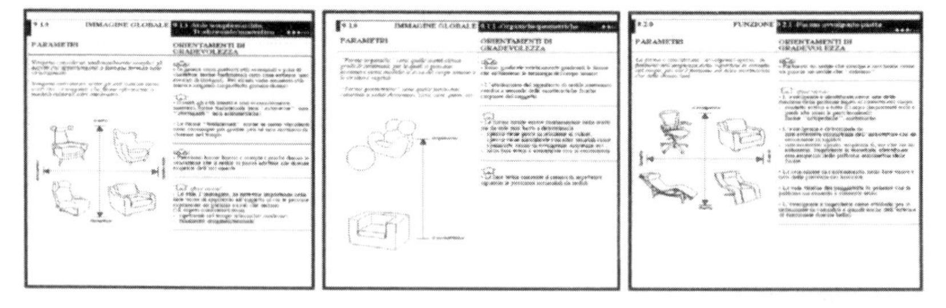

Figure 15.18 Examples of pleasurability guidelines for domestic armchairs and car seats

Steering wheels

The design of steering wheels for the Alfa Romeo 156 model(Figure 15.19 to 15.24)

Phase 1: subjective assessment with 18 users of 10 steering wheels available on the market. The following physical product properties were investigated: shapes and sizes of the crown, shapes and sizes of grip areas, shapes and sizes of central area, number of spokes (single, double, triple, etc.) overall size of steering wheel, softness, materials (plastics, wood, metal, fabric), colours.

Phase 2: subjective assessment with 18 users of 12 steering wheel mock-ups with varying sizes and shapes of the crown, heights and angles of grip areas and complete innovative design proposals.

Phase 3: development of complete design proposals with the design team. Dynamic testing of part prototypes (ring proposals, grips, sizes, etc.) with 10 users on internal race track. Dynamic testing of the complete and final proposal installed in the car prototype with 40 users during a 20 minute trial in which various driving tasks were executed.

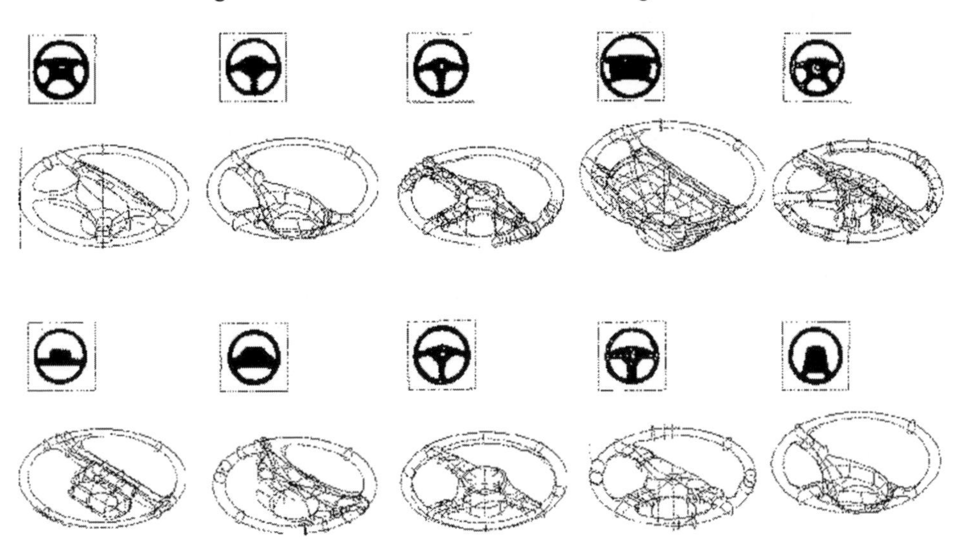

Figure 15.19 10 steering wheels of the phase 1 subjective investigation

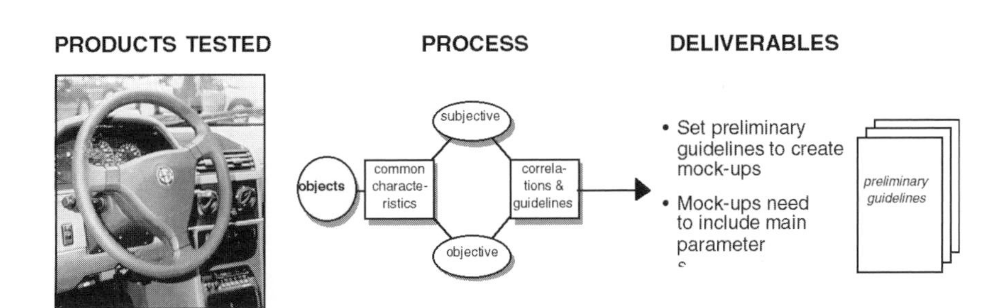

Figure 15.20 Flow chart of phase 1 steering wheel project

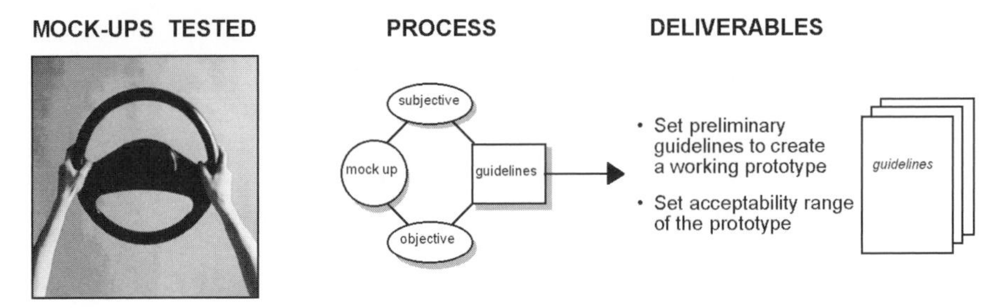

Figure 15.21 Flow chart of phase 2 steering wheel project

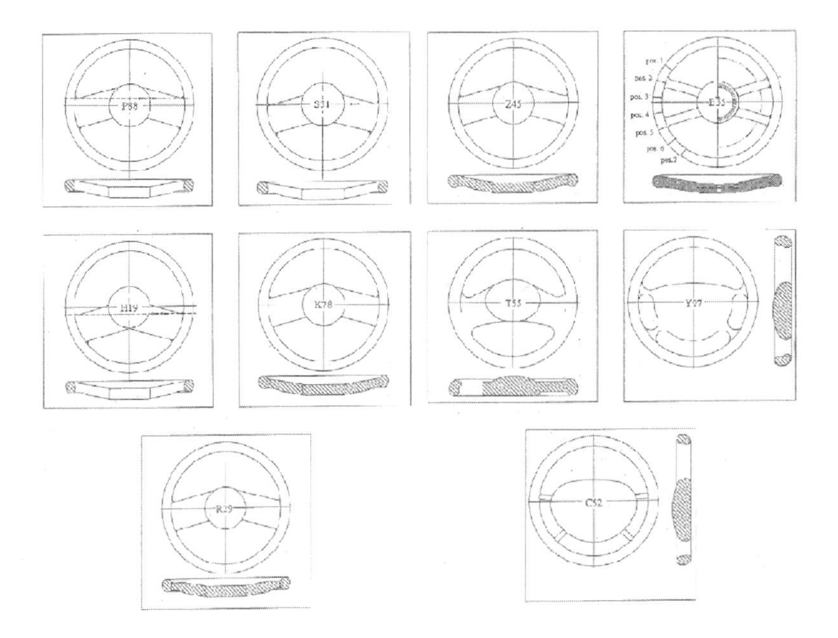

Figure 15.22 Mock-ups developed for the phase 2 subjective investigation

Figure 15.23 Mock-ups developed for the phase 3 subjective investigation

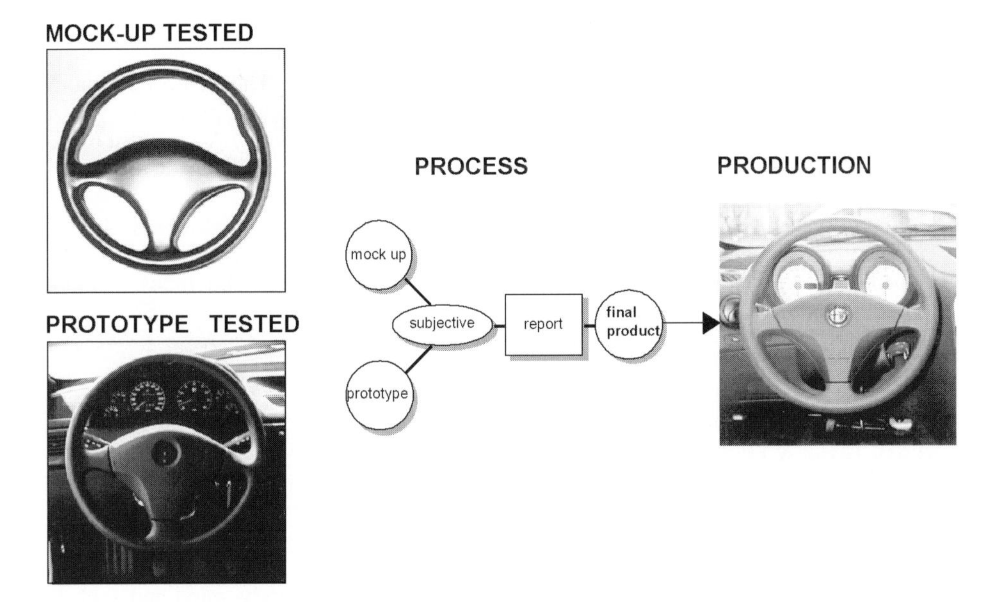

Figure 15.24 Flow chart of phase 3 steering wheel project

15.10 CONCLUSIONS

The challenge today for product manufacturers and all the members of the inter-disciplinary design team is to move beyond the stage of functionality and usability towards a more complete pleasure-based approach to product design.

SEQUAM is a technique which systematically provides concise data regarding the linking of the products' physical properties to subjective user responses. In this way people's likes and dislikes are clearly stated and linked to precise physical parts of the product.

SEQUAM has been applied in the investigation of various products. It has helped designers, through pleasurability guidelines and direct assistance in product design, by providing them with findings that have been integrated into their design proposals, and product manufacturers in providing users with truly pleasurable products to experience and use.

15.11 REFERENCES

Bandini Buti, L., Bonapace, L. and Tarzia, A., 1995. Analysis of car interior components using the Sensorial Quality Assessment Method SEQUAM. In *Proceedings of I.E.A. World Conference 1995*, Rio de Janeiro.

Bandini Buti, L., Bonapace, L. and Tarzia, A., 1997. Sensorial Quality Assessment: A method to incorporate perceived user sensations in product design. Applications in the field of automobiles. In *Proceedings of I.E.A. World Conference 1997 International*

Ergonomics Association 13th Triennial Congress, Taylor & Francis, London.
Bandini Buti, L., 1998. *Ergonomia e progetto: dell'utile e del piacevole.* Maggioli Editore, Rimini.
Bonapace, L., 1999. The ergonomics of pleasure. In *Human factors in product design. Current practice and future trends*. Green, W.S. and Jordan, P.W. (Eds), Taylor & Francis, London.
Bonapace, L. and Bandini Buti, L., 1997. The pleasant object: The Sensorial Quality Assessment Method. In *Ergonomia*, **5**, 10, Moretti & Vitali, Bergamo, pp.11-14.
Bonapace, L., Bandini Buti, L. and Tarzia, A., 1997. Sensorial Quality Assessment of car interiors. In *Proceedings of IBEC International Body Engineering Conference*, Stuttgart.
Buurman, R. den, 1997. User-centred design of smart products. In *Ergonomics*, **40**, 10, pp. 1159-1169.
Chapanis, A., 1959. *Research techniques in human engineering.* The Johns Hopkins Press, Baltimore.
Fulton, J., 1993. Physiology and design new human factors. In *American Center for Design Journal*, **7**, 1.
Green, W.S. and Jordan, P.W. 1999. Discussion and conclusion. In *Human factors in product design. Current practice and future trends*. Green, W.S. and Jordan, P.W. (Eds), Taylor & Francis, London.
Hofstede, G., 1991. *Cultures and organisations.* McGraw-Hill International, Maidenhead.
Jordan, P.W., 1998. Human factors for pleasure seekers. In *Ergonomia* **5**, 11. Moretti & Vitali, Bergamo, pp.14-19.
Jordan, P.W., 1999. Pleasure with products: Human factors for body, mind and soul. In *Human factors in product design. Current practice and future trends*. Green, W.S. and Jordan, P.W. (Eds), Taylor & Francis, London.
Jordan, P.W. and Servaes, M., 1995. Pleasure in product use: Beyond usability. In *Contemporary ergonomics 1995*. Robertson, S. (Ed.), Taylor & Francis, London, pp. 341-346.
Jordan, P.W., Thomas, B. and Taylor, B., 1998. Enhancing the quality of use: Human factors at Philips. In *Human factors in consumer products*. Stanton, N. (Ed.), Taylor & Francis, London.
Kahmann, R. and Henze, L., 1999. Usability testing under time-pressure in design practice. In *Human factors in product design. Current practice and future trends*. Green, W.S. and Jordan, P.W. (Eds), Taylor & Francis, London.
Kanis, H., 1999. Design centered research into user activities. In *Human factors in product design. Current practice and future trends*. Green, W.S. and Jordan, P.W. (Eds), Taylor & Francis, London.
Macdonald, A., 1998. Developing an qualitative sense. In *Human factors in consumer products*. Stanton, N. (Ed.), Taylor & Francis, London.
McClelland, I., 1995. Product assessment and user trials. In *Evaluation of human work. A practical ergonomics methodology*. Corlett, N. and Wilson, J.R. (Eds), Taylor & Francis, London.
Maslow, A., 1970. *Motivation and personality*. 2nd edition, Harper and Row, New York.
Nagamachi, M., 1986. *Kansei engineering*. Kaibundo Publishers, Tokyo.
Nagamachi, M., 1997. Kansei Engineering as consumer-oriented ergonomic technology of product development. In *Proceedings of I.E.A. World Conference 1997 International Ergonomics Association 13th Triennial Congress*. Taylor & Francis, London.
Pepermans, R.G. and Corlett, E.N., 1983. Cross-modality matching as a subjective assessment technique. In *Applied Ergonomics*, **14**, pp.169-176.
Pirkl, J.J. and Babic, A.L., 1988. *Guidelines and strategies for designing transgeneratio-*

nal products: A resource manual for industrial design professionals. Copley Publishing Group, Acton, Mass.

Rooden, M.J., 1999. Prototypes on trial. In *Human factors in product design. Current practice and future trends*. Green, W.S. and Jordan, P.W. (Eds), Taylor & Francis, London.

Smith, J., 1989. *Senses & sensibilities.* John Wiley & Sons, New York.

Taylor, A.J., Roberts, P.H. and Hall, M.J.D., 1999. Understanding person product relationships – a design perspective. In *Human factors in product design. Current practice and future trends*. Green, W.S. and Jordan, P.W. (Eds), Taylor & Francis, London.

CHAPTER SIXTEEN

Design Based on Kansei

SEUNGHEE LEE

Delft University of Technology, Department of Industrial Design Engineering, Jaffalaan 9, 2628 BX Delft, The Netherlands

Institute of Art & Design, University of Tsukuba,

AKIRA HARADA

Institute of Art & Design, University of Tsukuba,

1-1-1 Tennodai, Tsukuba, Ibaraki, Japan

and

PIETER JAN STAPPERS

Delft University of Technology, Department of Industrial Design Engineering, Jaffalaan 9, 2628 BX Delft, The Netherlands

16.1 INTRODUCTION

When a designer starts to design the form of a new product, she needs to integrate many demands and wishes that the prospective users of the product may have. Not only technical and objective demands are important, but also aesthetic, emotional, and other experiential factors, some of which are hard or impossible to express objectively. In design practice, the designer has to balance between objective and subjective properties, between functional technology and emotional expressiveness, between information and inspiration. Design development by 'Kansei (subjective criteria) science' or 'Kansei engineering' is a new approach which originated in these conditions.

When a design approach uses Kansei engineering, we give our attention to the behaviours of people when they perceive images or objects including products, and study how their personal preferences or cultural bases work to their feelings. 'Pleasure' would be one of the major feelings from the impression occured by Kansei. However, when designing products, designers are not only concerned with the visual appearances but also the other properties of the product. Objects are not only looked at in isolation, but are seen in a context, are handled, touched, sometimes also heard or even tasted. Therefore it is important that a wider range of experience, a fuller integrated Kansei appreciation, is incorporated into the design approach.

In this chapter, we introduce the basic ideas and the movement of figuring out the structure of Kansei in Japan, and how to apply Kansei to design approaches.

16.2 THE DEFINITION OF KANSEI

What is Kansei?

In Japan, the terminology of Kansei draws back on the German philosopher, Baumgarten. His work AESTHETICA (1750) was the first study that influenced Kansei engineering (Harada 1997). The aim of Kansei study is to seek the structure of emotions which exists beneath human behaviours. This structure is referred to as a person's Kansei. In the art and design field, Kansei is one of the most important elements which brings the willing or power of creation. In research by Harada, it was found that the attitude of a person in front of art work and design is not based on logic but on Kansei (Harada 1998a).

Kan Sei

sensitivity, sensibility — feel, touch, tactile sensation, emotion impression, appreciation — heart, mind, soul — be born, alive, dynamic

知性 → 矢 + 口 + (心 + 生)

Chi Sei

understanding, intellectual — arrow acute — mouth — heart, mind, soul — be born, alive, dynamic

Figure 16.1 The etymology of *Kansei*

The word 'Kansei' is interpreted variously and has been used in many researches related with not only design but also other research fields. It is a word which inclusively involves the meaning of words such as sensitivity, sense, sensibility, feeling, aesthetics, emotion, affection and intuition. Figure 16.1 shows the etymology of Kansei and Chisei interpreted from Chinese characters, both of which are processed in human minds when they receive the information from the external world. As you see in the figure, Chisei works to increase the knowledge or understanding which is matured by verbal descriptions of logical facts. And Kansei works to increase the creativity through images with feelings or emotions.

But we cannot doubt the fact that both Chisei and Kansei have the same level of power to stimulate human behaviours (Figure 16.2). So far, the practice of most designers has focused on Chisei. Kansei has been regarded as a totally subjective phenomenon so that anyone in the world has their own individual way of absorbing and presenting. In front of a painting, we appreciate it without thinking of any rules but 'just' feel a pleasure.

Kansei has an effect to create more various feelings in the human mind which appear as individualised emotions. But then in the history of product design, the emphasis on mass production caused a disregard for the individual's preferences and feelings.

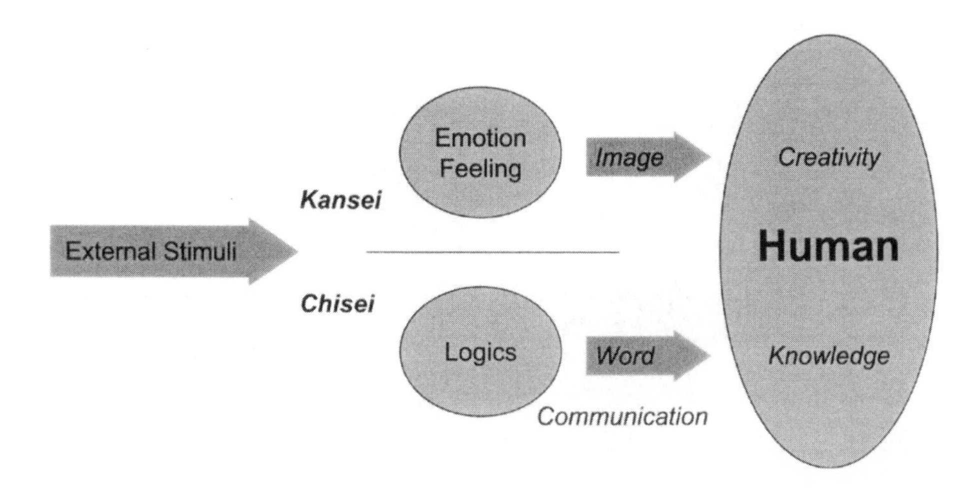

Figure 16.2 The effect of *Kansei* and *Chisei*

That's the reason why it has been hard for people to realise the relevance of Kansei for design. Nowadays, people's need for emotional satisfaction is growing and being acknowdged by manufacturers. The functional requirements can often be solved by Chisei, logical knowledge of technology. But the fulfilling of emotional requirements, including pleasure, requires attention for Kansei. Regarding those changes, we need to know the structure of Kansei and how to apply it to product design.

In 1997, the Kansei evaluation special project started as a five-year interdisciplinary project at University of Tsukuba in Japan. Because it was found that the term 'Kansei' was used in different meanings by different participating researchers, an initial study mapped their meanings (Harada 1998b). The researchers in the project were asked to give their definition of Kansei. These statements were analysed, and key words were clustered to five main aspects as follows:

- Kansei is a subjective effect which cannot be described by words alone.
- Kansei is a cognitive concept, influenced by a person's knowledge, experience, and character.
- Kansei is a mutual interaction between the intuition and intellectual activity.
- Kansei entails a sensitivity to aspects such as beauty or pleasure.
- Kansei is an effect for creating the images often accompanied by the human mind.

Most of all, it's important to understand Kansei implies that human behaviours can change dynamically, and indicates flexible and dynamic approaches are needed in the various fields of study.

Why is it important to develop modeling of the structure of Kansei ?

Before Kansei can be applied to product design, its structure should first be mapped. For design, this requires a human centered approach.

The Kansei evaluation special project includes various research fields divided into three research groups (Figure 16.3 to 16.6).

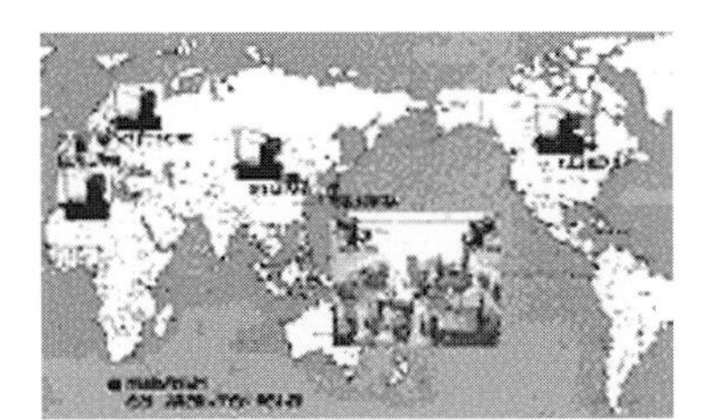

Figure 16.3 International network of controlling the robot

Figure 16.4 Robot in the art museum

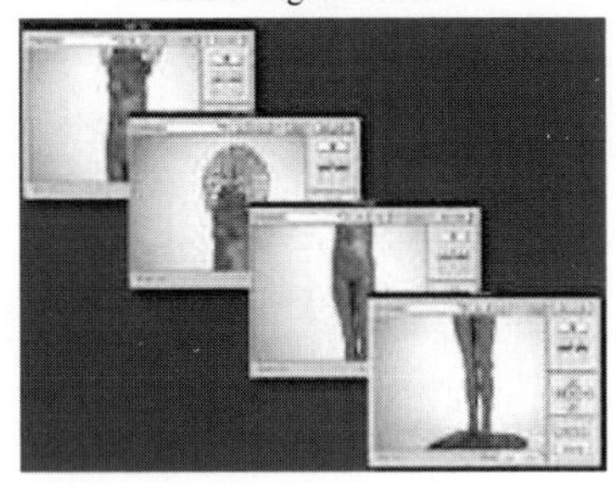

Figure 16.5 Example of tracking robot views

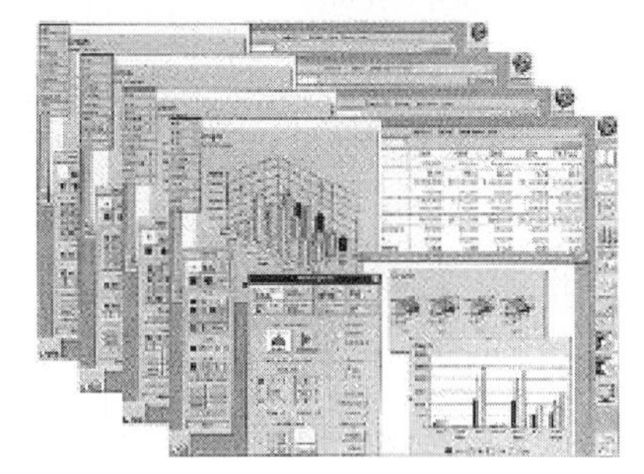

Figure 16.6 Analysis of appreciating behaviours

The first group is working to figure out the Kansei evaluation using the art and design works. They approach from the art critics, design history and cognitive science to know how people interact with art through their appreciation behaviours. The second group is developing network programming and user interface which is used to make these artworks accessible over the internet by studying. And the third group is designing the remote control robot which allows remote viewers to virtually more through a museum. This robot was made to let people explore an art museum over the internet.

The remote control robot is utilised as a tool to get the experimental data of human behaviours. By tracking the robot, these three groups can measure the viewing behaviours. These behaviors will be analysed on the basis of Kansei evaluation. We concentrate to find out not only the differences of behaviours but also the characteristic common points of people with different cutural backgrounds.

Based on the aim of Kansei study, many researchers not only from Japan but also from overseas are involved to uncover its structure.

16.3 HOW TO APPLY KANSEI TO DESIGN AND HOW TO EVALUATE KANSEI?

In a design context, Kansei emphasised the designers imagery skills: the power to produce a mental image and to use this in the creative process. It means, when a designer creates a new shape, all the impressions gathered in the process must be expressed. Generally, people hope to get a pleasant feeling from the appearance of the products. Here we introduce explorative design studies which attempt a new approach concerning Kansei in the early phase of design which would be one of the solution to improve the feeling of pleasure into design.

Design approach with abstracted images

Many designers use image collages at the first step of the design process. They try to communicate or confirm what the users/clients wish for a new design. Actually, the image collages often give the essential concepts of clients' wish to the designers (Figure 16.7). But in practice, these collage are rarely used after that point, when the designer is conceiving the form of the product.

From a point-of-view of Kansei, images are more effective than words to deliver feelings and concepts of formgiving in general. In this research, we developed a new method of formgiving based on the images that express the users' positive feelings including pleasure (Lee, Kato and Harada 1997).

Image abstraction was suggested as a method of creating new shapes of products. When some images are chosen by users/clients subjectively, this choice expresses their preferences of shapes, colours or atmospheres.

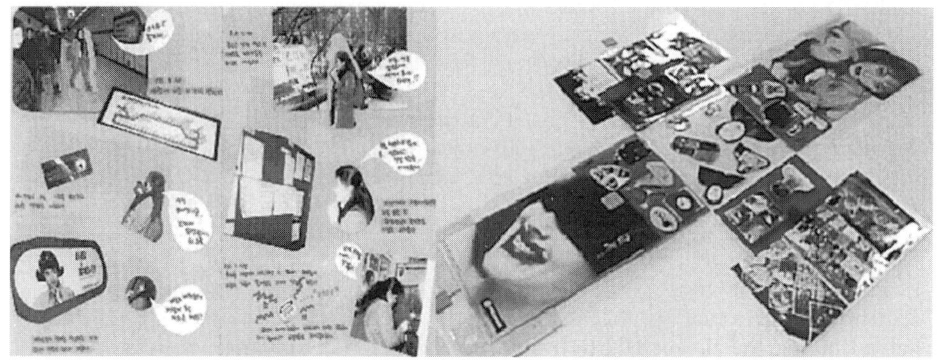

Figure 16.7 Image collages based on users/clients' preferences

But each of us may get different impression from the images. How then can we catch the subjective impressions? Here we introduce the 'image-icon' which shows a characteristic shape from the picture image. The users/clients and designers can communicate about the subjective, experiential Kansei qualities of the new product form through the inspirational visual materials and the abstracted image-icons. Designers seek for impressive parts on the image and draw this as a simple abstracted sketch. The user examines these image-icon sketches and selects those icons that best express their preferences for the new design. This feedback provides the designer with focused, nonverbal information on which she can base the new form of the product. The abstracted 2D image-icons carry the 'visual language' of the inspirational material as it is experienced by the designer and user. In creating the new form, the designer uses the images for visual elements. The 2D image-icons produced by the designers reveal the way in which they treat the images. They integrate the shapes of the image-icons and capture the essential impression in the 3D form of the new design. This process is shown in Figure 16.7.

In an experiment, we recorded image-icons made by designers and analysed which form aspects the designers chose to present in the icons. Figure 16.5 shows examples of image-icons abstracted from different image pictures. The designer who made these icons concentrated on visible edges in the images and suggested a linear shape of recombination for a new design. We found designers used different strategies in this process, some drew very geometric icons and designed a geometric form. Others enlarged a small part of the visual materials and used this in their design. Other designers transform the icons in more complex ways, for instance by combining and rotating several icons into a single form. These differences and similarities between designers' behavior showed more study is needed to find out how the designers' Kansei influence their creativity (Lee 1999a).

Extending of design approach based on Kansei by dynamic manipulation of 3D object

In a current cooperative project, we are exploring how the Kansei design approach can be extended from 2D images into richer product experiences, by introducing spatial materials and 3D icons. In this research, we describe the theoretical background of this project, and describe the experiment that was conducted at Delft University of Technology (Lee 1999b).

In the Kansei design approach developed by Lee and Harada (Lee, Kato and Harada 1997; Lee and Harada 1998), visual imagery is used to carry the subjective meanings, and the designer extracts form elements from inspirational material into abstracted image-icons. 2D images may be too limited to capture the dimensions of product appreciation, and the experiment sought to uncover ways in which the approach can be extended. Using 3D materials and 3D icons is an important first step. This experiment compares 2D and 3D viewing techniques and full tactile-visual exploration for different product types; the effect of these interaction modes on the richness of the 3D and 2D image-icons that designers produce is analysed using both objective (image processing parameters) and subjective (expert judgement) measures.

Tools and techniques for inspiration

External stimuli produce subjective responses in the observer. The impression often contains emotions such as favour, pleasure or wonder and these feelings reflect the individuals and the situations they face. The problems in design arise when the products reflecting the designer's feelings are not in accordance with the users, or when the analysis of the users' preferences are not accurately modeled in the designer's planning stage resulting in inappropriate products. Here, objective preference analysis becomes necessary.

The Kansei design approach is to let the client and designer communicate through the icons which visually represent the favourite shapes of both parties. To make a more creative shape and dynamic impression of the products, they need such a mutual supported communication.

The 'image' was introduced as a tool to narrow the gap of concept cognition between designers and users in the planning of product design, and to effectively reflect the sensibility each held in the design process. The evaluation or impressions of a form sometimes involves feelings that evade expression. This research also employed 'images' as a main information for the design support system, focusing on the shapes that were formerly regarded as difficult to apply so far.

Including 3D in the Kansei design approach

The way in which three-dimensional objects play a role in the design approach is different for the inspirational materials and for the image-icons. The inspirational materials are only evaluated or inspected, but not created, by the designer and client; the objects are also created by the designer. Moreover, inspirational materials, even for 3D forms, often come in the form of a visual representation rather than the real object. With the developments in the Internet, it can be expected that both 2D and 3D catalogues and databases of forms become a prominent part of the designer's world; a part from which inspiration can be taken. For this reason it is interesting to study in how far the relevant form information can be supplied through 3D or 2D spatial visualisations of a form, or what information is only supplied through the designer actually touching or handling the real product (Figure 16.6). Therefore we distinguish (at least) three levels of inspecting or evaluating the inspirational 3D materials: direct handling, visual inspection from many sides but without touching, and visual inspection through a virtual representation (such as QuickTimeVR or QuickDraw 3D).

The 3D icons preferably are real 3D objects, sculpted from clay. Current computer modelling techniques are still too clumsy to assist the designer in the very explorative type of shape modelling that is involved in creating the abstracted 3D icons. However, because many designers create some 3D sketches before they start modelling, we also let the designers create sketches (half the designers are asked to create the sketches before the clay icons, the other half in the other order).

About the experiment

The conditions included in the present experiment, where we make distinctions between the types of products being designed, requires a separation between products that are only looked at, such as speakers and clocks, versus products for which use implies touch, such as cassette radios. One can expect these different ways of using the product to have important implications for how their forms are experienced.

In this research, we chose a loudspeaker as a product which does not involve a touch, a cassette radio as a product that is frequently touched by its user. As for the subjects, we invited 24 students from 3rd to 5th grade in Industrial Design Engineering of TU Delft. The first session was a training session, to let the subjects get used to the method. Subjects were shown a picture of a camera, their task was first to sketch the most impressive parts using pen and paper, then to model impressive parts in clay. This resembled the earlier 2D version of the method. In the second session, the subject could view the loudspeaker (or cassette radio) in one of three observation conditions (see Figure 16.8 and 16.9). Either the product was displayed on a computer (different views were provided using a Quick Time VR-like display). In a second condition, the product was shown on a turn table and the subject was instructed not to touch it. Finally, subject in the third condition could look at and touch the product freely.

In the end of the second session, all the subjects filled in a personality test. This personality test was used earlier to show the strong relationship between the impressions on images (Lee and Harada 1998). It is the Maudsley Personality Inventory (by Eysenck) which is known for a high reliability and general validity in psychology and medical science (Eysenck 1964). In our earlier research, it showed clear results of impressions on images according to the personalities, such as substantial vs. unsubstantial and emotional vs. tactile. It provided the results based on two dimensions, one is extraversion-introversion (X) and the other is neurotic-stable (Y).

Measurement

Given the explorative nature of this research, and the great amount of variability involved in creative processes of form generation, we could not expect a clear-cut single objective measurement to work. First, we put all the outputs of image-icons and clay-icons from each subject on the personality space, as was done in the previous study. Also, the image-icons and clay-icons were classified on the expressiveness of their shapes by a panel of experts (design teachers), who are shown the original materials and the icons produced by the designers.

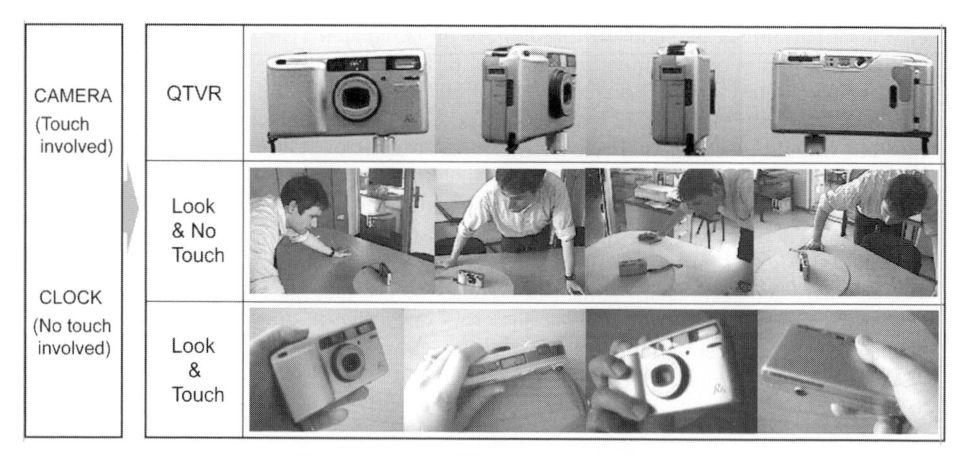

Figure 16.8 Three different conditions of observation

1. Influence of different conditions of observation

One of the aims of this study is to seek the differences of unconscious perception or impression when user or designer face various use of products in different situations.

Especially for the look-and-touch condition we expected the icons to reflect those shape elements which play a part in holding or operating the object, such as the scale and texture of grips. This effect should be stronger for the products that are meant to be handled than for the visual-only products, which would be dominated by the visual, geometrical shape alone. We also expect virtual and no-touch viewing not to differ much for objects of the latter category; this result might be of great practical importance, as it supports the credibility of the Internet VR-catalogues of products mentioned in the introduction.

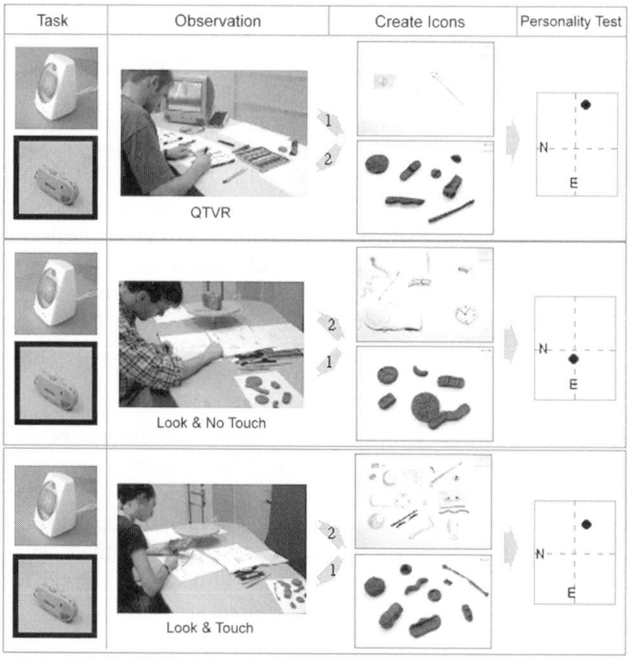

Figure 16.9 Abstracted impressions by sketching and modelling in the second session

2. Relationship between frequency of views and output of impressions

In case of QTVR condition, the frequency of views were stored automatically while the subjects produced the sketches and clay-icons. The model display programme used in the QTVR viewing condition kept a log of the total time that each view was on the screen. These times were converted to percentages of viewing times and averaged over the subjects, for each product class.

3. Parallel way of creations

Designers can approach to grasp the impressions or ideas of product by easier way of their own, such as 'sketching --> modelling' or 'modelling --> sketching' in the design process. We expect which impressions or elements of design could be easily created in each formgiving behaviour, sketching or modelling.

4. Relationship between the personality and design impressions

As far as noticed in the previous research, designers often scored high on extraversion. In this experiment, we found the same tendency for the design students. The clay-icons were placed in the personality space (see Figure 16.10). The experts (design teachers) remarked mostly about the variations along the neurotic/stable axis (note: the axis were not labeled). We got some dimensions of evaluation, such as simplicity (stable) vs. complexity (neurotic) and abstract (stable) vs. realistic (neurotic). These proved mostly the same tendency with the previous research which have collected impressions in words from the images.

Figure 16.10 Clay image-icons on the personality space

Discussion

When designers begin creating a product, they need to explore new shapes and elements which represent the preferences of the target users. Increasingly, users meet the products first through 2D images in Internet catalogues but they use them in real conditions. Because designers also change to use VR tools, it is important to study and improve the validity and workability of these VR conditions.

In this research, we compared the conditions of 2D, 3D viewing and handling in appreciations and observed how designers develop their shape ideas. We registered how people looked at the object and where they concentrated on the objects through a photograph, 2D image sketches and 3D clay-icons. To find patterns in the appreciation behaviour based on Kansei, the QTVR tool is an efficient method to register viewing of the forms by subjects (10). Building the Kansei appreciation structure on 3D objects, we study the dimensions such as forms (appearance), scale, surface textures and colours from 2D sketches, and tactile textures, physical emphasis from 3D clay-icons. Extending approach from 2D image-icons to 3D image-icons for manipulating in virtual space, we expect the 3D clay-icons play a role of rapid formgiving tool showing the essential elements of design for new products in searching among the 3D object database.

16.4 CONCLUSION

In this chapter, we introduced the basic meaning of Kansei and the movement of structuring appreciation model based on Kansei of people from all over the world. With this trial, we can easily find out people's ways to seek for pleasure through the art works. For the evaluation based on Kansei, we needed a new tool to communicate with other people in an emotional way.

We could develop a new method of design approach of communicating with users/clients using images to convey Kansei information. Furthermore, images give us many messages as impressive forms, colours, textures or atmospheres. Designers can interpret those characteristic impressions into abstracted simple drawings or models which we call 'image-icons'. With the image-icons, we've done several experiments so far to find a way to evaluate Kansei in design.

We believe that Kansei design which has a power of interpretation of people's emotion and feeling, makes it possible to create pleasure with products in our lives, because it focuses on human behaviours including emotional actions and responses. Finding the structure of Kansei related with art works and design will give us important ideas how to create pleasant products for human beings.

16.5 REFERENCES

Eysenck, H.J., 1964. *Eysenck Personality Inventory*. University of London Press, pp. 7-54.

Harada, A., 1997. The Framework of Kansei Engineering. In *Report of Modelling the Evaluation Structure of Kansei*, pp. 49-55.

Harada, A., 1998a. Modelling the Evaluation Structure of Kansei Using Network Robot. In *Report of Modelling the Evaluation Structure of Kansei II*, pp. 15-19.

Harada, A., 1998b. Modelling the Evaluation Structure of Kansei Using Network Robot. In *Report of Modelling the Evaluation Structure of Kansei II*, p. 2.

Lee, S.H., Kato, T. and Harada, A., 1997. Kansei Evaluation of Subjective Image by Iconic Abstraction. *Report of Modelling the Evaluation Structure of Kansei*, pp. 127-

135.
Lee, S.H., 1999a. *A Study of Design Approach by an Evaluation Based on Kansei Information 'Images'*. University of Tsukuba, pp. 133-139.
Lee, S.H., 1999b. Extending of Design Approach Based on Kansei by Dynamic Manipulation of 3D Objects. In *Bulletin of 4th Asian Design Conference*, pp. 686-693.
Lee, S.H. and Harada, A., 1998. A Mutual Supported Design Approach by Objective and Subjective Evaluation of Kansei Information In *Report of Modelling the Evaluation Structure of Kansei II*, pp. 359-374.

CHAPTER SEVENTEEN

Participative Image-Based Research as a Basis for New Product Development

STELLA BOESS and DR. DAVID DURLING

Research Scholar and Director of Research, Advanced Research Institute,

School of Art and Design, Staffordshire University, College Road, Stoke On Trent, ST4 2XN, Staffordshire, UK

CHERIE LEBBON

Research Fellow, Helen Hamlyn Research Center,

Royal College of Art, Kensington Gore London SW7 2EU,UK

and

PROF. CHRISTOPHER MAGGS

Director of Research, School of Health,

De Montfort University, The Gateway, Leicester LE1 9BH, UK

ABSTRACT

The paper discusses the concept of pleasure within the wider context of well-being. A qualitative study was conducted by researchers at Staffordshire University, UK, to gain insight into the strategies older people use in adapting their bathroom to their needs. The aim was to point towards qualities of use of products that would better correspond with participants' needs and wishes than currently available solutions. The study consisted of in-depth interviews in participants' homes and focus group sessions. Visual materials were made available for the participants to use during the group discussions.

Findings included that participants were variously engaged in achieving well-being through pleasure derived from product use. It is suggested that the participative approach taken in this study is particularly suited to eliciting such information. That is why it can be useful to designers and ergonomists in defining starting-points for new uses of environments and new types of products.

17.1 INTRODUCTION

This paper argues the case for consumer research as a method for re-contextualising during the design process. The study that will be used to illustrate this was carried out within the academic field. Nonetheless, it is suggested that the approach can also be useful in commercial product development. The paper first looks at a number of fields in which user needs have been researched, such as health research, architecture, gerontology and product design. Then our research methods are described, and sample results are reported to illustrate the benefits of conducting consumer research at the beginning of new product development. Based on the known problems that older people can face in retaining autonomy in daily life, the study was designed to gain insight into the strategies they normally use in adapting their bathroom to their needs. A goal was to derive qualities of use of products that would better correspond with participants' wishes and needs than currently available solutions. To enhance communication, visual materials were made available for the study participants to use during group discussions. The results suggest that participants were variously engaged for achieving well-being through pleasure derived from product use. It is argued that the participative approach used in this study is particularly suited to eliciting information about this aspect of consumers' lives.

17.2 DISCOURSES ON AGEING

Blaikie (1999) has reviewed various approaches to the study of ageing with regard to how concepts of ageing are formulated. These change along with societal processes, such as demographic change and the changing circumstances of retirement. Blaikie (ibid., 14) suggests comparing the (Foucauldian) discourses of the various expert fields and popular fields, looking at 'all items contained within a discourse, be these people, bodies, utterances, memories, images, or policy documents (...) as 'texts' or 'narratives'. The research task then becomes one of searching for internal consistencies or coherence within and between texts, rather than measuring these against some objective external truth.' In doing so, the social scientist's role is to 'elucidate complexity, not to simplify life'. We start with an exploration of the expert fields, as a basis for our elicitation of older people's wishes for bathing and personal care products.

Assistive equipment, emancipation and guidelines

One of the expert fields that produces descriptions of ageing is health research. As people age and their bodies change, their accustomed bathing environment may come to be disabling for them in daily life. In a recent study, 16% of participants over 70 years of age said they needed help with bathing or having showers, which means dependence on care or assistive equipment (Mayer and Baltes 1996: 169). Since the 1960s, architects and planners have addressed these problems by establishing criteria for supportive housing, and especially for bathrooms (reported e.g. by Goldsmith 1976; Kira 1976). Goldsmith (1997) tells the story of the disability movement's struggle in the UK and the US for access to public amenities. They succeeded in getting equal access legislation introduced (e.g. ADAAG 1991). By now, many types of equipment have been developed in industry in accordance with guidelines, to assist disabled and/or elderly people in their daily lives.

Further guidelines have been produced since, for example by Juul-Andersen and Jensen (1997). For accessible bathrooms they recommend the provision of grab-bars, level showers, seats in showers, and space for a wheelchair to turn. Older citizens' initiatives, for example the Dutch Senior Citizens Label (ANBO 2000), have formulated similar sets of prescriptions.

Recommendations and guidelines are being developed for the purpose of guaranteeing equal access, and their (ever evolving) contents are ideally something of a 'bottom line' for consumer products.

Discourses in gerontology

Sociological studies (e.g. Mollenkopf 1993), however, have highlighted the fact that assistive devices are often disliked and underused by older people in their daily lives. The devices are mostly provided through the mediation of medical professionals. Outlets are discretely tucked away, because ownership and use of such products is perceived to be stigmatising for their user.

In response to the many stigmas that society attached to older people, Laslett (1989) and others argued for the recognition of the Third Age as a phase of enjoyment, individual fulfilment and activity. Initiatives like DesignAge at the RCA in London (founded in 1991) raised the issue that older people are a target market waiting to be discovered, and stimulated designers' interest in products for older people (Coleman 1997).

The change towards a more positive view is now very evident in Western societies. However, the concept of the Third Age has also been criticised as having an implicit desire to negate the "Fourth Age of decrepitude" (see for example Blaikie, 1999: 186). So is there *one* appropriate way of thinking about older people? Tinker (1995: 45) warned that 'All research on older people concludes that it is unwise to generalise about elderly people. It is not so much age that distinguishes older people from the rest of the population but the particular contingent circumstances which tend to be associated with later life.'

The field of environmental gerontology researches the individual contingent circumstances of life in older age. The concept of 'environmental press' (Lawton 1982) was proposed as a theoretical model for the relationship between an ageing individual and their environment. Since it was proposed, its emphasis has shifted towards the opportunities for proactivity of individuals towards their environment (for example, Slangen-de Kort 1999; Steinfeld and Danford 1999). In the interest of reliable measurement, the scope of research in this field is often limited to a set of existing environmental options for research subjects. That limits its direct applicability to new avenues for design. It does, however, inform us that people have various psychological 'styles' of acting and coping as they age.

Discourses on consumer culture

In consumer culture, individual style is a more general term. Consumers, it is said, shape and display their lifestyles through acts of consumption (Oswald 1996; Chaney 1996). At the major bathroom fairs, such as the ISH in Cologne and the KBB in Birmingham, it can be observed that the bathroom is being promoted by the industry as a place for physical recreation and enjoyment, in tune with the trend towards healthy living and fitness of recent years (see also Hansgrohe 1993; Hoesch 1997). Do the wishes of older consumers match those of other target groups? Manufacturers have discovered the Third Agers, and this has created a demand for research into their lifestyles and preferences. Where are they homogeneous in their wishes, and where heterogeneous? Recent research on the topic of bathrooms has therefore focused on the preferences (Berlo 1996) and experiences of individual older consumers (Mullick 1999; Boess and Lebbon 1997).

The Individual in consumer culture

Both Blaikie (1999: 7) and Mayer and Baltes (1996: 503) remark that individuals are shaped throughout life: by their biological ageing, but also by all the contingent circumstances that they live through. Given that an 80-year-old was an adult before there were bathrooms in every house, let alone acrylic bathtubs, it can be supposed that older people's attitude to bathing is far from exhaustively known. Blaikie (ibid.: 25) further points out that an individual's perception of herself may differ from that of others: 'If, as (...) detailed interviews suggest, older people do not feel old, then what are the implications of having a young mind trapped within an ageing body?'

Our study on consumer preferences needs to elicit their own perspective, as well as the context of their lives; in other words, the culture they live in. Kay Milton, an anthropologist, has compared definitions of 'culture' and has suggested that: 'It consists of two complementary processes: that whereby culture generates actions, and that whereby culture is sustained, reinforced or modified through actions (Milton 1996: 20).'

The question then arises, who modifies culture and why? Douglas (1996) has described 'protest shopping' as a strategy of members of the consumer society for introducing cultural changes. They express their protest through visual attributes, for example clothing, which they make or buy, and display. Through its visibility, the values which it embodies are introduced into and absorbed by society. McCracken (1988a: 97) has suggested that disadvantaged groups try to claim equal status with a hegemonic group, an example being women adopting male dress codes in the professional world. Milton (1996: 21) proposes a less uni-directional model: individuals come from and move within a variety of systems, and: 'Social interaction becomes an arena in which the participants each assert their particular way of knowing the world, and in which they try to make their knowledge count [...] in the process through which culture is continually recreated.'

From consumer behaviour research, it is known that an individual's involvement with a product, which includes their knowledge about it, is a predictor for their *'predisposition to perform'* (Laaksonen 1994: 176), to act in some way towards it (usually purchase). The *actual* process of owning, using and consuming products, however, is also influenced by the cultural values and the general situation of an individual. That is why Laaksonen called for a more comprehensive approach to the study of consumption (ibid.: 179).

The concept of 'quality of life' is an attempt to look at life in a comprehensive way. In the field of health research, Mayer and Baltes (1996: 497) and Bowling (1995) have discussed quality of life, remarking that it is too broad and too idealistic to be readily applicable. Kristensen (1999) attempts an application of the concept to product design, but does not relate it to previous definitions. For the purposes of multidisciplinary approaches, such as our focus on the individual within culture, Mayer and Baltes (ibid.: 497) suggest using the term 'well-being', which is easier to limit and avoids the discussion of definitions.

Mayer and Baltes (ibid.: 518) report that the subjective assessment of well-being and objectively 'measurable' living conditions do not necessarily correlate, because of internal compensation processes on the part of the individual. Financial security, health and participation (family and friends) were named as more important than a good house by respondents in their study. However, the authors argue that the home strongly influences all that, though indirectly. In our research, therefore, subjective assessments of well-being in relation to the home are elicited.

Jordan and Macdonald (1998a) have proposed a framework for pleasure in human factors for design. In the following, the methods and results from our study are reported, with the aim of connecting that framework to the contexts of relationships, life course and engagement for change. The concept of pleasure is looked at within the wider one of 'well-being'. Research questions are: where and how are people 'trying to make their knowledge count'; are they 'engaged' for changes that will enhance their well-being?

17.3 THE WELLBATHING STUDY

Research design

The task of the Wellbathing study was to research points of departure for the design of bathrooms for older people. Several collections of new and not-so-new methods for design research have recently been published, for example by Poulson, Ashby and Richardson (1996); Jordan (1998); Aldersley-Williams (1998); and Hom (1996). Recent methods include the use of visual materials (Jordan and MacDonald 1998b) and 'make tools' (Sanders and Dandavate 1999) to aid communication and transfer of tacit knowledge. Sanders and Dandavate (ibid.: 89) explains that: 'Understanding how people feel gives us the ability to empathise with them. This way of knowing provides tacit knowledge, i.e., knowledge that can't readily be expressed in words (quoting Polanyi 1983).'

By witnessing and participating in the consumers' experience, the designer/ researcher can connect it to their own way of experiencing the world.

The Wellbathing study aimed to provide designers with knowledge they can make sense of and act on, both about participants' relationships with their present environment, and about the ways they have adapted their environment for future use. Elicitation methods were chosen that would allow participants to produce descriptions, explanations and artefacts, under the assumption that these methods would elicit tacit as well as explicit knowledge.

Research methods were therefore chosen that would address:

- Patterns of daily life

In daily life, many products are available to us to perform tasks in a routine manner; they are 'ready-to-hand' (Verbeek and Kockelkoren 1998, quoting Heidegger). The products are 'embodied': they are merely an extension of our body in carrying out the task. When a tool breaks or an activity is cumbersome, the object becomes 'present-at-hand'. Conscious action, 'engagement' is required to return it to a state of functioning as an extension of the body. But is this process merely reactive on an individual's part, or how and to what extent is maintaining and proactively changing the environment a part of daily life? To answer these questions, we looked for descriptions and traces (Zeisel 1981: 89) of active engagement with objects, as well as routine patterns of behaviour. The main interest was in descriptions related positively or negatively to well-being.

- Knowledge of environment and products

What knowledge is available to people in assessing their home environment and in deciding what needs to be done with it? Bound up in this question is the cultural 'meaning' of home. The home, throughout life, becomes increasingly connected to a person's identity, storing cherished objects and memories (Blaikie 1999: 200; McCracken 1988a: 124; Csikszentmihalyi 1995: 121). Do people have strong feelings about their home, privacy, and about assistive products? Do they protest, against whom or what, and what knowledge do they then bring to bear? We looked for explanations of wishes or aversions, and for descriptions of ideas to initiate change.

- Experience of environment and products

Diane Willcocks (1984: 13) found visual tools to be helpful when she conducted consumer research in old people's homes, because participants 'have difficulty in making choices' when the products that would be right don't exist or are not available, i.e. when 'these are at all hypothetical and involve them considering something they do not currently experience.'

To elicit experiential qualities attributed to products, we designed a visual research element for our study. It was hoped that this element, in particular, would be a vehicle for the elicitation of tacit knowledge, which might be lost in the translation to verbal reports. This was considered particularly important in the light of the intended usefulness of the study for the design of consumer products. The visual element needed to be

- Open-ended. It was not to present already-made solution proposals, but to allow for discovery by the participants.
- Appropriable, as a tool, so that something could be said about HOW participants appropriated it.
- Enjoyable and reality-related, to generate empathy between the researchers and the participants.

Data collection

Two academic researchers carried out the study. A combination of data collection strategies was employed, comprising three focus group interviews, fifteen individual interviews and ten photographic recordings. All three methods have been used in design-related research for decades (see e.g. Zeisel 1981).

The 36 participants were between 50 and 90 years of age and were residents of sheltered housing. They were invited to take part through the mediation of two housing associations, in Staffordshire and Wiltshire, UK.

Focus groups

The focus group methodology is now frequently used for design research purposes because it is an economical means of eliciting a broad range of consumer responses on products. See for example McDonagh-Philip and Denton (1999), Strickler (1997), and Jordan (1994). The Royal College of Art's user forums (Coleman 1997) and the Glasgow School of Art study for furniture design (Macdonald 1996) have used group discussions in collaboration with older people. For useful guidance see for example Krueger (1994).

The focus group interviews were conducted in each building complex's common lounge, and many of the participants in each group knew each other well. These interviews served to open the topic and bring out as many facets of it as possible. They were followed up by interviews in participants' own homes if they expressed interest.

The visual element: A tool-kit

In the group interviews, participants were offered a tool-kit comprising magazine cut-outs, line drawings, and an A3 board on which to arrange them. The method is based on 'moodboards' or 'lifestyle boards' as commonly made up by designers to visualise the target market for a product (e.g. Baxter 1995). What was different about our tool-kits was that they were used by the participants *themselves* to communicate their idea of their 'ideal bathroom'. Similar tools have previously been tried out in marketing research, with the reported benefit of engaging consumers' imagination (Sampson and Bhaduri 1987). Given the interest of our study in present and future product use, such a tool-kit was likely to be an enhancement to the discussions. There is a limit to the amount of pictures that can be offered, but participants were encouraged to expand and modify where desired. The choice was meant to be wide enough to be perceived as open. Each participant was asked to show and explain their collage to the others.

The visual exercise made the session fun. It was valuable in opening sensitive topics for discussion and broadened the scope of it, because images could be referred to.

Individual interviews

These were semi-structured, in-depth interviews of about one and a half hours duration (McCracken 1988b), conducted in participants' own dwellings. Generally, the participants were happy to show the researcher their bathroom and to explain about its contents, sometimes demonstrating their ways of using them.

Photographic documentation

The third element, photographic documentation, served to record what participants' bathrooms looked like, and to make an approximate assessment of their accessibility and usability, with reference to guideline data and the person's own descriptions. With the respondent's permission, a researcher took about three photographs of the bathroom.

17.4 ANALYSIS

The audio-recorded group sessions and individual interviews were transcribed and the data were categorised and coded by looking for instances of conflicts, trade-offs, motivations, goals, aspirations, strategies and fears – everything that was connected to a person's interaction with their environment (For an accessible introduction to qualitative data analysis using e.g. grounded theory, see Polit and Hungler (1997: 375-98)). Comparisons between statements made in the group interviews, the individual interviews, the photographic documentation and the researchers' background knowledge of usability problems were made. Statements on issues such as cleanliness were carefully monitored because of the likelihood of their being biased towards social desirability.

The photographs, the picture boards and the descriptions were compared for salient differences and similarities, such as degree of decoration, amount and type of assistive and other functions present, and the care invested in adapting the space to individual preferences and requirements.

The validity (here: meaningfulness) of qualitative research results can be gauged by examining the data for their consistency within and across individuals and data sources. For example, a person's verbal *account* of what their bathroom looks like is compared with a *photograph* of the bathroom: what has the person left out; what have they emphasised. The various types of information are triangulated (compared) until a plausible picture emerges. Inter-researcher comparisons of interpretations and conflicting explanations of the same phenomenon by different participants are also brought in. As a further test of validity, summaries of the group discussion protocols were sent to all participants to check whether they agreed on the content (member check).

The reliability (here: generalisability to the target group) of the results is basically a matter of confidence. Social scientists have indicated that analysis of qualitative data from a small group that is typical of the target group, e.g. demographically, produces a reliable indication of many of the salient issues for a larger group. For the purposes of product development, this is often enough to get started. For the purposes of scientific inquiry, it can inform the design of empirical studies.

Care was taken to try not to oversimplify the complexities of participants' lives. The findings, grounded as they are in participants' life circumstances (Strauss and Corbin 1990), were used to formulate hypotheses which constitute steps towards theory.

17.5 FINDINGS AND HYPOTHESES

Life over time: Routine and changes

- Daily life

Practically all participants occasionally made small changes to their bathrooms, such as moving a stool in and out of the room, or using objects for purposes not intended by their design, such as using grab rails as clothes hangers. Participants did not even describe these as changes at all. Yet they could have a significant effect on the way activities would be carried out. There was a low threshold to effecting such changes. This also applied to items for functional compensation such as small grab-bars, or bath-boards (the latter are laid across the rim of a bathtub for sitting on to wash). Small, reversible changes were regarded by participants as 'little tricks to make life easier'.

- Major changes

Items that were too large for changing or moving just by hand, and which had to be fixed using tools, were seen as major changes. Participants would actively reject assistive items of this kind if they did not think they needed them in the long term. Wall-fixed shower seats were an example for this, as were fold-down toilet grab bars which were regarded as being 'for them, the disabled'. In the event of provision with such equipment, a change in self-perception followed: 'Now that I'm disabled, I prefer a level entry shower. I do miss a bath though …' Though disability is age-related statistically, participants generally thought of it as a potential but not inevitable future. Even when in the process of adjustment to long-term care, participants strove to create and preserve opportunities for (shared) intimacy and informal assistance.

Minor changes have small effect on the self-perception of an individual; major changes have great effect. People would reject products which they saw as putting them permanently in the 'sick role' (Stimson and Webb 1975), which means being released from responsibilities and rights, in order to have the space to recuperate. Participants did not equate their aged state with being 'sick'. Their wishes remained directed towards maintaining or enhancing competence, health and enjoyment. At the same time, they were selecting their environment in such a way as to limit the demands it makes on them.

Hypothesis

Time with products is not uniform. The experience (of a product) is pleasurable when it adapts itself to a person's fluctuating state of health: while being in the 'sick role', a person makes different demands on their environment than in the course of daily life. And one partner of a couple may have different health-related needs from the other.

A conclusion for design is that supportive objects/functions have to be easily removable or ignorable, adaptable for various ways of handling, and have to make low initial demands on the physical and psychological strength of an individual. A major change, on the other hand, makes demands on resources, as will be further discussed below (5.3). It follows that fluctuations in health and fitness should preferably be cushioned by 'designing in' minor changes in the course of daily life.

Knowledge: New impulses and life experience

The conventional style of the bathroom in Britain today is a bathtub, a washbasin and a toilet, all in a matching pastel colour. The bathrooms of the study participants had all been equipped this way before they moved into their flats.

With the tool-kit, all participants were presented with approximately the same impulses. On their collages, they all chose new or unfamiliar things from the cuttings when these were thought to offer desirable functions. Some used them to describe things they knew or had known in the past, and to describe desired functions for which they knew of no available objects. Some were more adept than others at designing an 'ideal bathroom'.

In participants' own bathrooms, however, interpersonal differences became visible in another way. New developments such as level entry showers were implemented only by those participants who had trust in technology and were familiar with the means of procuring it; i.e. who could approach it pragmatically. Others acted and thought 'romantically': the practicalities of changing things were not readily accessible to them. Their collages showed that they had fantasies, for example for bubble baths, but that they were more likely to reject level-entry shower cabins and, in the extreme, electrical equipment such as a thermostat.

There was implicit pressure on the participants to give up the aspiration of pleasurable bathing for independent functioning. Some resisted this pressure to be 'sensible' and acted subversively in various ways: by obtaining informal care, negating physical obstacles, and adapting objects to uses that these were not originally meant for. Sophisticated assistive devices available and known to the participants were perceived as belonging to the organisational context of professional care. When that prospect appeared, it was resisted because it represented a loss of autonomy in the organisation of one's private life.

Hypothesis

For the participants of this study, only a pragmatic approach to providing themselves with assistive equipment, based on experience with technology provision, was a successful one. The 'pragmatist' brought no ideals into the process, but just adopted what the supply system offered and operated within its organisational framework.

The 'romantic', conversely, was left to negate the problems the material environment poses, to resort to fantasies, and to maintain makeshift situations. For her to be able to choose assistive functions they might have to be available as part of a trend, as a means of enhancing self-perception, with opportunities for hedonistic pleasure, and with a low threshold to implementation, for example through visual presentation, easy access, opportunity of comparison and trustworthy assurances as to technical harmlessness.

The challenge is to design assistive functions in such a way that they enable the individual to start off from where they stand, with their varying needs for relationships with their environment. Objects could then be used by a consumer alone, preserving her autonomy, but also as mediating objects in a social relationship, for example with an informal or formal carer.

Level of gratification and resources

- Physical pleasure

For some of the participants, bathing was *central* to pleasure. Others preferred showering, some both. Those who were more focused on bathing tended also to be opposed to solutions such as baths with hydraulic chairs, because these were cumbersome and offered only half the pleasure.

In many old people's residences, level-entry showers are currently being installed to replace bathtubs. This is done out of responsibility and care towards the welfare and functional autonomy of residents, and is being recommended by guidelines. The preference of some participants for a bathtub is at odds with that, however. They miss enjoyment and regeneration, which would most likely benefit their well-being.

- Personal expression

Most participants rejected clutter and untidiness in the bathroom. Some, however, liked to create a visually dense environment, for example through opulent displays of toiletries. Women tended to decorate their bathrooms more than men, and expressed more elaborate visions for them in their descriptions and on their collages. In their bathrooms the ready-designed shape of functional objects took backstage to their own activities of decoration. Participants' styles were quite diverse. Pieces of equipment which would seem alien in the room were undesirable.

- Habituation and sustained engagement

Whatever their ideal bathroom, in reality participants all settled for a certain status quo for long periods of time. In this 'snapshot' study, this finding is inferred from observation of traces and from participants' own accounts. Two alternating phases are distinguished, which depend on the resources available to an individual at a given point in time:

A. Sustained engagement

When a person has sufficient resources to keep on devoting attention to something, with the goal of realising aspirations. Examples:

- Defiance: sustained investment of intellectual resources.
  'They wouldn't let me change my bath to a shower, so now I won't invest anything. But one day I'll get a jacuzzi.'

- Subversion: sustained investment of social resources.
  Asking a friend to help every day, instead of buying assistive equipment or accepting formal care. 'My friend brings me a bath-board ... I don't want to use steps and age yet.'

- Confrontation: sustained investment of physical resources.
  'I've fixed a horizontal bar to the wall behind my bathtub. I use it to get in and out from my wheelchair' (This last example is taken from literature (Walker 1997). No such clear instance was found in our study.).

B. Habituation

Where basic functioning is at issue, aspirations are not necessarily abandoned (as seen in the collages), but not actively pursued either. Examples:

- Compromise: 'We bought a bath-lift you know ... it turned out we couldn't use it, so now we have a bath-board. Oh, we learn to make do.'

- Trade-off: pragmatic provision.
  'Now that I'm disabled, I prefer the shower. I do miss a bath though.'

- Resignation
  'I can't use the bathroom at all now ... I'm not normal. They take me to the assisted bathroom once a week.'

Hypothesis

The ability to present a 'civilised' bathroom is a source of self-worth, the inability to do so a source of shame. Equipment that is designed to have an assistive or supporting function must address the need for personal expression if it is to be acceptable to consumers.

The qualities of use of the bathtub – relaxation, warmth, containment, rest, and weightlessness – and the qualities of use of the shower – refreshment, activation, invigoration, cleansing – were both desired, enjoyable physical experiences for participants of this study.

What is needed are new forms that offer these use qualities in accessible ways; objects which avoid cluttering the bathroom and which make low demands on resources. The door bath is a solution which attempts to offer these qualities, but it was financially beyond the reach of most participants.

A briefing for a design developed from these considerations might describe, for example, a piece of low-tech-looking, removable, comfortable seating furniture which can be placed in a level shower space, and has a lifting element integrated into the body of the product, not apart, which can be operated by the user. The piece of furniture might contain massaging functions and softly running or still water.

17.6 USE OF THE STUDY METHOD FOR ERGONOMICS AND DESIGN

This study has looked at the concept of pleasure as a part of well-being. The idea was to capture many relevant facets of consumers' experience of their environment. It was done to provide points of departure for design which go beyond the improvement of existing product types.

What does the method contribute to ergonomics and design? A contribution that is hoped for is that measurements or other research methods can be further developed to include the following aspects, which emerged as central to the way participants of this study used things:

- Life over time: Routine and change

Participants moved in their environment and used things in such a way as to produce preferred daily life patterns, which varied between individuals. For example, they preferred to stand rather than sit under the shower and vice versa, employed little tricks to make life easier, or helped a partner carry out personal care activities. These are private ways of acting. Eliciting them requires a permissive yet probing approach. Participants might hesitate to reveal them, on the assumption that they are not normal, or because they are not actually aware of them as adaptation processes. Major adaptations fall outside the normal course of life and are perceived as changes. Desirable products were those which allowed participants to provide themselves with assistive functions without having to adjust their self-perception too radically.

- Knowledge: Impulses and life experience

The visual tools were useful in providing impulses. Participants were able to connect new or unfamiliar things with their life experience and could use them to describe desirable qualities of use.

Comparison between the image-based work and participants' actual situations brought the observation that not all approaches to (or styles of) environment adaptation resulted in a well-adapted living environment. Some participants put pleasure (bath) above safety and autonomy, in order to preserve at least potential enjoyment. The strength of this preference could be read from their protest or subversion against available solutions. This suggests that there is a need for assistive products which offer more opportunities for enjoyment and are obtainable in a more pleasurable context than is now the case.

- Level of gratification and resources

Physical pleasure and personal expression were important to many of the participants, so much so that they would become actively engaged for them. The level of gratification achieved depended on the resources of an individual. A person would became disengaged when they had to concentrate their resources on just managing daily life, though they would still be able to 'dream up' an ideal state. Objects have to be obtainable and usable without making too many demands on resources.

17.7 CONCLUSIONS AND FURTHER WORK

The paper has looked at pleasure in products within the broader context of well-being, illustrating the approach with a study of bathrooms for older people. Concepts that were investigated in particular were patterns of daily use, individual knowledge of the world from experience (e.g. about expected qualities of use), impulses, and levels of engagement in relation to resources.

It is hoped that investigations of this kind will become part of design methodology. The results can be used as guiding ideas by design teams, and provide criteria for the evaluation of subsequent design solutions. Of course, they do not replace a further process of testing and re-testing with consumers in the subsequent stages of design.

The core of the method is the elicitation of consumers' engagement towards environments in which they can make their knowledge count.

In this study, paper-based research tools were used. How such tools could be enhanced, technologically or otherwise, to be more efficient but without loss of depth and ease of handling, is a challenge for further research.

17.8 REFERENCES

ADAAG, 1991. *Americans with Disabilities Act Accessibility Guidelines.* Checklist for Buildings and Facilities, Architectural and Transportation Barriers Compliance Board, Washington, D.C., U.S.

Aldersley-Williams, H., 1998. *The Presence Forum.* Discussion forum: MethodsLab. URL: <http://www.presenceweb.org>. Last accessed February 1, 2000.

ANBO, 2000. The 'Seniorenlabel'. http://www.seniorweb.nl/anbo/ (the senior citizens' organization), Weerdsingel WZ 18a, 3513 BB Utrecht. Last accessed May 1, 2000.

Baxter, M., 1995. *Product Design. Practical Methods for the Systematic Development of New Products.* Chapman & Hall, London, pp. 221-227.

Berlo, A. van, 1996. *Ontwerp van veilige en comfortabele badkamerprodukten voor alle leeftijden. (Designing Safe and Comfortable Bathing Products for all Ages).* Centrum Techniek Ouderen, Eindhoven.

Blaikie, A., 1999. *Ageing and Popular Culture.* Cambridge University Press, Cambridge.

Boess, S. and Lebbon, C., 1997. Wellbathing, a Study for Design. In *Improving the Quality of Life for the European Citizen.* Placencia Porrero, I. and Ballabio, E. (Eds), Proceedings of TIDE (Technology for Inclusive Design and Equality), held 23-25 June 1998, Helsinki, Finland. IOS Press, Amsterdam.

Bowling, A., 1995. *Measuring Disease.* Open University Press, Buckingham.

Chaney, D., 1996. *Lifestyles.* Routledge, London.

Coleman, R., 1997. *Working Together: A New Approach to Design.* Coleman, R. (Ed.), Royal College of Art, London.

Csikszentmihalyi, M., 1995. Design and order in everyday life. In *The Idea of Design.* Margolin, V. and Buchanan, R. (Eds). MIT Press. First appeared in *Design Issues*, **8**, 1, (Fall 1991), Cambridge, Massachusetts, pp. 24-34.

Douglas, M., 1996. *Thought Styles.* Sage, London.

Goldsmith, S., 1976. *Designing for the Disabled* (3rd ed.). RIBA, London.

Goldsmith, S., 1997. *Designing for the Disabled, The New Paradigm.* Butterworth-Heinemann, Oxford.

Hansgrohe, 1993. *Badewonnen (The Pleasures of Bathing).* Hansgrohe (Ed.), DuMont, Cologne.

Hoesch, 1997. *Crossing the Waters.* Hoesch (Ed.), Plitt Druck und Verlag, Düren.

Hom, J., 1996. *The Usability Methods Toolbox.* US Industrial and Systems Engineering Department, San Jose State University, URL: <http://www.best.com/~jthom/usability/usahome.htm>. Accessed 1 February 2000.

Jordan, P. W., 1994. Focus Groups in Usability Evaluation and Requirements Capture: A Case Study. In *Contemporary Ergonomics 1994.* Taylor & Francis, London, pp. 449-453.

Jordan, P.W., 1998. *An Introduction to Usability.* Taylor & Francis, London.

Jordan, P.W. and Macdonald, A., 1998a. Pleasure and Product Semantics. In *Contemporary Ergonomics 1998.* Taylor & Francis, London, pp. 265-268.

Jordan, P.W. and Macdonald, A., 1998b. Human Factors and Design: Bridging the Communication Gap. In *Contemporary Ergonomics 1998.* Taylor & Francis, London, pp. 551-555.

Juul-Andersen, K. and Jensen, E.M., 1997. *Design Guidelines for Elderly and Disabled Persons' Housing.* Danish Centre for technical aids for rehabilitation and education, Copenhagen, and, Royal College of Art, London.

Kira, A., 1976. *The bathroom.* Viking Press, NY.

Kristensen, T., 1999. Quality of Life: A Challenge to Design. In *The Design Journal*, **2**, 2, pp. 10-19.

Krueger, R.A., 1994. *Focus Groups: A Practical Guide for Applied Research.* Sage, London.

Laaksonen, P., 1994. *Consumer Involvement.* Routledge, London.

Laslett, P., 1989. *A Fresh Map of Life: The Emergence of the Third Age.* Weidenfeld London.

Lawton, M.P., 1982. Competence, environmental press, and the adaption of older people. In *Gerontological Monograph No. 7 of the Gerontological Society.* Aging and the Environment: Theoretical Approaches. Lawton, M. P., Windley, P. G. and Byerts, T. O. (Eds), Springer, New York.

Macdonald, A., 1996. *The Challenge of Age.* Macdonald, A. (Ed.), Glasgow School of Art. Foulis Press, Glasgow.

Mayer, K.U. and Baltes, P.B., 1996. *Die Berliner Altersstudie* (The Berlin Study of Ageing). Mayer, K. U. and Baltes, P. B. (Eds), Akademie Verlag, Berlin. (quotations are translated from the orig. German by the authors).

McCracken, G., 1988a. *Culture and Consumption. New Approaches to the Symbolic Character of Consumer Goods and Activities.* Indiana University Press, Bloomington and Indianapolis.

McCracken, G., 1988b. The Long Interview. In *Qualitative Research Methods Series,* **13**. Sage, London.

McDonagh-Philip, D. and Denton, H., 1999. Using Focus Groups to Support the Designer in the Evaluation of Existing Products: A Case Study. In *The Design Journal*, **2**, 2, pp. 20-31.

Milton, K., 1996. *Environmentalism and Cultural Theory: Exploring the Role of Anthropology in Environmental Discourse.* Routledge, London.

Mollenkopf, H., 1993. Technical Aids in Old Age. In *Report of the Arbeitsgruppe Sozialberichterstattung.* Wissenschaftszentrum Berlin für Sozialforschung, Berlin, pp. 93-106.

Mullick, A., 1999. *Bathing For Older People With Disabilities.* The Idea Center at State University of New York at Buffalo,
URL: http://www.ap.buffalo.edu/~idea/publications/free_pubs/pubs_bathing.html>
Last accessed 1 February 2000.

Oswald, L., 1996. The Place and Space of Consumption in a Material World. Review Essay. In *Design Issues*, **12**, 1, pp. 48-62.

Polanyi, M., 1983. *The Tacit Dimension.* Peter Smith, Gloucester, MA.

Polit, D. and Hungler, B., 1997. *Essentials of Nursing Research.* Lippincott, Philadelphia.

Poulson, D., Ashby, M. and Richardson, S., 1996. *Userfit: A Practical Handbook on User-Centered Design for Assistive Technology.* TIDE, ECSC-EC-EAEC, Brussels-Luxembourg.

Sampson, P. and Bhaduri, M, 1987. Getting the Basics Right: Qualitative Data, Interpretation or Misinterpretation? In *Qualitative Research: The 'New', The 'Old' And A Question Mark.* E.S.O.M.A.R. Sampson, P. (Ed.), Marketing Research Monograph Series, **2**, E.S.O.M.A.R., Amsterdam, pp. 71-108.

Sanders, L. and Dandavate, U., 1999. Design for Experiencing: New Tools. In *Proceedings of the First International Conference on Design and Emotion.* School of Industrial Design Engineering, Department of Industrial Design, Delft, pp. 87-92.

Slangen-de Kort, Y., 1999. *A Tale of Two Adaptions.* Dissertation. Eindhoven University of Technology, Eindhoven.

Steinfeld, E. and Danford, G.S., 1999. *Enabling Environments. Measuring the Impact of Environment on Disability and Rehabilitation.* Kluwer, New York.

Stimson, G. and Webb, B., 1975. *Going to See the Doctor.* Routledge, London.

Strauss, A. and Corbin, J., 1990. *Basics of Qualitative Research: Grounded Theory Procedures and Techniques.* Sage, London.

Strickler, Z., 1997. The Question of Validity in Data Collection. In *User-Centred Graphic Design.* Frascara, J. (Ed.), Taylor & Francis, London, pp. 43-59.

Tinker, A., 1995. Preventing Accidents to Elderly People: A Strategy for Managing Risks. In *Promotion of Safety for Older People at Home.* Rogmans, W. and Illing, B. (Eds), Proceedings of a European conference held in Stavanger, 14th-16th May 1995, European Consumer Safety Association, Amsterdam, pp. 41-50.

Verbeek, P. and Kockelkoren, P., 1998. The Things That Matter. In *Design Issues,* **14**, 3, pp. 28-42.

Walker, A., 1997. Zimmer mit Aussicht. In *Design für die Zukunft (design for the future).* Coleman, R. (Ed.), DuMont, Cologne (quotation translated from the orig. German by the authors), pp. 117-120.

Willcocks, D., 1984. The ideal home visual game, a method of consumer research. In *Research, Policy and Planning,* **2**, 1, pp. 13-18.

Zeisel, J., 1981. *Inquiry by Design: Tools for Environment-Behaviour Research.* Cambridge University Press, Cambridge.

CHAPTER EIGHTEEN

Emotional Responses to Virtual Prototypes in the Design Evaluation of a Product Concept

KATJA BATTARBEE and SIMO SÄDE

Smart Products Research Group, Department of Product and Strategy Design, University of Art and Design, Helsinki UIAH, Hämeentie 135 C, FIN-00560 Helsinki, Finland

18.1 INTRODUCTION

The topic of this paper is a spin-off from a study in a much larger 'VIRPI – Research and Development Project on Virtual Prototyping Services for the Electronics and Telecommunications Industries' project. A number of Finnish universities and companies participated in VIRPI, which was funded by the companies and the National Technology Agency TEKES. The goal of the research was to develop virtual prototyping techniques for use in product development.

The aim of one of the subprojects of VIRPI was to study the use of Internet-based virtual prototypes for usability testing. The result was that virtual prototypes can be used in usability evaluation, that task-based tests seem more beneficial than the free exploration approach, and that emphasis and further research must be directed to the design of the data logging (Kuutti, Säde and Battarbee 1999; Battarbee *et al.* 1999). In the study, a usability test with a virtual reality modelling language (VRML) model was simulated. An interactive VRML model of a rather simple conceptual mobile phone was placed in a web browser window, where it could be manipulated, and its buttons could be pressed to navigate in its user interface with the according feedback on its display. The fictive product concept was a very small mobile phone with no numeric keypad, to be used as an extension for a communicator. The VRML model was not a very high fidelity prototype aesthetically, which is typical for virtual reality models, because they must be light enough, in terms of computer file size to be manipulated through the Internet. A VRML model is not as precise and detailed as a Computer Aided Design (CAD) model, but it can represent the interactivity of the product. 'Virtual reality prototyping is a process in which a product or a product concept, its behaviour and usage are simulated as realistically as possible using computer models and virtual reality techniques' (Kerttula 1999).

In the course of comparing the task-based approach to free exploration as the usability testing strategy, users were shown a physical model of the concept at the end of the evaluations. They gave negative emotional evaluations of the virtual prototype and positive ones of the physical model, while still remarking that they looked the same. This was a puzzling and interesting aspect and it led to a new study described in this paper, to test which factors with the virtual prototype might have affected the users' opinions in such a negative way. If virtual prototypes are to be used for evaluating concepts, the possible negative influence of the prototyping platform should be recognised and minimised.

The reasons for negative evaluations of the virtual prototype were assumed to be related to three issues: the fidelity of the prototype, the number of layers between the prototype and the user, and the look and style of the test environment.

Nielsen (1993) notes that it is financially reasonable to reduce the fidelity of a prototype for evaluations at the early stages of design when the concept still is undergoing changes. This can be done by cutting down on the number of features (vertical prototype) or by reducing the level of functionality (horizontal prototype) or both. Virzi, Sokolov and Karis (1996) add two dimensions to the idea of fidelity by stating that it depends on the breadth of features, degree of functionality, similarity of interaction, and aesthetic refinement.

In the VIRPI study the virtual prototype for the usability tests had the full breadth of features and degree of functionality, it had the interaction of the phone implemented into it: the microphone could be pulled out, the roller turned and buttons reacted visibly when pressed. It lacked fidelity concerning the aesthetic refinement, because it had been constructed to make the file small enough to be used over the web.

Another factor lowering the fidelity – the similarity of interaction – might have been the many layers of user interface between the prototype itself and the user. The prototype concept was manipulated through the computer hardware (the mouse and the display), the operating system, the web browser page with instructions, and the virtual environment (CosmoPlayer ™ window).

In addition, from the way the users talked, it was suspected that the technical and complex appearance of the prototype environment described above might have caused feelings of techno-coldness, which might be transferred to the evaluation of the concept itself.

18.2 AIMS

The aim in the spin-off study was therefore to study the design evaluation of the virtual prototype over the Internet and to find out which factors had caused the negativity in the assessment. The test was set up by varying two likely aspects of the virtual prototype:

- the fidelity of the virtual prototype itself;
- the look and feel of the testing environment and web page.

It was hoped to better understand the requirements and limitations of evaluating design with virtual prototypes in order to broaden the possibilities of using virtual prototyping in product design and development.

18.3 METHODS

The comparative qualitative study was conducted with three different combinations of background and prototype, simulating a test situation taking place over the Internet.

Test setup

A new higher fidelity VRML model of the mobile phone concept was built. The technique of building virtual prototypes involves having the surface of the model constructed of a mesh of triangles. The finer the mesh, the finer the surface detail and the larger the file size. The surface materials are then created by projecting or mapping an image onto the mesh, where it will appear smooth except at the edges. The higher-quality virtual prototype had a much finer surface with no jaggy edges, and the same image as on the original prototype mapped onto it. The only functional part was the opening microphone, and there was no interaction implemented, as it was not seen as essential for the design evaluation.

The original background was the web browser window with the black, space-style CosmoPlayer window and its controls on the left and the web page's help and feedback options on the right. The second, and less technically-styled version was a centred, neutral white window, with the prototype rotating in a pale grey space. The CosmoPlayer controls were hidden, and the virtual tools were text buttons instead.

The background and the prototype were tested in the following combinations:

A. the space-style background with the original low-fidelity prototype (Figure 18.1)
B. the space-style background with the higher fidelity prototype (Figure 18.2)
C. the neutral background with the higher fidelity prototype (Figure 18.3)

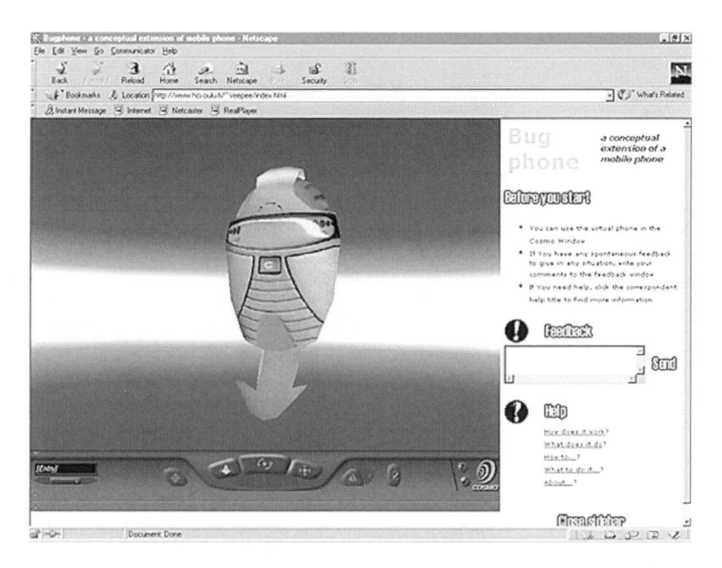

Figure 18.1 Test setting A

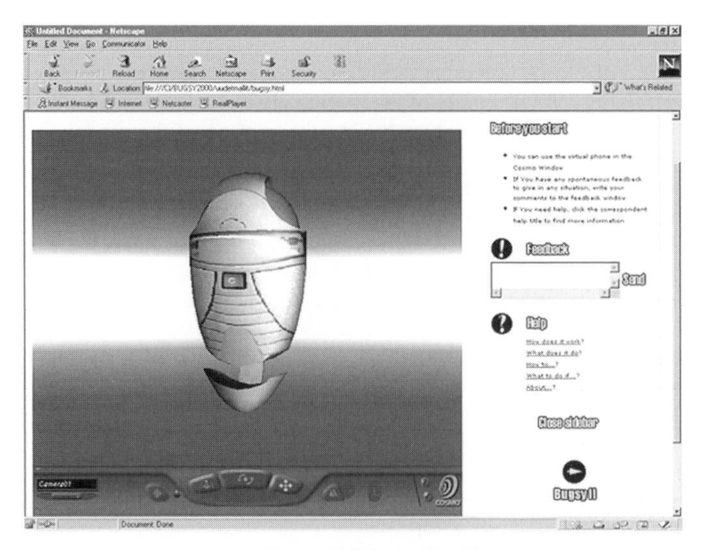

Figure 18.2 Test setting B

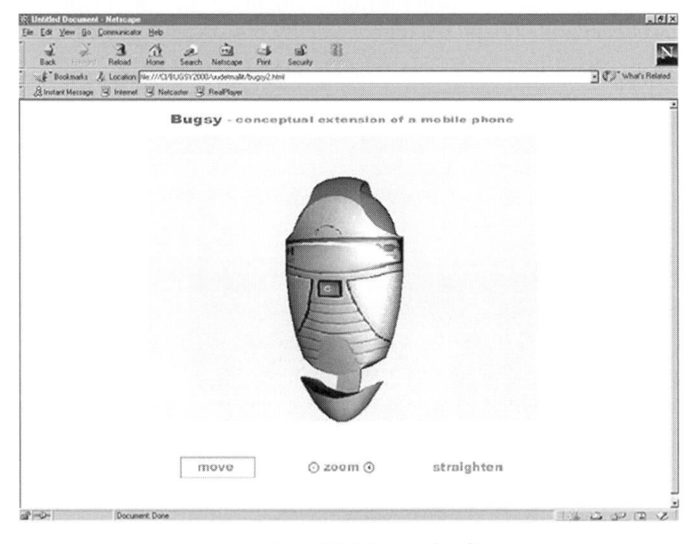

Figure 18.3 Test setting C

Process of the test

The users were 18 students of industrial design at the University of Art and Design Helsinki, representing a wide range of nationalities and equally both genders. Most of them have good skills in 3D CAD modelling but were not very or at all familiar with virtual reality prototypes. They observed the virtual prototype on the computer screen in front of them, operating it with a mouse. They evaluated the prototype in pairs, each evaluation lasting about 30 minutes. Each combination (A, B and C) was evaluated by three pairs of users. First the evaluators were presented and explained the concept, what the product was for, and encouraged to use and explore the virtual environment. They were asked for some first impressions and then to comment and discuss the overall design and details. Finally they were given the physical model, and the two prototypes were compared and discussed. The evaluations were videotaped for analysis.

18.4 RESULTS

Analysis of results

The videotaped interviews were transcribed, and from the text the comments regarding the design evaluation were isolated. It was found that the evaluation of the design consisted of different kinds of evaluations, which seemed to fall into three wide categories. The categories were titled concept (Figure 18.4), character and design. This division is different from design evaluation of finished products, which, according to e.g. Ulrich and Eppinger (1995) consists of the quality of the user interface, emotional appeal, ability for maintenance and repair, appropriate use of resources and product differentiation.

A product concept is an approximate description of the technology, working principles, and form of the product (ibid.). The concept category was for comments about the purpose and use of the product and its functionality.

The character category consisted of comments about the style of the product, about the intended user group and about things that the product reminded the users of. Characters are high-level attributes that help us understand people and things (Janlert and Stolterman 1997). They are a widely understood set of attributes and characteristics that apply to the appearance and behaviour of things in a comprehensive way. If some of the characteristics are known, reliable guesses can be made of the missing ones. This virtual prototype was a somewhat unusually strongly styled product, eliciting a high number of quite strong comments about the character.

The design comments were both ergonomically and aesthetically grounded comments about the overall design and the parts and details of the product.

The comments were classified under the three categories, and by comparing these different 'modes' of evaluation, the process could be understood better and the sources and effects of the negativity placed more accurately.

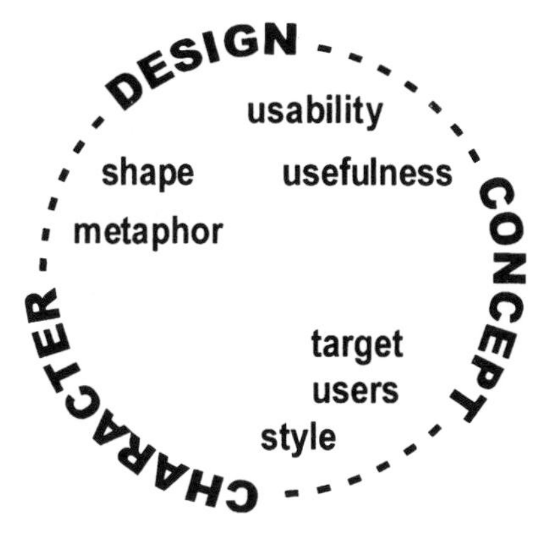

Figure 18.4 The categories of concept, character and design form a seamless continuum. Design is about issues of the product and physicality. It opens up to the concept through practical usability, the user interface, and general usefulness. Concept is about the product related to use, users, and context. It is linked to character via the target users and the style suitable for them particularly. Character is about spirit and personality. It is expressed through metaphors and shapes in the concrete design.

As a reliability inspection, the comments were analysed using content analysis (see e.g. Coolican 1994), a method originally used for analysing messages from the media and other recorded material. In this case a list of the comments was given to four users, two men and two women, who had taken part in the evaluation, and they were asked to judge how positive or negative the comments were on the scale from 1 (negative) to 5 (positive). These were compared to the evaluations of the researchers. The ratings of design details and usefulness were often almost identical, especially when the user had often stated a clear opinion. Many comments reagarding the character such as '*this is like a martian*' were more vague, and as there was no predefined target group or character as a baseline for evaluation, they were evaluated less consistently. Some users had made strong comments, and some had made quite careful comments, which were also more difficult to grade consistently. It was also a value judgement of the user, whether for example playfulness or seriousness in the character was more favourable. In the later analysis of the negativity of comments, the most unreliable comments were disregarded.

18.5 COMPARISON OF VIRTUAL PROTOTYPE FIDELITY, SETTING A VS. SETTINGS B AND C

The differences between the two virtual prototypes were in the solidness of the objects and in the jaggedness of the surfaces. The image mapped onto the surfaces was the same, as were the overall dimensions. For example, in the lower fidelity prototype, the support of microphone was paper thin if seen from the side, and on the round parts the jagged edge of the polygon shape was easily seen, especially when zooming in. The better prototype, however, had all its parts solidly filled and it looked smooth and round even from close up.

The concept evaluation had no clear prototype-dependent differences: in all settings A, B and C users pointed out it did not look like a phone, criticised somewhat the need for the phone to be connected to another computer-like device, and mostly liked the idea of the physically rolling menu selector.

In the design evaluation, as could be expected, setting A did not prompt as many detail level comments, because the details were not as clearly depicted as in B and C. The similar evaluations were that the general shape was nice and pleasing, very holdable and portable, but there were too many design details where they were not needed, the colours were not particularly nice, and especially the pull-out microphone should go, as it seemed quite breakable. The design-related comments were more negatively toned in A, speaking of aggressiveness, alien heads and fish hooks when describing the microphone, and of anchors and a robot's smiling and talking mouth in B and C.

The character evaluation in A described the product as belonging to a space science fiction film, with a clear tendency towards masculine action and aggression (Starship Troopers, Star Wars, Darth Vader) and a small thought that it might be something children could enjoy. In B and C the discussion was if the product is serious enough for businessmen or not, in most cases not. It was compared to a number of hand-held products (e.g. dictaphone, stop watch) and to toys and insects (tamagotchi, scarab) and it was experienced as more like a fashion accessory product for young people or gadget-loving businessmen.

18.6 COMPARISON OF BACKGROUND, SETTINGS B VS. C

It was suspected that the rather complex original background had partly influenced the response to the concept. This was tested by presenting the new high fidelity virtual prototype to group B mounted on the original background and to group C on the new neutral background. It was supposed that there might be a slightly more negative tone in the comments of group B. The result, however, was that no such impact could be identified. The overall tone of the evaluations of concept and design with both alternatives B and C were quite neutral.

The character evaluation was where the differences were expected to arise. The tone of evaluations in both groups was, however, equal in being tentatively positive. Actually, contrary to the expectations, the C evaluators mentioned slightly more phrases related to 'space' or 'futuristic techno' than the B evaluators, but this is most likely due to differences in the evaluators' personalities.

The background in C was slightly easier to use, as the controls were labelled clearly and there was nothing else in the interface of the virtual environment. Ease of use is important for independent user testing over the Internet.

18.7 RESPONSES TO THE PHYSICAL MODEL

In all cases, the comments criticising the design were repeated, with much more confidence, when the physical model was shown. But their attitudes in comparing the model to the virtual prototype were different. In case A, they were almost relieved to see and feel something concrete, and thought the model was much nicer, funny, toyish, and harmless.

In cases B and C, users were comparing really detail level differences between the virtual prototype and the physical model. Again, the mock-up focussed the character assessments more on the idea of toys. When evaluating the virtual prototype alone, only four out of 37 remarks about the character included the idea of it being toy(ish). After the mock-up was shown 11 remarks out of 26 included such expressions. Moreover, some of the evaluators thought the virtual prototype seemed more finalised, more serious and even somewhat nicer than the obviously toyish physical model.

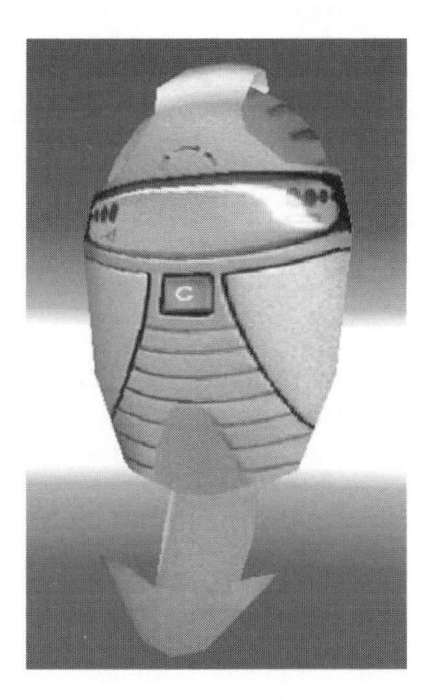

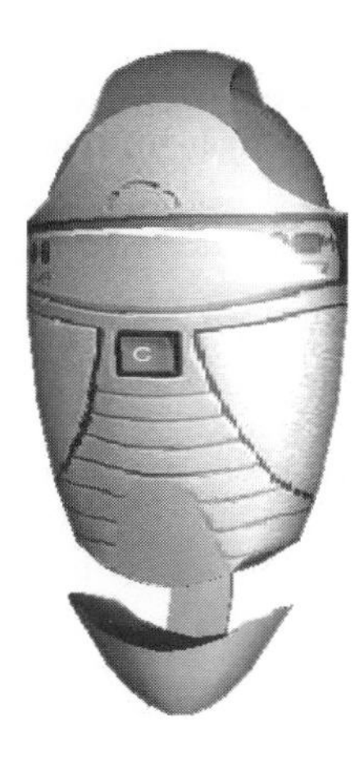

Figure 18.5 Comparison of prototypes A, C and the physical model

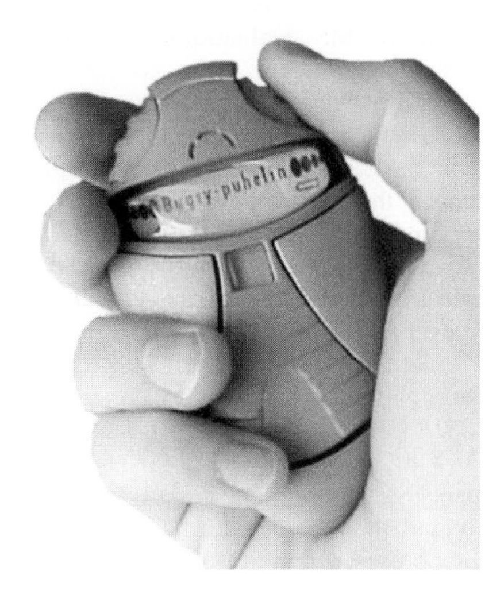

Figure 18.5 Comparison of prototypes A, C and the physical model (continued)

18.8 VIRTUAL PROTOTYPES CONFUSED WITH CAD MODELS

As the users were not very familiar with virtual prototypes, and because the virtual prototypes had not been utilised to their fullest potential, there were some comments that a CAD model would be much better for evaluating the design. This was said in all of the settings. However, users also said that they would be happy with the better quality virtual prototype in B and C, if one could explore the physical functionality of the product more: open and close parts, put the battery in and take it out, show the parts exploded and assembled, as well as explore the functionality of the interface. As one user pointed out: 'if I was the manager, I would be able to say that you have a nice idea here, but it needs more work.'

18.9 SCALE INDICATION

Almost all the users wished to see the virtual prototype compared to real products (such as computer disks, match boxes or coffee cups) or held in the hand to understand the proper size. However, with small clues e.g. in the display, some of them were able to guess roughly what kind of size the prototype was supposed to be.

18.10 CONCLUSIONS

The evaluations of users consisted of three kinds of comments: comments regarding the concept (purpose and functionality), character (style, likeness, metaphor) and design (overall shape and details).

Virtual prototypes work well in presenting the product concept – general idea, functionality, interface concept and also the overall shape. The credibility of the prototypes could further be enhanced with multimodality, more specifically in the Internet environment audio feedback, such as key sounds, sounds of a connecting call, etc. According to experiences from Johnson and van Vianen (1993) the added value of having sounds implemented into demos and prototypes cannot be overestimated.

The structural and aesthetic fidelity of the virtual prototypes is not high enough for the evaluation of the actual design and details of the product with VRML models over the Internet. The higher fidelity prototype of B and C could only be run on a high-powered computer and directly from the hard drive.

The fidelity of the virtual prototype may also affect the character evaluation of the design and the emotional tone of the comments. For example, round forms and warm colours suggest that the object is warm, friendly and protective. Therefore it is not very surprising that a jaggy-edged representation of a round form is not so well received, and some users even pointed out that probably the sharp corners were giving an impression of aggressiveness. This evaluation of the character typically changed or became more complete as users were evaluating the details and sorting the impressions they were getting. The concept they were evaluating seemed to have a collection of characteristics which did not quite fit together, but instead were suggesting both on the one hand something funny and toyish, and on the other hand something more serious and useful, thus confusing the users. As one user said: 'It looks like a bug for sure, but it's not ...... buggy.' It also seemed that when the users in the end decided which kind of character they felt it should be, they were more happy and confident in saying how the design should be changed to fit that character.

As was expected in the beginning of the study, the fidelity of the prototype affects several dimensions of the design evaluation of the virtual prototype. However, the background did not have any noticeable effect.

18.11 ACKNOWLEDGEMENTS

We would like to thank the National Technology Agency TEKES, the Academy of Finland, and the Foundation for Economic Education for funding our research. We would also like to thank our project partners from the VIRPI project, design students Anne-Mari Pöyhtäri for the model and Topias Teirikko at Cyber Design for the new virtual prototype, professor Ilpo Koskinen for good advice, and our colleagues Anu Mäkelä and Tuuli Mattelmäki for support and good suggestions. Thanks also to all of our participants at UIAH.

18.12 REFERENCES

Battarbee, K., Säde, S., Kuutti, K., Mattelmäki, T., Keinonen, T., Teirikko, T. and Pöyhtäri, A.-M., 1999. Internet-Based Usability Testing Using 3D Virtual Prototypes. In *Bulletin of the 4th Asian Design Conference, International Symposium on Design Science.* Nagaoka, Japan 30th–31st Oct, ISBN 4-9980776-0-0.

Coolican, H., 1994. *Research Methods and Statistics in Psychology.* Second edition, Hodder and Stoughton.

Janlert, L.-E. and Stolterman, E., 1997. The Character of Things. In *Design Studies*, **18**, 3, July 1997, Elsevier, pp. 297-314.

Johnson, G.I. and van Vianen, E.P.G., 1993. Practical Experiences with Consumer Products, Users and Prototyping. In *Contemporary Ergonomics.* Lovesey, E. J. (Ed.), Taylor & Francis, London.

Kerttula, M., 1999. Virtual Design and Virtual Reality Prototyping. In *New Product Development Based on Virtual Reality Prototyping.* Kerttula, M., Battarbee, K., Kuutti,

K., Palo, J., Pulli, P., Pärnänen, P., Salmela, M., Säde, S., Tokkonen, T. and Tuikka, T. (Eds), MET Publications, Helsinki, ISBN 951-817-719-8

Kuutti, K., Säde, S., and Battarbee, K., 1999. Virtual Prototypes in Usability Testing. In *New Product Development Based on Virtual Reality Prototyping*. Kerttula, M., Battarbee, K., Kuutti, K., Palo, J., Pulli, P., Pärnänen, P., Salmela, M., Säde, S., Tokkonen, T. and Tuikka, T. (Eds), MET Publications, Helsinki ISBN 951-817-719-8.

Nielsen, J., 1993. *Usability Engineering*. Academic Press, Inc., Boston, ISBN 0-12-518405-0.

Ulrich, K.T. and Eppinger, S.D., 1995. *Product Design and Development*. McGraw-Hill International Editions.

Virzi, R.A., Sokolov, J.L. and Karis, D., 1996. Usability Problem Identification Using Both Low- and High Fidelity Prototypes. In *Proceedings of ACM CHI'96 Conference*, pp. 236-243.

CHAPTER NINETEEN

Understanding Attributes that Contribute to Pleasure in Product Use

ANNE-LISE HAUGE-NILSEN

Master in Product Design Engineering,

Godalsvei 14, 0871 Oslo, Norway

and

MARGARET GALER FLYTE

Professor, Head Ergonomics and Design Group,

Department of Human Sciences, Loughborough University

Loughborough, Leicester LE11 3TU, UK

ABSTRACT

In recent years the concept of usability has been broadened from its original concerns with comfort, convenience and ease of use to include notions of pleasure and delight. Specifically these notions of pleasure and delight have been applied to the experience of actual use rather than to that attributed to the ownership or outcome of the use of a product, e.g. the pleasure of listening to music produced by an operated audio system. There is now interest among ergonomists and human factors practitioners concerned with evaluating the usability of products in including criteria which relate to pleasure and its opposite 'displeasure' in the experience of using a product.

But what is 'pleasure' or 'delight' in product use? How can it be described and hence included in product evaluation?

A number of studies have been carried out in the Ergonomics and Design Group at Loughborough University to develop methods for understanding what is meant by pleasure in the use of products. The research has looked at how people describe the attributes that contribute to the experience of pleasure or displeasure in using a product. It has also looked at whether there are generic concepts which can be applied to different products, or whether the concepts are product specific. Ways of assessing the strength of the association of the concepts generated with pleasure have also been investigated. Products have included such things as pepper grinders, nut crackers and bottle openers. A variety of methods have been used to investigate what is meant by pleasure in product use, including focus groups, questionnaires, user trials and semantic differentials. Research results so far have not conclusively shown generic attributes that relate to pleasure and displeasure in product use. The product attributes or features which have been identified as being most closely linked with pleasure are aesthetics, effectiveness, grip, ease of use and control of the product. The features thought to be most important in judging displeasure are related to uncomfortable grip, unacceptable force, ineffectiveness and safety issues.

Certainly this is a start in our attempts to understand what contributes to pleasure or lack of pleasure in using products. There is a long way to go yet.

19.1 INTRODUCTION

In recent years, the concept of usability has been broadened from its original concerns with comfort, convenience and ease of use to include notions of pleasure and delight. Specifically these notions of pleasure and delight have been applied to the experience of actual use rather than to that attributed to the ownership or outcome of the use of a product; for example, the pleasure of listening to the music produced by operating an audiosystem. There is an increasing interest among ergonomists and human factors practitioners concerned with evaluating the usability of products including criteria which relate to pleasure and its opposite 'displeasure' in the experience of using a product.

There seems to be a trend in society to focus upon aspects of product use other than 'plain' usability. An example is a report from ISTAG (Information Society Technologies Advisory Group) Working Group 3 which is advising the EU Commissions 5th work programme (Ayre 1999), which suggests '(...) that there needs to be a greater recognition that the devices carried by users or the products, content and services that they interact with, have values other than a mainly practical one (...).'

There is clearly a commercial interest in investigating pleasure in the use of products (e.g. electronic games). User friendliness alone is not enough in a competitive market place. Products need to have other qualities in order to sell and be characterised as a good product in our commercialised world. One of these qualities is probably the feeling of pleasure in use.

Human factors research into ease of use has shown that the fact that something is easy to use is due to several attributes of a product and how these attributes are arranged. It is therefore not unlikely that the feeling of pleasure is created by several product attributes. But what are these attributes and how can they be measured and described and hence included in product evaluation and design? Jordan (1996) defines pleasure in product use: 'The emotional and hedonic benefits associated with product use'. And the opposite, 'displeasure' in product use: 'The emotional and hedonic penalties associated with product use'. These definitions were used as the basis for the studies presented here.

In this chapter, three studies carried out at the Ergonomics and Design Group at Loughborough University are presented. The studies discuss how traditional human factors methods can be used for the investigation of pleasure in the use of products. Further, we discuss how the results from the studies can be applied to the design of new products. Finally, there is a discussion on generic principles that describe the feeling of pleasure in product use.

19.2 WHAT SHOULD THE STUDIES INVESTIGATE?

Traditionally, human factors research has sought to identify generic principles for user friendliness, including concepts of learnability, memorability, efficiency, intuitiveness, error tolerance and ease of use. Today one can say that there are some generic overall principles for user-friendliness that can be applied to most products and settings as, for example, the importance of consistency in a user interface. However, studies performed by Grundin (1989), discuss that the concept of consistent user interfaces is both context and product specific. The question is whether pleasure in product use is specific or generic. Thus, it is important to investigate if there are some overall principles for pleasure in the use of products that can be applied in product design and development.

The aim of most usability studies is either to develop methods or to use these methods to gather user specific information for the development of more user-friendly products and systems. The results from these studies must be applicable for those who design products, namely designers. For the methods and results to be applicable for designers they must be put into the context of the design process. The design process consists of several more or less defined phases. The name, number and content of the phases vary between designers, organisations and the products to be developed. In this chapter we will refer to a design process consisting of five phases, including an analytical phase, specification phase, concept development, detailed design and, finally, production. The studies presented in this chapter end at the specification phase.

If methods for the investigation of pleasure are to be of any use to designers, they must be easy to perform and require few resources. This is a problem for some traditional human factors methods for usability investigation. However, there are some generic principles of usability that designers have been able to take advantage of. Thus it has not always been necessary to conduct a complete study for every new developed product. Therefore, we also strive towards some generic principles concerning pleasure in the use of products and the development of 'quick and dirty' methods that can be used by designers directly.

The three studies presented in this chapter apply traditional human factors methods for the investigation of pleasure in the use of products. Three methods that should be well known by most designers were chosen: Focus group, questionnaire and user trial. The aims of the studies were both to evaluate the applicability of the methods to the investigation of pleasure and to identify attributes that contribute to pleasure in the use of products.

19.3 DESCRIPTION OF THE STUDIES

The three studies described below addressed a number of issues for increasing our understanding of the concept of pleasure in product use. The first issue was 'How can the concept of pleasure in product use be described – what are the key attributes?' This was addressed in word generation exercises. The second issue was 'How strongly are those words and concepts associated with pleasure in product use?' This was addressed in questionnaire exercises. The third issue was whether any of the words and concepts generated could be considered generic and hence applicable to other products. This was addressed in studies using two different products. A fourth issue was how the information gathered can be translated into design characteristics and information of value to designers, with the intention of designing products that are pleasurable to use. A more pervading issue addressed in the studies was whether current practice methods in human factors could be used to successfully investigate the topic of pleasure in product use. Although the studies were limited in scope they have produced evidence that leads us towards some positive conclusions.

All three studies where carried out at Loughborough University and the participants were all students at the same university. The products used as stimuli were nutcrackers and pepper grinders.

In our study, all three methods were interdependent. This means that data from the first focus group was used further in the questionnaire and data from the latter was used further in the user trial. Figure 19.1 below shows a simplistic view of the data flow between the three methods.

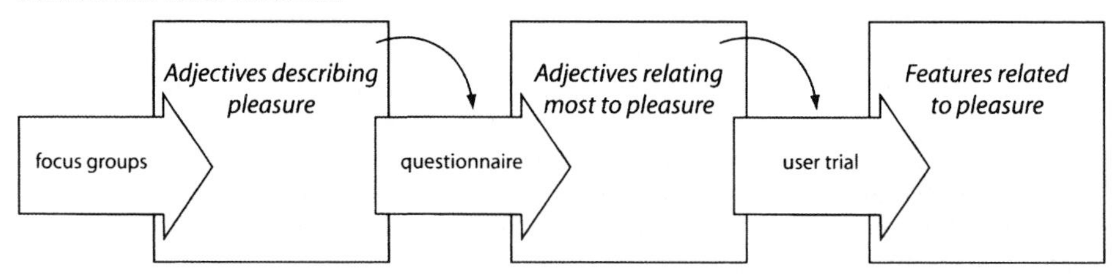

Figure 19.1 Data flow between the methods (studies)

19.4 STUDY ONE: FOCUS GROUPS

The aim of this study was to generate and study words that people use to describe pleasure and displeasure in the use of products. Four focus groups with five to six participants were involved. All four sessions were audiotaped. During the session the participants were first asked to use the three nutcrackers placed on the table in front of them, while individually filling in a form. The form asked them to write down as many words as possible that they related to pleasure and displeasure in the use of each of the nutcrackers. After completing this the participants discussed what they had each written down and tried to agree on three words that described pleasure and displeasure in the use of all three nutcrackers.

The focus groups generated many words (mainly adjectives) and phrases, which the participants used in describing pleasure and displeasure in the use of nutcrackers. These words were sorted into nine groups, depending on which attributes of the product they described. For example, if a participant wrote that the nutcracker had a 'nice colour' this would be regarded as applying to 'good aesthetics'. The nine-word groups are shown in Figure 19.2.

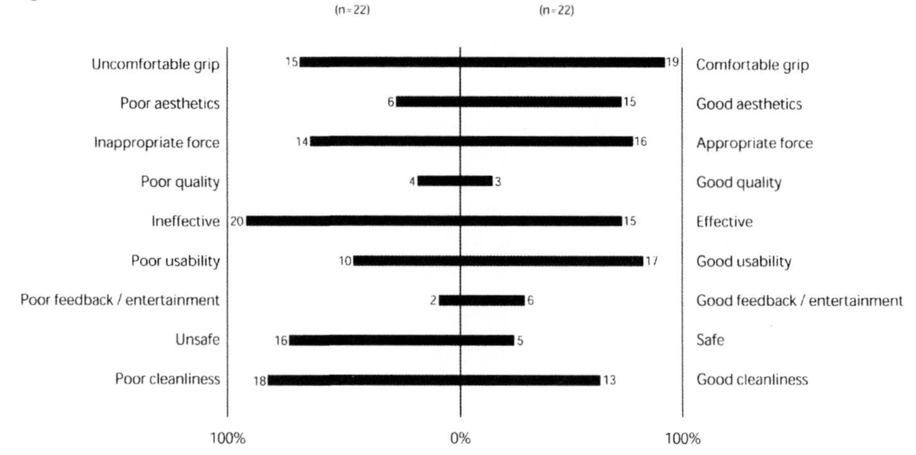

Figure 19.2 Word groups that relate to pleasure and displeasure
The number of participants mentioning one word or more in the respective word groups are shown beside the bars

19.5 STUDY TWO: QUESTIONNAIRE

The aim of the study was to investigate the strength of association of words generated in the focus groups to do with pleasure and displeasure in product use. The words generated in the focus groups were all given a suitable opposite and distributed randomly. This means that all the opposite pairs were split and randomly mixed with all the other words. The participants were asked to think about the words as product properties and rank each word on a scale from 1 (least pleasurable) to 7 (most pleasurable) in respect to how this word would relate to pleasure when using the product. Before the questionnaire was distributed to the participants it was pilot-tested by potential participants, and words that were difficult to understand were removed. At the end there were 29 pairs of words left (all together 58 words) (Figure 19.3).

	1 least pleasure	2	3	4 neutral	5	6	7 most pleasure
comfortable							
difficult							
smooth							
cold							

Figure 19.3 An example of how the questionnaire was constructed

The mean score of each word was calculated and opposites tested for consistency. This was done by counting the number of times each pair scored on the opposite sides of the scale. Figure 19.4 shows the words found to be most and least strongly associated with pleasure in the use of a product. The words shown in Figure 19.4 were also found to be the most consistent opposites.

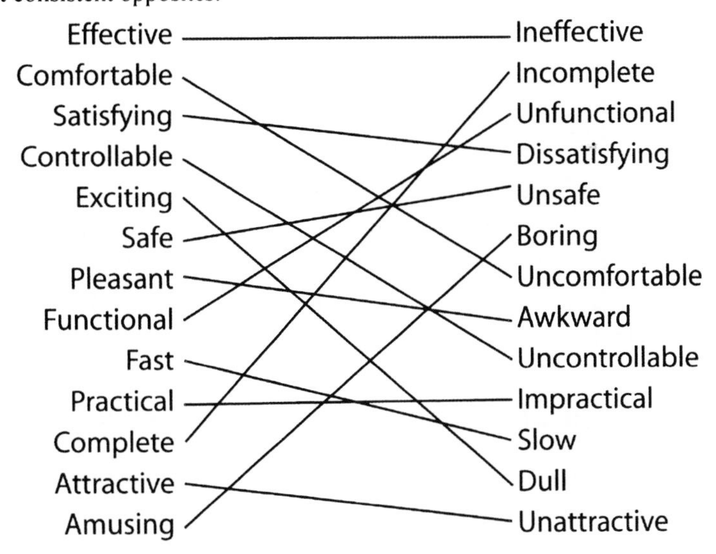

Figure 19.4 The words found to be most and least related to pleasure sorted with the most extreme first (by mean). The lines between the words connect the pre-defined opposites

19.6 STUDY THREE: USER TRIALS

The aim of this study was to investigate which features people relate to pleasure in the use of products and if the words found to relate to pleasure in the focus groups with the use of nutcrackers are also applicable to other products. Pepper grinders were chosen as the new product to be tested. There were ten participants in the pepper grinder trial and nine in the nutcracker trial. The participants took part one at a time and were asked to use the products while talking about the pleasure or otherwise of using them. After using the products for 5-10 minutes they were asked to pick out the product they found to be the most pleasurable to use and to fill in a form based on the 18 words most strongly related to pleasure from the questionnaire study. The participants were asked to write down the product properties of the chosen product that they found related most to the 'pleasure' word given.

The properties generated in the user trials were sorted into nine groups based on the attributes of the product the property described. Frequency counts were done to see how many participants mentioned each word under each of the 18 given words from the questionnaire study, and in total independent of the given words. Figure 19.5 shows the frequency of attribute groups in total.

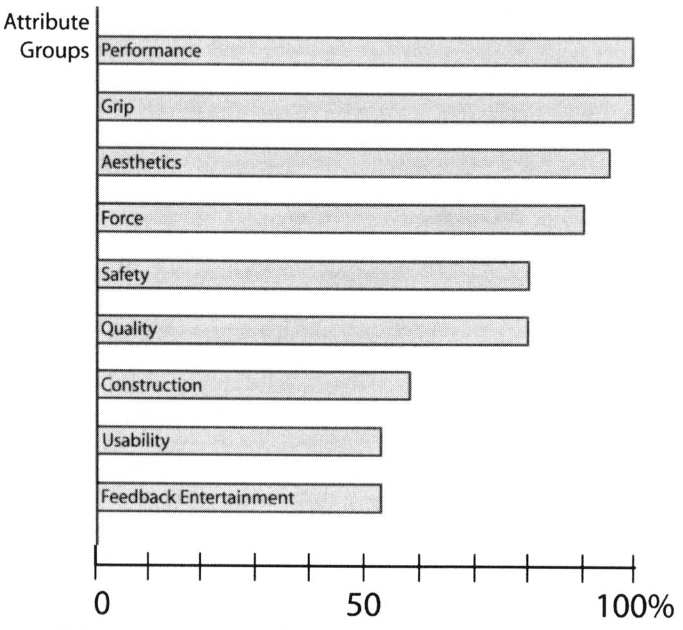

Figure 19.5 Total frequency of attribute groups (in percent) mentioned under the 18 given words

19.7 HOW CAN THE RESULT BE APPLICABLE TO DESIGN?

What do these three studies tell us about pleasure in the use of products and how can the results be applicable to designers? Let us start with the applicability for design. As mentioned before, the studies generated several words that were found to be important for the feeling of pleasure in the use of nutcrackers and pepper grinders. These words can be used to formulate a product specification for the pleasurable nutcracker and pepper grinder, which can be used directly in the design of new, more pleasurable products.

Figure 19.6 shows the product specifications generated for the pleasurable pepper grinder and nutcracker.

The pleasurable pepper grinder	The pleasurable nutcracker
Grip * Nice, smooth, warm and rounded touch. * Should give a firm and secure grip, so user feels that it will not slip out of their hands, and so that it is possible to control where the pepper ends up on the plate. * Should fit into any hand. * Wood was preferred to touch. *Aesthetics* * Should be a nice thing to look at, should blend into the kitchen, should not stick out too much. * Most participants preferred wood or see-through plastic. *Force* * Easy to turn, that is should not require too much force, but should feel resistance. *Quality* * Should have no wobbly parts, should feel solid. * Should give a sturdy impression, smooth and solid. * Should not easily tip over. * Must withstand being dropped or tipped over in the kitchen. *Feedback/entertainment* * Should give a nice grinding sound, and resistance so one can feel the mechanism working. *Safety* * Should be no small loose part that could be lost, or eaten by small children. *No chance of getting fingers into the grinding mechanism. *Usability* * Should be obvious how to fill it (most participants thought it to be from the bottom, and not the top). * Should be easy to get the pepper corns into the container. *Performance* * Should give enough pepper, not too much or too little. * The pieces ground should be small, no big shucks allowed. * Should be fast, should not have to turn many times. *Cleanliness* * Should not spill too much when putting it back on the table after use. *Construction* * All parts should fit perfectly together, so they slide naturally together when assembling. * The opening mechanism should stay fastened when grinding, should not be necessary to re-tighten it. * Should not be too small so it gets lost, but not too big either. * Should hold enough pepper so do not have to fill it very often.	*Grip* * Long (10-15 cm), smooth, rounded handles, which are not too thin. * Should fit any handsize, and should not open so wide that small hands cannot grasp it. * Wood was preferred to touch. *Aesthetics* * Should have some aesthetic value, should be able to be left on the table with the nuts. * Wood is said to blend in with the nuts, natural look. * Should look sturdy, smooth surfaces *Force* * Should be able to control the force, so the nut does not crack into pieces. * Should not require too much force, anybody should be able to use it. *Quality* * Should give a sturdy impression, no wobbly parts. * Metal was preferred when thinking of durability. * Should not break when dropped. * Must withstand the pressure from the nut. *Feedback/entertainment* * Should be able to watch the nut being cracked, and see when it is ready cracked. * Should give a satisfying crunching sound. * Should be fun and not a hassle *Safety* * Should not shoot out nutshells. * It should not hurt to use the nutcracker, due to too much force or uncomfortable handles. * Should not be any hazard, or threat of hazard, to pinch fingers, skin, or palm of hand. *Usability* * Should be obvious where to put the nut, and how to use it *Performance* * Should crack the nut on first attempt. * Give a whole nut. * The nutcracker should hold the nut safely. * Should be adaptable to different nuts. *Cleanliness* * Should not shoot out nutshells, but keep them contained. *Construction* * Should not be too big, so could fit into the bowl of nuts.

Figure 19.6 Product specifications for the pleasurable pepper grinder and nutcracker

These product specifications are rather vague. This means that the specification can result in very different solutions, depending on the designer. However, this does not mean that the specification cannot result in a pleasurable product. It might be that the specification is detailed enough. Thus a recommendation for future studies would be to carry through the whole design process. This study ends at the specification phase. However, it would be very useful to test prototypes based on the specification for pleasure in the use of products to establish if the new products are more pleasurable to use than the original ones.

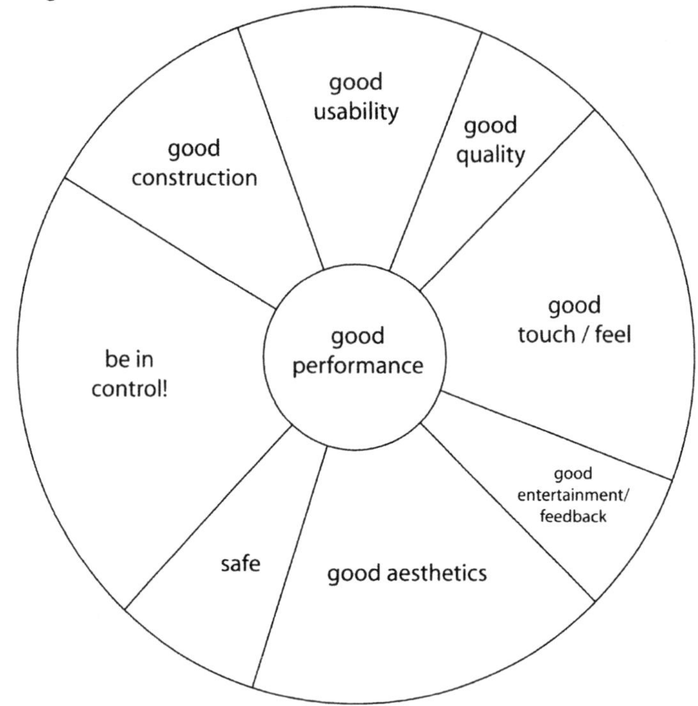

Figure 19.7 The pleasurable cake

The figure shows the attributes that contribute to pleasure in short-term use. The most important attribute for the feeling of pleasure seems to be good performance and it is therefore placed in the middle (the best slice of a cake).

So the results can be used to design a pleasurable nutcracker and pepper grinder, but does it tell us anything about pleasure in use in general? The results from our studies suggest that there are some attributes that contribute more strongly to pleasure in the use of products than others. It also suggests that there are attributes that must be present to create the feeling of pleasure in the use of the product. The 'pleasurable cake' (Figure 19.7) illustrates the results by the size of a cake slice. The bigger the slice, the more important for the feeling of pleasure. Our study only looked at short-term use. Most of the users had never used a product exactly like the ones in the trial. This means that the participants did not have the time to develop a relationship to the products. When a product is used and owned over months and years, other attributes like pride and nostalgia could be present (Jordan and Servaes 1995; Jordan 1996; 1998a).

19.8 IS THE FEELING OF PLEASURE PRODUCT SPECIFIC OR GENERIC?

The next important question to ask is whether the feeling of pleasure in the use of products is generic or product specific. Our study shows clearly that there are several attributes that contribute to pleasure in product use and that some contribute more than others. Thus it is interesting to investigate if the importance of the attributes is the same, independent of the products. Our last study attempted to investigate this by including a second product, pepper grinders. Pepper grinders and nutcrackers are very similar. They are both hand held, are operated by muscle force and are used to grind or crunch something to be eaten. This might suggest that in the case of these two products the feeling of pleasure should depend on the same attributes.

In the user trial, a frequency count was performed of the number of times the different words most associated with pleasure (from the questionnaire study used in the user trial) were related to the different attribute groups. The results from this frequency count show that there are similarities, but also some slight differences between nutcrackers and pepper grinders when it comes to the feeling of pleasure. Attributes that seem to be generic (at least for pepper grinders and nutcrackers) are comfortable grip, controllable force, performance, safety, aesthetics, ease of use and quality. Some of the attributes that show differences between the products are firm grip, little force and construction. These three attributes all seemed to be more important for the nutcrackers than the pepper grinders. This might be explained by the fact that to crack a nut requires more force than grinding pepper. Thus a firm grip, little force and a sturdy construction are important. However, these are speculations. What it does show is that one can hardly conclude that the feeling of pleasure is completely generic. There are probably some attributes that will always contribute to the feeling of pleasure in product use. However, the importance of each attribute is likely to vary between products. We anticipate that all the attributes in Figure 19.7 (the pleasurable cake) will contribute to pleasure to some extent even though the importance will not always be distributed as in this study. A good example would be the 'good feel/touch' which was very important in our study. That the product is nice to touch and feel is possibly only of importance in the case of hand held products. It would not be so important for a TV but rather more important for the remote control.

Although it is possible to conclude that some attributes will contribute to pleasure in the use of product, it is not clear what it is that makes an attribute pleasurable. As mentioned earlier, this has also been a topic of discussion for the concept of usability, as with consistency, which is context and task specific. The feeling of pleasure is probably task, context and product specific. In addition, we have to deal with attributes concerning aesthetics and entertainment which depends on individual taste and feelings. Jordan (Jordan and Servaes 1995; Jordan 1996; 1998a) also concludes that in long-term use attributes like pride and nostalgia contribute to the feeling of pleasure, which again are probably individual. For less individual attributes like usability, safety, performance, control, construction, touch/feel and quality it should be possible to find some general principles as to what makes it pleasurable. Today there exist several guidelines on how to make a user-friendly product, which can be used also for the purpose of pleasure. To make a product nice to touch and feel it should, for example, not have any sharp edges or be too cold or too hot. The performance of a product has to be seen in relation to the task which is to be performed. These examples suggest that it should be possible to create a list of issues that should be considered when a new product is to be developed, in order to make it pleasurable. However, the importance and details of each attribute must be investigated in each case.

19.9 APPLICABILITY OF THE METHODS

Study One: Focus groups

The focus group seemed to work very well in inspiring the participants. All participants had a lot of fun during these sessions. The concept of pleasure in the use of products can be difficult for users to understand and even more difficult to talk about. Focus groups, where several people come together to discuss a topic, can create a good group dynamic and can therefore make it easier to talk about difficult topics such as pleasure. Our experience from these studies was that it was easier to make a person talk about pleasure in a group than alone in the user trial.

The amount of output was also greater in the focus group than in the user trial. The forms that the participants were asked to fill in seemed to help to keep the focus on pleasure, and it also gave the participant some time to think through the topic on their own and to prepare before they had to talk about it in public. The results from our study show that a focus group is useful if the goal is to generate a lot of words or phrases that participants use in describing pleasure in the use of products. Because the focus group allows the participants to discuss a topic it probably generates more words than a user trial with only one participant. However, this requires that the topic of the focus group is made very clear to the participants through the whole study. Well-structured focus groups can give a lot of useful and important information that can be used in the development of a new pleasurable product. However, for the purpose of more empirical studies, other methods should be considered.

Study Two: Questionnaire

Several respondents said that they found the questionnaire difficult to complete. The reason for this was that they were not asked to think about a specific product. Some respondents mentioned that they would have filled in different answers, depending on the product they were thinking about, because some characteristics can be seen as pleasurable in some products and not in others. This might have resulted in more neutral scores than would have been the case with specific products. It would be interesting to test the questionnaire with specific products, to see if the answers would differ between products. Questionnaires are useful for empirical studies where a lot of data is required and it is also not a very expensive method, although this depends on the type and the extent of the questionnaire.

Study Three: User trial

As already mentioned it was more difficult to make the participant talk about pleasure in the user trial than in the focus group. It was also more difficult to keep the focus on pleasure because the participant quickly ran out of things to say about it. For this study to be of any use it is necessary to include some kind of stimuli (as the form in our study) to give the participant hints about what he/she could talk about. A user trial is fairly costly and time-consuming in comparison with the other two studies mentioned in this chapter. Thus it is probably not the most applicable method for the purpose of gathering requirements on pleasure, when developing a new product. However, if the goal is to gather quantitative data about how people see pleasure in the use of products, then user trials should be considered. It would not, for example, be possible to deduce much about individual differences in responses to product attributes from a focus group in which participants are free to influence each other.

19.10 APPLICABILITY TO THE WHOLE POPULATION

It must be noted that all the participants in our studies were students and that the results would not necessarily apply to the rest of the population. For a broader understanding of pleasure in the use of products, studies including groups from the whole population must be included. For example, it is reasonable to assume that elderly people will feel differently about pleasure in product use than students. Pleasure depends on different product attributes which depend on individual taste and meaning as, for example, aesthetics. Some people would like to use a product more or less only because of its appearance and others would prefer a product that worked well even though the user would consider it ugly. In addition, the task setting is important when dealing with what people find pleasurable. For example, a professional chef would be likely to find it more important for the feeling of pleasure that a knife was functional, while this would be less important to a fourteen-year-old boy. Social and cultural settings are probably also important, especially when it comes to long-time use where attributes like pride and nostalgia are present. Figure 19.8 suggests that there are at least three settings that influence what makes us feel pleasure in the use of products.

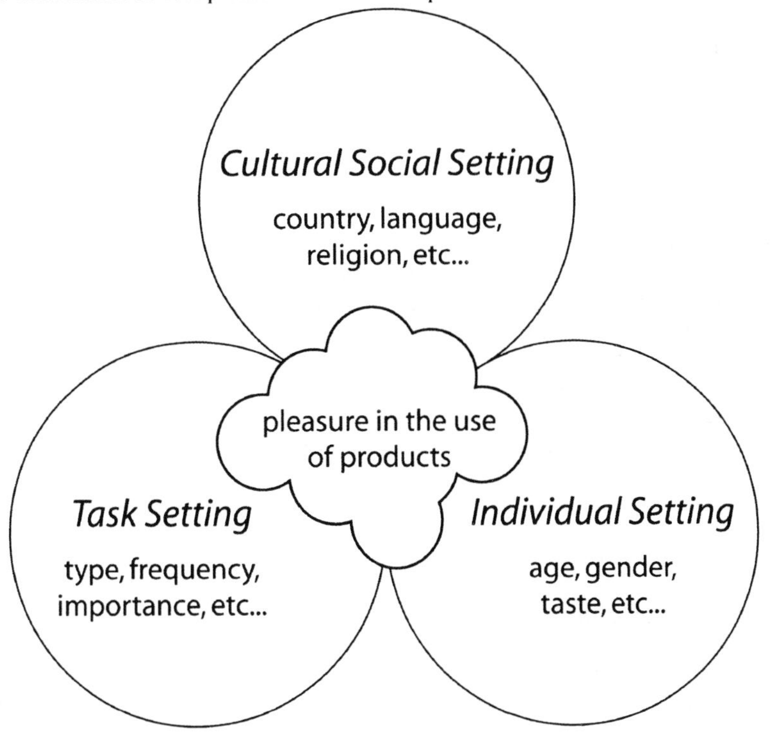

Figure 19.8 The figure shows three settings that probably influence the feeling of pleasure in the use of products

19.11 CONCLUSIONS

From the studies presented in this chapter we can conclude that the most important attributes that contribute to pleasure in short-term product use are good performance, pleasing aesthetics, good feel/touch, control of the product, good quality, safety, good construction, good feedback entertainment and good usability.

It is probably not possible to find a complete set of generic principles for the feeling of pleasure in product use that can be used directly in product development. However, it should be possible to find overall principles describing attributes, like those found in our studies, that must be considered if the goal is to create a pleasurable product. The details concerning the attribute design and importance must be investigated for each product and context.

The traditional human factors methods used had different applicability to the investigation of pleasure in the use of products. Focus groups were probably the most applicable method for use in product development when the right stimuli are given. However, both questionnaire and user trials are applicable when more quantitative data are required.

A recommendation for future studies would be to carry through the whole design process, from analysis to prototyping and testing. Testing one or more prototypes based on the results from the analysis and the specification based on it (see Figure 19.6) would give the ergonomist more confidence.

It is clear that the concept of pleasure in the use of products is complex. It is not due simply to one or two attributes of a product, and it can probably not be distinguished entirely from personal taste and values, cultural and task setting. Human factors experts do have a long way to go to find a valid and robust theory of pleasure in the use of products.

19.12 REFERENCES

Ayre, J., 1999. Working group report on user-friendly IST applications and services. ISTAG Working Groups, Draft of 25th May 1999.

Grundin, J., 1989. The case against user interface consistency. Communications of the ACM. In *The Association for Computing Machinery*, **32**, 10, pp. 1164-1173.

Jordan, P.W. and Servaes, M., 1995. Pleasure in product use: Beyond usability. In *Contemporary Ergonomics*. Robertson, S.A. (Ed.), Taylor & Francis, London, pp. 341-346.

Jordan, P.W., 1996. Displeasure and how to avoid it. In *Contemporary Ergonomics*. Robertson, S.A. (Ed.), Taylor & Francis, London, pp. 56-61.

Jordan, P.W., 1998a. Human factors for pleasure in product use. *Applied Ergonomics,* **1**, pp. 25-33.

19.13 BIBLIOGRAPHY

Baber, C., 1996. Repertory grid theory and its application to product evaluation. In *Usability evaluation in industry*. Jordan, P.W., Thomas, B., Weerdmeester, B.A. and McClelland, I.L. (Eds), Taylor & Francis, London.

Baxter, M. 1995. *Product design*. Chapman & Hall, London.

Chin, J.P., Diehl, V.A. and Norman, K.L., 1988. Development of an instrument measuring user satisfaction of the Human-Computer Interface. In *Proceedings of the CHI'88, Conference on Human Factors in Computing*, ACM, New York, pp. 213-218.

Collier, G.L., 1996. Affective Synthesia: Extracting emotion space from simple perceptual stimuli. In *Motivation and Emotion*, **20**, pp. 1-32.
Cushman, W.H. and Rosenberg, D.J., 1991. *Human factors in product design*. Elsevier, Amsterdam.
Gould, J.D. and Lewis, C., 1985. Designing for usability: Key principles and what designers think. In *Communications of the ACM*, **28**, pp. 300-311.
Hekkert, P. and Snelders, D., 1996. Empirical studies of the arts. *In Journal of the International Association of Empirical Aesthetics*. Martindale, C. (Ed.), **14,** 1, pp. 105-106.
Hirschman, E.C. and Holbrook, M.B., 1982. Hedonic consumption: emerging concepts, methods and propositions. In *Journal of Marketing*, **46**, pp. 92-101.
Hofmeester, G.H., Kemp, J.A.M. and Blankendaal, A.C.M., 1996. Sensuality in product design: A structured approach. In *Proceedings of HCI 96*, Vancouver, pp. 428-435.
Holbrook, M.B. and Hirschman, E.C., 1982. The experiential aspects of consumption: consumer fantasies, feelings and fun. In *Journal of Consumer Research*, **9**, pp. 132-140.
Hutchinson, R.D., 1981. *New horizons for human factors in design*. McGraw-Hill, New York.
Jeffrey, R., 1994. Handbook of usability testing. Wiley & Sons, New York.
Jordan, P.W., 1997a. Products as personalities. In *Contemporary ergonomics*. Robertson, S.A. (Ed.), Taylor & Francis, London, pp. 73-78.
Jordan, P.W., 1997b. Usability evaluation in industry: Gaining the competitive advantage. IEA'97, From experience to innovation 13th triennial congress of the International Ergonomics Association. In *Proceedings*, **2**, pp. 150-152.
Jordan, P.W., 1997c. Putting the pleasure into products. In *IEE Review*, November.
Jordan, P.W., 1998b. *An introduction to usability*. Taylor & Francis, London.
Jordan, P.W., 1998c. Pleasure made in Japan: Kansei Engineering and Design, In *Global ergonomics*. Scott, P.A., Bridger, R.S. and Charteris J. (Eds), pp. 467-470 and *Proceedings of the Ergonomics Conference*, 9-11 September, Elsevier Ltd. Oxford, ISBN 0-08-04333340, Cape Town, South Africa, p. 947.
Jordan, P.W., Thomas, B. and Taylor, B.C., 1998. Enhancing the quality of use: Human factors at Philips. In *Human factors in consumer products*. Stanton, N. (Ed.), Taylor & Francis, London.
Jordan, P.W., Thomas, B., Weerdmeester, B.A. and McClelland, I.L., 1996. *Usability evaluation in industry*. Taylor & Francis, London.
Kasteren, J. van, 1995. The paradox of beauty. Delft Outlook, **4**, pp. 3-7.
Keinonen, T., 1997. Does usability influence product preference? In *Proceedings of the IEA CHECK*.
Kelly, G.A., 1955. *The psychology of personal constructs*. Norton, New York.
Kent, J.C., 1999. What features of a product cause people to feel 'delight' when in use and do they differ if the product is used for working purposes as opposed to recreationally?, BSc Thesis, Department of Human Sciences, Loughborough University.
Logan, R.J., Augaitis, S. and Renk, T., 1994. Design of simplified television remote controls: A case for behavioral and emotional usability. In *Proceedings of the Human Factors and Ergonomics Society 38th annual meeting*.
Loudon, D.L. and Della Bitta, A.J., 1993. *Consumer behaviour: Concepts and applications*, 4th Edition, McGraw-Hill, New York.
Macdonald, A.S., 1998. Developing a qualitative sense. In *Human factors in consumer products*. Stanton, N. (Ed.), Taylor & Francis, London.
McCallum, M., 1985. A methodology for identifying the psychological dimensions of product preference. In *Interface 85 proceedings of the 4th symposium on human factors and industrial design in consumer products*. Human Factors Society, Santa Monica, CA, pp. 53-56.

Morgan, D.L., 1988. *Focus groups as qualitative research.* Sage Publications Inc., London

Nagamachi, M., 1995. *A story of Kansei engineeing.* Kaibundo Publishing, Tokyo.

Oliver, R.L., 1996. Satisfaction: a behavioural perspective on the consumer. McGraw-Hill, New York.

Pugh, S., 1990. *Total design.* Addison-Wesley Publishing Company, England.

Sinclair, M.A., 1990. Subjective assessment. In *Evaluation of human work.* Wilson, J.R. and Corlett, E.N. (Eds), Taylor & Francis, London.

Stanton, N., 1998. *Human factors in consumer products.* Stanton, N. (Ed.), Taylor & Francis, London.

Stewart, D.W. and Shemadasani, P.N., 1990. *Focus groups: Theory and practice.* Sage Publications Inc., London.

Takahashi, S., 1995. Aesthetic properties of pictorial perception. In *Psychological Review*, **4**, pp. 671-683.

Taylor, A., 1999. The relationship between ergonomics and industrial design in new product development. In *Contemporary ergonomics 1999.* Hanson, M.A., Lovesey, E.J. and Robertson, S.A. (Eds).

Ziesel, J., 1984. *Inquiry by design.* Cambridge University Press, Cambridge, UK.

CHAPTER TWENTY

Measuring Experience of Interactive Characters

KRISTINA HÖÖK, PER PERSSON and MARIE SJÖLINDER

SICS, Box 1263, 164 29 Kista, Sweden

ABSTRACT

We created the Agneta and Frida system with the aim of strengthening and encouraging exploration of information spaces. On the user's personal desktop we placed two animated females (mother and daughter), sitting in their living-room chairs, watching the browser (more or less like watching television). In a user study of Agneta and Frida, we attempted to capture their effect on the subjects' experience of the space. We measured facial expressions, reactions to Agneta and Frida in a questionnaire, mood after using them, and actual time spent versus subjects' estimation of time spent. As it turns out, none of the parameters is correlated with the others. We discuss the implications of this.

20.1 INTRODUCTION

In the PERSONA-project* we have tried to develop a richer picture of what it means to navigate in and interact with information spaces. One important distinction was made between *wayfinding* and *exploration* (Benyon and Höök 1997). In wayfinding activities, users have a clear understanding of what they want to find. In exploration, on the other hand, people are not trying to get anywhere, they are not trying to find their way. In exploration, it is difficult to apply the traditional usability criteria such as efficiency, ease of learning and task completion. Instead, it is the overall experience that matters: whether the space encourages curiosity, feels interesting or entertaining, provides for a socially rich environment, or appeals to our emotions and aesthetic values (Munro, Höök and Benyon 1999).

We created the Agneta and Frida system with the intention of strengthening and encouraging exploration. In the ensuing study we wanted to capture some aspects of exploration. Earlier studies of anthropomorphic interfaces did not really address these issues. In the complicated process of developing our own methods and measurements, we found that the influence of the prevailing interface culture is particularly hard to cater for and factor out. Some of the results of the study, we believe, can be accounted for only if we take into consideration the difference between the interface of Agneta and Frida and prevailing web interface culture. We also arrive at a more complex picture of what it means to measure experience and pleasure in a system. We start by describing the system.

* PERSONA: PERsonal and SOcial NAvigation, an EC-funded i3 project. http://www.sics.se/humle/projects/persona/web/index.html

20.2 AGNETA AND FRIDA

The web is mostly conceived as a spatial environment (Höök *et al.* 1999b). We 'move up and down', we 'enter sites', we 'surf', and we 'jump between pages'. By adding Agneta and Frida and their commentaries, ironic remarks and gossip to a collection of websites, we tried to mix this spatial experience with a more 'narrative' one. On the user's personal desktop we placed two animated females (mother and daughter), sitting in their living-room chairs, watching the browser (more or less like watching television) (Figure 20.1). On top of a collection of approximately 40 film production company websites we hard-coded the comments and behaviours of Agneta and Frida. Basically these were short ready-made commentaries and/or animations which were triggered by three sorts of cues.

First, behaviours were connected to the user's activity (e.g. loading sites, moving and clicking the mouse) and document content (e.g. text, imagery, sound files, error messages, browser malfunctions). Loading a page sometimes triggered general everyday speculations from Agneta and Frida on what something on a site could mean, what the purpose of the site was, or if the design was likeable or not (see Figure 20.1). Moving the mouse over an image might trigger a comment on the content of that image. The comments were often humorous and reflexive in nature.

Another set of behaviours/comments were of a more general nature, unrelated to content or user's activity. This included blinking, picking noses, going to the toilet/kitchen, drinking coffee, or general gossiping about uncle Harry and Miss

Figure 20.1 Agneta and Frida reacting to the site of a film production company

Andersson (the owner of a repulsive poodle that often enters Agneta and Frida's back yard, and about which they occasionally fantasise killing).

Thirdly, the user could choose to let small narratives run parallel to the browsing session, interweaving with the two behaviour types above. Clicking on the right mouse button meant the user would be presented with a menu in which he or she could make the choice between different genres and titles. In the experiment, the system provided one comedy ('Poodles: Cute Fluff or Ambassadors of Evil?') and one melodrama. Each of these plots involved about ten scenes, which at certain time intervals were played out on the desktop, mingling with the more 'content-related' behaviours.

We implemented these ideas in the prototype via JavaScript and Microsoft Agent Tool.

20.3 AGNETA AND FRIDA STUDY

With Agneta and Frida we wanted to investigate explorative aspects of navigation. In the study, we wanted to capture four different issues:

- Are they entertaining?
- Do they create affective responses in users?
- Do users perceive them as human-like (life-like, believable) characters?
- Can they tie together the pages in a site to a 'narrative' coherence?

18 subjects used Agneta and Frida (the 'withA&F' group) and 20 subjects explored the websites without the characters (the 'withoutA&F' group). Subjects were given the following instruction: 'Imagine the following situation: you are at home one evening, and you have nothing in particular to do and you are not especially tired. You have a fast and efficient computer at home. A friend has suggested some cool websites on the net that you should have a look at. I'll be in my room over here, and you can come and get me when you are finished'. Subjects explored the system for as long as they wanted, and afterwards were asked a set of questions. Their interactions with the system and facial expressions were video-recorded.

In short summary, the study showed that Agneta and Frida did raise strong affective responses – both positive and negative. They were perceived as human-like, they made users feel more relaxed while browsing, and users described their experience with Agneta and Frida in more narrative terms than did users who were not using Agneta and Frida.

While the entire study of Agneta and Frida is described elsewhere (Höök *et al.* 1999a; Höök *et al.* 1999b), we are going to revisit a set of problems with the study related to experience and interface culture.

20.4 VARIABLES MEASURING EXPERIENCE

We took several measurements related to subjects' experience of Agneta and Frida:

1. How much time subjects spent in the system. We expected longer usage time if they found the system interesting/richer in experience.
2. Subjects' estimation of how much time they spent in the system versus how much time they actually spent. We assumed that the more they overestimated time spent, the more they disliked the system (with or without Agneta and Frida), while the more they underestimated time spent, the more they liked the system. When having fun – 'time flies'!
3. Subjects' evaluation of Agneta and Frida after using them captured in a set of Likert-scale questions (7-grade-scale from 'a lot' to 'not at all'):
    - 'Agneta and Frida's commentaries are fun'
    - 'The animated characters make the browsing situation nicer'
    - 'It was fun to surf together with the animated characters beneath the browser'
    - 'I would like to use Agneta and Frida again' ('Yes, often' – 'No, never')

These questions were mingled with other questions so that subjects would repeatedly have to make judgements of their experience:

1. Subjects' facial expressions were videotaped during the use of Agneta and Frida. Here we counted smiles – proper smiles, giggles and outright laughs. We constructed a variable, smiles per 10 minutes, consisting of the total number of smiles, subtracting the number of negative smiles (laughing *at* Agneta and Frida) and unrelated smiles (arising from the study situation or the webpage content *per se*). The experimenters made this categorisation purely intuitively. The number of smiles was supposed to capture some of the non-conscious affective reactions.
2. Subjects' estimation of their mood after using the system (with or without Agneta and Frida). Measurements 3 and 4 were expected to capture subjects' retrospective and conscious experience and attitude towards Agneta and Frida.

Table 20.1 Smiles/minute and the four 7-grade Likert-scale statements displayed in the same diagram for each subject, 1 - 18. The subjects are sorted by how many smiles they made, from person 1 who only smiles 1.2 times per 10 minutes to subject 18 who smiled 13.3 times per 10 minutes

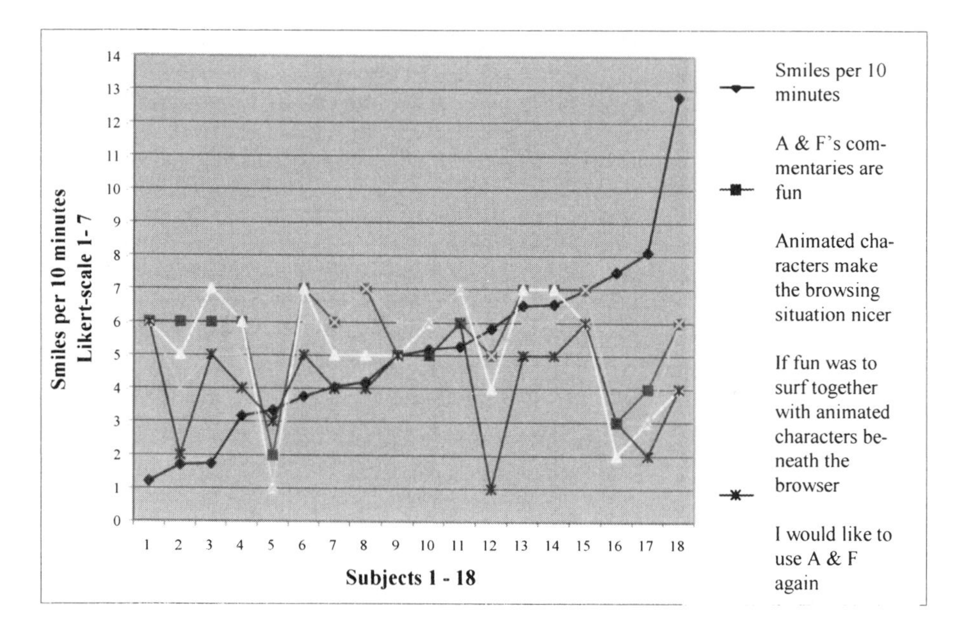

20.5 RESULTS

One might think that it is fairly straightforward to divide the group of subjects exposed to Agneta and Frida into two groups: one that liked them and one that did not, especially given that Agneta and Frida raised strong emotional reactions in the subjects (Höök *et al.* 1999a). But from the five different measurements outlined above we arrive at a more complex picture.

Time spent versus perceived time

Time measurement (1) showed that the subjects in the 'withA&F' group spent on average more time than the 'withoutA&F' group (27.4 minutes/subject versus 20.7 minutes/subject). This might indicate that the experience with Agneta and Frida was richer and more interesting, and therefore made users stay longer.

Table 20.2 Correlation matrix for the questionnaire queries: 'Agneta and Frida's commentaries are fun' (fun), 'The animated characters make the browsing situation nicer' (nice), 'It was fun to surf together with the animated characters beneath the browser' (surf), and 'I would like to use Agneta and Frida again' (again). $P < .05$

Correlation Matrix

	again ok	nice ok	surf ok	fun ok
again ok	1.000	.699	.669	.468
nice ok	.699	1.000	.804	.815
surf ok	.669	.804	1.000	.790
fun ok	.468	.815	.790	1.000

This result was consistent with the second time measurement (2). The subjects in the 'withA&F' group *under*estimated time spent by 2.5 minutes on average, while the 'withoutA&F' group *over*estimated time spent by 3.1 minutes. This result seems to be in line with our expectation that Agneta and Frida creates a richer experience. The difference in estimated time is, however, too small to judge.

Questionnaire replies and number of smiles

While the time measurements were encouraging, the questionnaire replies (3) complicated things. As seen in Table 20.1 and 20.2, there was a consistency among the conscious judgements in the 'withA&F' group, which was expected. However, as Table 20.1 also shows, the number of positive smiles per 10 minutes (measurement 4) did not vary with users' conscious judgements (not significant correlation value). Subjects who claimed to like Agneta and Frida did not necessarily smile a lot (as for e.g. subjects 3, 4, and 6), while those who smiled a lot sometimes claimed not to like them at all (as for e.g. subjects 12, 16, 17).

Table 20.3 Smiles per 10 minutes, one of the Likert-scale questions 'Agneta and Frida's commentaries are fun', and the difference between subjectively estimated time spent and the actual time spent depicted in the same table. The 18 subjects are sorted by the number of smiles per 10 minutes. None of the measurements follow one another

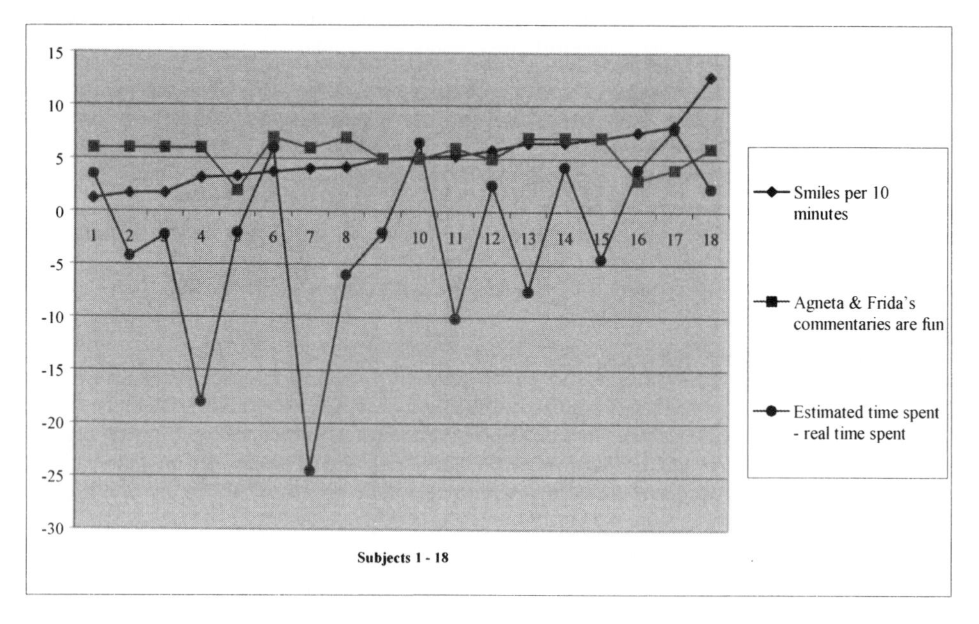

nfortunately, the time measurement (2) did not follow either the questionnaire replies nor the smiles measurement, see Table 20.3 (again a non-significant correlation). This was also the case in respect to the absolute time measurement (1) (a non-signification correlation value). Subject 16, for instance, smiled as often as 7.5 times per 10 minutes and spent 36 minutes (9 minutes above average) with the system, which would indicate that he had a good time. However, his post-usage view on Agneta and Frida's commentaries was only 3 on the 7-grade scale.

Table 20.4 Number of subjects who were in a better, equal or worse mood after using the Agneta & Frida system, versus only using the web pages

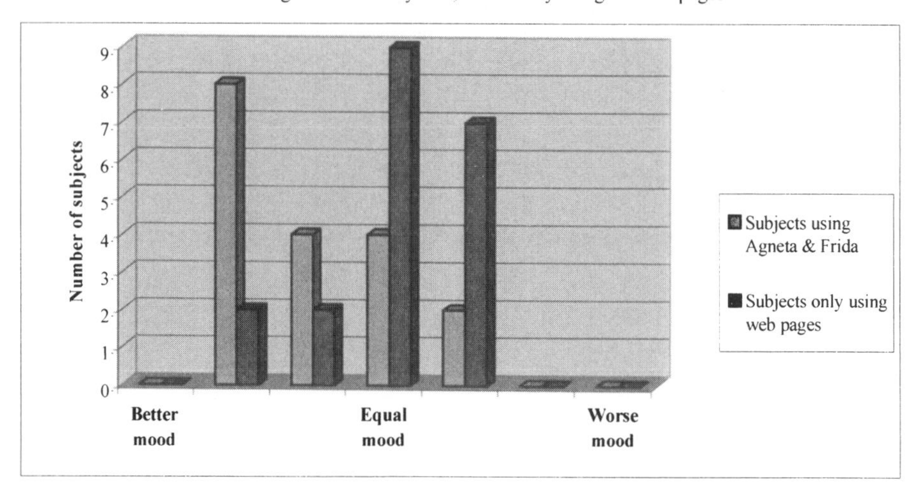

Mood

The mood measurement (5), depicted in Table 20.3, shows that the 'withA&F' group claimed to be in the same or better mood afterwards, while the 'withoutA&F' group was in the same or worse mood. This indicates that Agneta and Frida's humorous commentaries and behaviour did, for most subjects, provide a positive experience.

But again, the change in mood did not follow any of the other measurements, see Table 20.4 (a non-significant correlation).

20.6 DISCUSSION

Measuring different things

None of the measurements were correlated, which indicated the difficult nature of measuring experience. As there are five variables that do not consistently follow one another, these must be said to measure different things. It is not obvious that either one of them is more *truly reflecting the experience* than the other. Depending upon viewpoint, we might think that either the questionnaire replies, the mood after using the system or the facial expressions best reflect the overall experience. For example, we might argue that facial expression is not a good measurement of such a complex process as interpreting, enjoying, and afterwards evaluating a system like Agneta and Frida. Facial expression will 'only' show the instantaneous reactions to the jokes, which might be different from subjects' overall experience in retrospect. Also, as some people get nervous and 'giggly' in front of cameras, the experimental situation might have had an effect on the smiles measurement. In this view, the reflective experience would be a better judgement of the overall experience of Agneta and Frida.

On the other hand, the post-usage replies might not reflect the true picture either as they might be affected by 'after-thoughts' of moral and ethical nature. For instance, one might first laugh and then afterwards reconsider and judge ones own reaction. Subjects will then express their 'official' view on what humour should or should not be, whether humour should be added to web navigation, whether interactive characters belong in the interface at all, and other issues more to do with their attitude than the interactive experience they had of Agneta and Frida while using it. A result that possibly backs this up is that we found a correlation between how much subjects where disturbed by 'Agneta' and 'Frida' and their web- and computer experience.

Users who had a lot of web experience were more disturbed by Agneta and Frida ($r=.54$, $P < .05$), the same for computer experience ($r=.60$, $P < .05$). One possible explanation might be that computer-experienced users have a stronger model of how to explore the web, and given that Agneta and Frida break with this 'tradition', they are more disturbed than users who do not have such strong expectations or 'preconceptions'. Some users even said before, after, and even during the session that they generally disliked interactive characters in the interface. Experienced users sometimes take a definite stand for or against characters in the interface. Users who are accustomed to having complete control over the computer – from the inside of the operating system out – find it particularly hard to accept processes that run outside their control. Users who are less experienced and who have not 'grown up' with, for example, text-based command interfaces, but where their first experience is through the 'www', have another view on what computers are and how much insight they can have into what is going on.

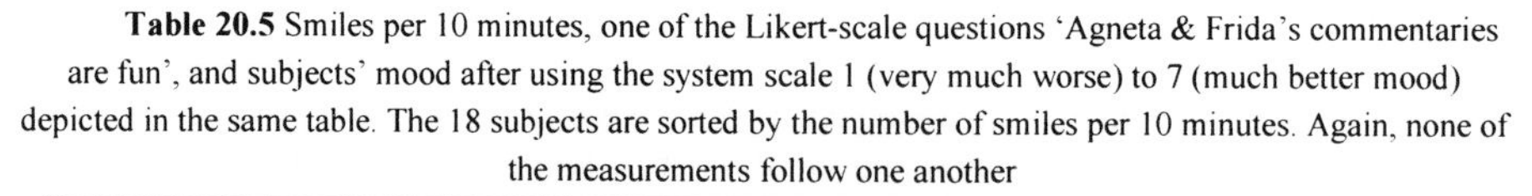

Table 20.5 Smiles per 10 minutes, one of the Likert-scale questions 'Agneta & Frida's commentaries are fun', and subjects' mood after using the system scale 1 (very much worse) to 7 (much better mood) depicted in the same table. The 18 subjects are sorted by the number of smiles per 10 minutes. Again, none of the measurements follow one another

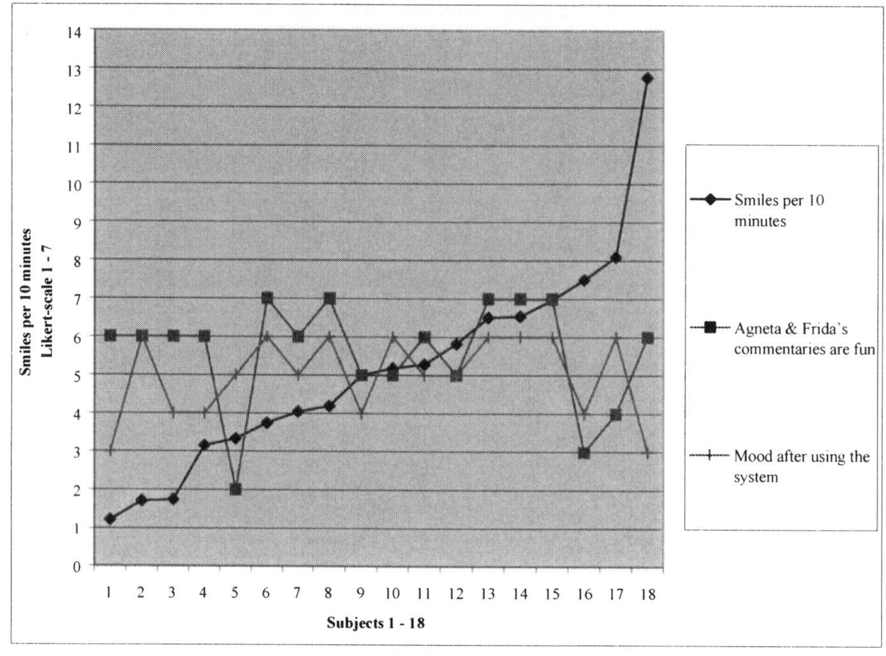

The mood measurement – which lands somewhere between the instantaneous reactions during use and the reflective replies afterwards – will again measure something else.

The relationship between design goal and measurements

The fact that subjects' facial expressions or time spent in the system did not predict their questionnaire evaluations is intriguing. We stated in the introduction that we wanted to measure subjects' experience. We assumed that a richer, more entertaining experience would prompt people to stay longer in the system, make users want to return to the system more often, trigger more smiles during use, and provide a higher appreciation of Agneta and Frida. As this was not true in any consistent way, we will have to decide which of these measurements most accurately captures the subjects' experience in relation to the design goal. If we aim to entertain for a one-time usage situation, then maybe it is more important that subjects smile a lot; if we want subjects to return to the space, then their post-usage evaluation should be emphasised. In both cases, we can identify a group of subjects who *did* have a rich experience of Agneta and Frida.

We may need to differentiate between a number of design goals related to experience. For some applications, the goal might be to make people stay for a long time in the space and smile a lot while they are there. For other applications, it is more critical that users will want to return to the space often – that their curiosity is aroused. The fact that many users were disturbed by Agneta and Frida – but still enjoyed their company – indicated that we failed to create a feeling of flow or relaxed relationship with the space. If that is our design goal, then other means should be used.

Ill-defined measurements

A third way of accounting for the difference between the measurements is to claim that they are in themselves lacking in definition. Counting the number of smiles is problematic since people may have different body language. The presence of a video-camera might make some smile more, others less. When analysing the videotapes, we saw that some users were more inclined to ignore the audio-based comments from Agneta and Frida, and focus instead on the content of the web pages, whilst others were more alert and attuned to the jokes. Smiles can also mean many different things: we smile because we like something, because we find it silly, or as a reflection of someone else's smile or laughter. Although we tried to filter out these variables they may still have influenced the data.

The views reflected in a Likert-scale question such as 'Agneta and Frida's commentaries are fun' might also be ill-formed and 'fluffy'. Subjects are generally known to behave in a socially desirable way, i.e. according to what they believe the experimenter desires. Since experimenters and designers were considered to be one and the same by some subjects, it is likely that such 'politeness' effects were present.

Providing a richer context for interpretation

On another level, our results also point at the difficulty of gathering facial expressions and using those as a means to measure subjects' affective reactions towards computer systems. Within the area of affective computing (Picard 1997), many attempts are made to measure users' physical reactions to interactions with systems. So far it has been difficult to find any good predictors of users' inner mental states from measurements such as blood pressure or transpiration. It is possible to see that the user gets aroused, but not whether it is a positive arousal or not, nor whether it reflects a mental state to which the system should adapt (Riseberg *et al.* 1998). In order to pinpoint finer distinctions in the emotional reactions, we have to consider the user's interpretation, understanding, attitudes and expectations of computer culture. These processes will be determined by personal expectations, but also by social and cultural context.

With Agneta and Frida it is important to remember that jokes and irony must be interpreted and understood with respect both to the prevailing culture and to the individual user's attitudes and taste. Our views on humour are reflections of our personality and who we want to be in the eyes of others. For instance, some of the jokes Agneta and Frida make are ironic towards the web interfaces:

Frida: They say that computers save so much time. But sometimes I wonder... At work I often feel like I'm spending 90% of the time getting the damned thing to work, and about 10% of the time actually accomplishing things with it....
Agneta: I don't really know... I'm not that experienced...
Frida: Maybe we should buy a home computer...? Just for the fun of it...
Agneta: Naa, I'd prefer a television instead... there are more stories on TV....

Some jokes concerned with the male dominance of the IT-world:

Frida: Stupid! Nothing works! Who would ever publish a page like this?
Agneta: A man?

Some users might approve or disapprove of the views and humour of Agneta and Frida. Subjects' instantaneous physical arousal will only provide *some* insight into their overall understanding and reactions towards them.

Lacking from our study is a series of questions to determine subjects' attitudes towards interface culture and the role of humour, irony and characters in the interface. We should also have checked their attitudes towards different genres of humour and irony in general. If we had known more about the subjects we would perhaps have been able to better understand the difference between facial expressions, mood, time spent, and post-usage views.

20.7 OVERALL EXPERIENCE OF AN INFORMATION SPACE

In recent years there has been an explosion in the number of usages and functions of information technology. In respect to entertaining, fun, social, affective and character based interfaces, we need to include not only traditional human factors measurements, but also overall experience, including emotion, understanding and interpretation. 'Goodness' can no longer only be defined in terms of efficiency.

This broadening of the notion of evaluation undoubtedly makes things more complex. Understanding and experience of a system involves extremely intricate mental processes, about which not much is known by either computer scientists nor psychologists. Furthermore, many of these processes are influenced by dispositions, expectations and everyday models of the world, which are determined not only by personal experience but also by the socio-cultural context. From our discussion above, it becomes obvious that there is no way we can capture such a complex understanding of user reactions unless we bring into the study a larger picture of the contextual factors pertinent to the interpretation of the system. In this respect, our study failed to take into account enough of the users' personal expectations and attitudes toward interface culture and culture in general.

What complicates matters in this case is the exposure of users to a completely new kind of interface: one with interactive, ironic, not very 'helpful' characters. When a type of computer interface has been around for a while and has been extensively used – e.g. point-and-click interfaces – users will learn how it works, start to expect certain behaviours, and act accordingly. We have yet to be exposed to many different kinds of interactive character interfaces – perhaps this will happen in the near future (see for example Microsoft Office Assistant (Horvitz *et al.* 1998), www.extempo.com (Hayes-Roth *et al.* 1998), Catz and Dogz at http://www.petz.com (Hayes-Roth *et al.* 1998)).

In conclusion, in spite of the complexity we should try studying overall experience of information spaces. We must, however, take into account a larger context. The effects of interface culture, expectations, personal taste and attitude must be investigated.

20.8 ACKNOWLEDGEMENTS

The work reported here was done as part of the EC- and SITI-funded i3 PERSONA (PERsonal and SOcial Navigation). We wish to thank all the members of the project for fruitful discussions.

20.9 REFERENCES

Benyon, D. and Höök, K., 1997. Navigation in Information Spaces: Supporting the Individual, Presented at *Interact'97*, Sydney, Australia.

Hayes-Roth, G.B., Ball, C.L., Picard, R.W. and Stern, A., 1998. *Panel Session: Affect and Emotion in the User Interface at the Intelligent User Interface (IUI) Conference*, San Francisco.

Horvitz, E., Breese, J., Heckerman, D., Hovel, D. and Rommelse, K., 1998. The Lumiere Project: Bayesian User Modeling for Inferring the Goals and Needs of Software Users. In *Proceedings of the Fourteenth Conference on Uncertainty in Artificial Intelligence*. Madison, WI. Morgan Kaufmann, San Francisco, pp. 256-265.

Höök, K., Sjölinder, M., Ereback, A.-L. and Persson, P., 1999a. Dealing with the Lurking Lutheran View on Interfaces: Evaluation of the Agneta and Frida system. In *The Workshop Behaviour Planning for Life-Like Synthetic Characters*, held during i3 Spring Days, Sitges, Spain.

Höök, K., Persson, P., Holm, J., Tullgren, K., Sjölinder, M. and Karlgren, J., 1999b. Spatial or Narrative: A study of the Agneta and Frida system. Workshop on Affect. In *Interactions: Towards a New Generation of Interfaces*. Held during i3 conference in October 1999, Siena.

Munro, A.J, Höök, K. and Benyon, D., 1999. *'Social Navigation of Information Space'*. Munro, A.J, Höök, K. and Benyon, D. (Eds), Springer Verlag, September 1999 ISBN 1-85233-090-2.

Picard, R., 1997. *Affective Computing*. MIT Press.

Riseberg, J., Klein, J., Fernandez, R. and Picard, R.W., 1998. *Frustrating the User on Purpose: Using Biosignals in a Pilot Study to Detect the User's Emotional State*. CHI 98 late-breaking results.

CHAPTER TWENTY-ONE

Understanding People and Pleasure-Based Human Factors

BIRGIT H. JEVNAKER

Assoc. Prof., Norwegian School of Management,

PO Box 580, N - 1302 Sandvika, Norway

ABSTRACT

If the late industrial society is being transformed by the search for new cultural and emotional experiences, industrial design needs to extend its knowledge and creativity in the direction of more culture-creative and pleasure-based human factors. To improve forthcoming designers' market-cultural understanding and imagination beyond their own subculture, a range of students' field-studies were conducted in an industrial design-school context. This chapter describes the benefits and challenges of combining an observational and a scenario-inspired approach in a time-compressed field-study. It focuses on the art of capturing actual codes and dynamics among stage-setters and users engaged in pleasurable practices. This qualitative design-research experimentation is relevant for design in business. Iterative tools combining elements of future-research with ethnography and socio-cultural studies provide rapid exposure to life-contexts of emerging groups, which may ground and inspire empathic and imaginative designing taking *Homo ludens* into account.

21.1 INTRODUCTION

In this chapter, we will look at how future designers – or other interested students of market cultures – can learn to understand people's pleasure. Pleasure is a fundamental human aspect of products, services, and living experience. But what that pleasurable product or experience is, few can say. Recent design historians point to the influence of *context* (Sparke 1987): what is desirable or pleasurable is changing over time and space. Context-rich information may thus be of help to the designers. Wisdom on how humans seek pleasure has always been around, Aristotle emphasised pleasure as a perfect whole following various practices (see Aristotle, in Stigen 1996). Yet the product success literature seldom explicates this holistic experience beyond the functional, coherence, and fit arguments (Bowen *et al.* 1994) or the more diffuse product superiority or uniqueness character (Cooper 1993). Nonetheless, since the pleasurable seems to be on the increase in the affluent society, firms may need to expand their capabilities in designing for pleasure, even cracking the complex code of cultural change.

Culture and cultural change are not new concepts within the social sciences. They have, however, received increased attention recently when internationally oriented firms want to ride the waves of culture (Trompenaar, Hampden-Turner and Trompenaars 1998). Yet there is still a considerable gap of knowledge as to how actual cultural practices are constituted and shared by groups of people, for example in an urban culturally segmented setting. Interesting similarities and particularities may be found among people engaged in shared practices whether in London, New York, Milan, Helsinki or Oslo. Not surprisingly, trend investigation firms are offering their services, claiming to be in touch with particular trendsetters. Also, a professional interest in social anthropology and ethnographic tools has been emerging during the 1990s, even among marketeers (see North 1995). The sociological or ethnographic field-study is one way of getting insight through going out in the field and collecting data first-hand, typically through year-long detailed studies. More rapid field-studies have been introduced recently, which may enrich and enable the user-informed practice of design professionals. This paper describes and analyses a learning experience from conducting rapid field-studies as the main part of a Market and Cultural Understanding course in an industrial design-school context. The aim was to improve design students' abilities to understand people and market cultures from a broad set of perspectives.

21.2 PERSPECTIVE AND THEORETICAL FOUNDATION

Hearing the voice of the market is not a new concern (Barabba and Zaltman 1991), albeit perhaps more challenging within the new rapidly shifting economy. Recent innovation literature has pointed to an increasing need to 'learn from the market' by adopting more explorative methods such as *field expeditions* that study actual people in their natural settings. Leonard-Barton (1995) argues that this approach is essential in order to create a more empathic design in relation to users. Achieving empathy with 'the others' is also an old concern within cultural studies (Klausen 1970). Yet the links between user cultures and design are still on an embryo level. Gains may be achieved by making more creative connections between design and cultural processes, and one opportunity is to enable designers to study peoples' life-contexts and experiences (see also Johnson and Ireland 1995).

This chapter is based on a conceptualisation of culture and part-culture in the modern social-anthropological tradition focusing on the everyday practices of groups of people (Gullestad 1989). This means that our interest is on the *micro* level of constituting practices rather than seeing culture as an all-encompassing life-form, or alternatively, as a separate (artistic) sector in society at large. The micro level allows for studies of live agents with intentions, actions and meanings.

From a design-in-business point-of-view, we need to gain insight into how cultural codes, expressions and interactions are continuously being created and interpreted among real groups of people. Socio-cultural processes going on even in the periphery of the society may, now or later on, drive cultural change in the mainstream, according to recent sociologically-based future-literature (Frønes and Brusdal 2000). It may thus be of critical interest for designers to study selective part-cultures as these studies may capture cues and signals that drive ongoing and future cultural change in consumer and user behaviour.

Language can be important. The benefit of using the concept *part-culture* rather than subculture is to avoid potential value-laden preconceptions, e.g. associating the term with standing against the mainstream. The focus on part-cultures is in line with the anthropologist Marianne Gullestad's (1989) conceptualisation.

The chapter is methodological in its focus delineating how a *field-study* approach may be used to explore selective part-cultures and how these cultivate certain themes/values, etc. In accordance with phenomenological or grounded theory perspectives of the field-study (see e.g. Giorgi 1975; Glaser and Strauss 1967), this is an inductive method to increase our ability to describe people's behaviour and experiences. *Thick description* (as coined by Geertz, see Czarniawska-Joerges 1992) is adopted as a perspective as well as a tool to gain a more in-depth understanding of people in their natural cultural context. This method aims to give a rich narrative and a multidimensional interpretive description; in short, a feeling of 'how the life is over there' (Czarniawska-Joerges 1992: 101). By contrast, 'thin description' would describe merely the more obvious data or stay on a factual level.

As design students, the art of seeing and visual skills are already being stressed, which may have its benefits as well as its challenges. The potential of thick descriptions may, on the one hand, be explored by using visual tools such as drawings, photography and video recording. Designers are also sensitive to notice artifacts, and may do digital mappings, storyboards, etc. Indeed, this is what some field-researchers from non-design disciplines currently are seeking, e.g. in visual anthropology and post-modern consumer behaviour. On the other hand, design students' descriptions may also be constrained by little research-based training in observational and interviewing skills.

One aim of the Market and Cultural Understanding course is, therefore, to initiate and help design students learn and practise a more solid, inductive exploration approach, which may be useful in their later professional practice. Seeing beyond the conspicuous signs may also be learnt. The initiation by the teacher reflects on how to immerse oneself in another culture in a way that brings you in touch with it while avoiding the traps of being too anchored in ones own culture (*ethnocentricity)* or alternatively, going native (Dahl and Habert 1986). Both anthropologists and intercultural communicators argue that it is essential, at least temporarily, to be *culture-relative* (a professional norm in anthropology, see Klausen 1984) and reducing one's natural inclination to favour one's own cultural values.

In addition to identification and description of cultural practices in micro by use of qualitative or ethnographic field-studies, a light version of a scenario-inspired approach was introduced. The fundamental point of departure in scenarios is to identify strategically interesting *themes* here-and-now that seem to relate to socio-economic and cultural forces that may affect the future. For example, sociological scenario building has focused on the two career families as one of its crucial themes leading to valuable reflections on the 'life cycle squeeze', whereas growth-oriented scenarios often take fluctuations in oil prices as a critical assumption. The choice of themes and framing perspective is thus critical. In the Market and Cultural Understanding course, students may themselves pick and choose, which is experienced as inspiring.

Relevant themes and related issues are selected and refined and provide the main building blocks for a future scenario. It is a portrayal of what *might* happen – rather than what will happen – within a selective set of driving forces and assumptions that create a certain internal consistency. From the multiple scenario-based perspectives and methods existing, we only used the fundamental first step approach of identifying and conducting a reflective dialogue around what students found to be interesting and meaningful cultural themes. Themes were selected from what can already be traced here-and-now making quick outlooks or *windows analysis*[*] focussing on certain trends in society, for instance the increase in entertaining games. It emerged as a theme some years ago when looking to multiple media and is cultivated mainly among boys (of various ages!). And the offerings seem to be heavily influenced by, for example, Sony's Playstation, Nintendo and the recent Sega's Dreamcast. Yet we know very little of how games are reframing children's play culture.

[*] I owe this term to Barbara Czarniawska-Joerges who used it in a presentation referring to quick outlooks of socio-economic trends using looking out of the window as analogy.

In fact, kids' play may change rapidly, such as families quickly adopting new rituals when television was introduced. According to recent management literature, the late industrial society is being transformed into new cultural codes of more individually creative conduct which even affect corporations (Ghoshal and Bartlett 1997). Business people need to understand the new *tribes* driven by frequent communication and exchange, affluent offerings, globalization and a search for new sensations (see Nordström and Ridderstraale 1999).

Since thorough cultural studies may take years to conduct, it suggests a dilemma that human factors researchers are grappling with: the way systematic field studies are to be organised, and the emerging practice, the way rapid design research actually gets done. In this chapter, we suggest evoking more empathic and pleasurable design through the practise of a combined modified approach. A scenario-inspired perspective on people's everyday practices helped to focus in on emerging cultural themes, which were then more rapidly tracked using what may be observable here-and-now with a visually somewhat trained design eye combined with an adapted anthropological-oriented lens. Furthermore, the futurist approach constituted an imaginative foundation for selecting strategically interesting groups in relation to the choice of meaningful themes.

21.3 METHOD

The field-study method draws primarily on real observations and face-to-face interviews in a particular setting. Yet the methods of gaining an increased knowledge of user cultures in relation to design are still on an embryo level. A notable exception is an article in Design Management Journal, "*Exploring the Future in the Present*", by Johnson and Ireland (1995). Working in a similar tradition with design students since 1992, through the course called Market and Cultural Understanding which includes a culture-focused field-study, I suggest that gains may be achieved by making more creative connections between design and cultural processes.

The material in this chapter is collected by second-year industrial design students (undergraduates of Oslo School of Architecture) who participated in a range of task assignments (see below). Since the framework for the cultural expedition in this case was merely two weeks within student and teaching limited budgets, the field-studies were brief and intensive and might suffer from failing to dig very deeply into the part-cultures. Although most data were collected in Oslo or its close surroundings, its relevance may not be limited to an urban Nordic setting since the field-studies and methodology may be of more general interest.

In presenting the material, I have chosen to draw particularly on exemplary cultural themes and communities of interest such as mountain climbers, base jumpers and urban 'beauties' that can illuminate more general learning points and give the reader some insight into a few contexts.

21.4 THE SCENARIO-INSPIRED RAPID FIELD STUDY

The task assignment consists of a set of exercises including a sketch of future-oriented themes related to contemporary culture and rapid fieldwork among part-cultures and related stage-setters (see Figure 21.1). Also, a presentation is to be made of the collected material with analytical and expressive reflections, both orally and in a course paper. The tasks are to be done by small teams (1-3 persons) and group-collaboration is encouraged. The set of exercises is organised to let the students be inspired by looking for the presence of future-oriented themes, and then exploring one or two of these themes through a qualitative field study, within a relevant cultural setting. Furthermore, the groups are encouraged to focus on autonomously selected themes and part-cultures that may be exciting and may trigger their further work on the assignment.

- **Exercise 1: Sketch of the Future**
- **Exercise 2: Identification and Sketch of Cultural Themes**
- **Exercise 3: Empirical Fieldwork 1 – Users**
- **Exercise 4: Empirical Fieldwork 2 – Stage-setters**
- **Exercise 5: Analytical Interpretation**
- **Exercise 6: Preparation of Exciting Presentation**

Figure 21.1 Overview over a set of exercises in the scenario-based field study

21.5 EXERCISE 1: SKETCH OF THE FUTURE

Starting with a reflection on past, present and future societies, the teacher encourages a dialogue around the Hunter-Gatherer, the Farmer's Era, the Age of the Industrial Worker, Mass Production and Mass Consumption, and the Late Industrial Era with Information and Communication Technology. Through this dialogue the design students are actively trying not merely to understand the past, but rather to sort out relevant themes among emerging new trends and 'what comes next' (Jensen 1999). A reflective approach is essential. Imaginative features of what *might* come next are delineated and discussed in class with reference to new terms such as *Knowledge-based Economy*, *Experience Economy, Edutainment* or even *Dream Society*. The discussion is based on an introductory lecture fueled by a broad reading including recent management and futurist literature such as the Americans Pine and Gilmore (1999) and the Danish futurist Rolf Jensen (1999). In particular, the potential scenario of a *blend* of multiple flexible media and Dream Society (Jensen 1999) is discussed, with its pleasures as well as its 'hard fun' (thinking of all the project-nomads for example). Vivid illustrations of this non-stop communicative society can be traced already here-and-now, as when Scandinavian teenagers cannot go to bed without their mobile phones (observation of own daughters). Students are then asked to *expand* on these or other themes, finding their own points of departure through an Internet expedition or using other relevant media.

This exercise typically provides sufficient food for thought and triggers rich meta-reflections. By using lots of comparative historical and cross-cultural experiences (first-hand or literature-based), the teacher tries to keep up a critical discussion and avoid being trapped in superficial or romantic visions of socio-economic 'trends'. Fads and fashions shift all the time. Rather, a critical-constructive attitude is encouraged which may enable a reflective exploration of how people actually follow interests or activities that may lead to pleasure.

21.6 EXERCISE 2: IDENTIFICATION AND SKETCH OF CULTURAL THEMES

Based on the initial more loosely scenario-based thinking, we move on to focusing on self-selected *cultural themes* that relate to the overall socio-economic trends grounding. In particular, the students were encouraged to trace what are already around us as potentially significant clues of a near future. It was a self-selection process on the group level that was found to be very stimulating. The students wanted to carry on their search and follow up the approach in new projects after the course was over, which is a strong indication of interest.

The cultural themes selected by students varied from focus on particular *part cultures* (e.g. Student groups, Rebels, etc.) to more common *shared interests* (Beauty) or *activities* (Skateboarding, Extreme Sports). For example, during the 2000 Spring course one group identified Base Jumping as a theme of particular interest for understanding values and norms of extraordinary sensational experiences:

'Öyvar (one of the students) eagerly turns up in the corridor after the lecture: Come and join in exploring the Base culture! Heidi is nodding consentiently: I was thinking of the same, it would be cool, albeit I was doubtful since you know...' (the three students were all engaged themselves in this sport).

'Öyvar and Morten understand what she means, – that's cool, then we will *know* what we are talking about! Yeah, and then we can take the climbing culture as well! Oh yeaaaahhh, it must have been Öyvar who said it!

– We may perhaps compare different extreme sports? Oh, yeah! It fits with the idea that people are searching for excitement, adventure and experiences, yet how to frame this? (...)

The three heads were thinking hard... in a café. They start to collect what they know about climbers and jumpers.'

21.7 EXERCISE 3: EMPIRICAL FIELDWORK 1/USERS

After initial review and reflection on what is known à priori about, for example, mountain climbing, it is necessary to put oneself in a more open inquiry mode in order to get close to the concrete practice. Phenomenologists recommend distancing oneself from any preconceptions through literally putting this knowledge temporarily out of one's mind (Giorgi 1975). The hard task now is to sense, collect and describe what is going on in the focused area among various groups of people and between people and artifacts in the setting selected.

Edgar Schein (see Hatch 1997: 211) has provided a three-level framework that may help to investigate culture and what are, in fact, its multiple levels. Layers at the top encompass concrete *Artifacts* that are visible but often undecipherable (objects, stories, visible symbols and signs, etc.). On the mid level, there are *Values* and norms which represent a greater level of awareness compared to the next one which is *Assumptions* or "*weltanschaung*", the most-taken-for-granted invisible and deep cultural level and thus the most difficult to describe. It may be useful to think of this as an iceberg where only the top is sticking up from the sea. However, one may also focus on processes rather than layers or elements of culture, which may highlight how images grounded in assumptions and values, for example, are made real or given tangible form. Mary Jo Hatch calls this process *realisation* (Hatch 1997: 363-364).

Since culture is a dynamic, fluid phenomenon, which springs from multiple sources, one needs to look closely at actions and interactions. It is important that the fieldworker tries to immerse herself or himself in what is actually going on in the field rather than merely collecting what people *say* is going on (cf. Argyris' and Schön's *theory-in-use* vs. espoused theories; see Argyris 1982: 85). In fact, what is not said may be as important as what is said, since cultural codes are often mostly implicit.

Though fieldwork is dependent on engaging in the selected real part of the world, doing one's homework may be of great help: Reading extensively before going into the field brings a repertoire of possible observation points. During this course exercise, most students tended to use Internet to search for existing literature of relevance to their fieldwork. Preparing a set of relevant issues of inquiry may also help to keep a soft focus with a sharp lens. My experience is that students are grappling with how to ask the right questions; What should we ask about? However, through their scenario-inspired preparation, students generated lists of triggering issues. In fact, what was most triggering was the observation that some of the best students were able to relate to multiple layers of the cultural practice in focus. Small details of behaviour, such as a dress code among the urban 'beauties', were captured without losing sight of the deeper cultural norms, such as expressing a blend of group and self-expressive identity that was sufficiently cool for the club insiders.

It is generally recommended that *how*-questions are asked, rather than *why*-questions, since causal interpretations are more demanding for the informant and may also bias the data given; e.g. through influencing the informant to giving a more 'proper' or improved-upon representation rather than telling what actually happened. Hence, it may be wise to start with a concrete incident and try to track actions by just letting the informants tell their stories and listening carefully to their storytelling. When ambiguous information emerges, a follow-up question or *probing* may tie naturally into the overall story sequence. Probing is a technique to further explore and check what respondents are telling you during in-depth interviews; Can you tell me something more about this? Did this mean that person X was the first to use chalk when climbing in this area or had it happened before? (Using chalk was originally seen as bad among groups of climbers, since it left marks on nature.)

Perhaps since some of the students were active practitioners themselves in the focused activity, vivid illustrations and engaged outlines of field visits were given. For example, the group investigating climbers and jumpers was able to capture rich descriptions of a variety of actual practices. Soon the group came up with the idea that climbing may have evolved through similar cultural dynamics that are experienced among base jumpers of today. Organising values among groups immersed in what was conceived as unregulated wild activity seemed to be paradoxical and thus worthwhile exploring.

Written rules were found among the mountain climbers of today. By contrast, the group found that climbing by the turn of last century was not following such strict safety rules; equipment was sparse, extra food might be missing and yet quite demanding mountain climbing tours were undertaken. It was perhaps even more dangerous than today's base jumping, the students argue, since high risks were taken among the pioneers with perhaps few safety concerns, for example, dr. Arentz and Theresa Berthau climbing peaks in the West Norwegian Alps with only a rope as support.

Other fieldworkers did have some initial problems with establishing rapport with their potential informants. The reasons varied, one problem was to deviate too sharply from the group in focus. For example, the students investigating the modern elderly could not mingle into the part culture of interest without being conspicuous. However, these students were able to overcome this initial obstacle.

21.8 EXERCISE 4: EMPIRICAL FIELDWORK 2 / STAGE-SETTERS

The idea is to find people and organisers that influence or affect the particular culture, target group or focused theme through their designing or otherwise organising behaviour. Using the theatre metaphor we call them *stage-setters* since these people may affect the users' appearances, dos and don'ts (norms) and how values are communicated.

A group studied the lifestyle culture associated with customers of Diesel, the clothing manufacturers. The group studying the Diesel lifestyle culture among teenagers did a range of interviews at the regional Diesel marketing department, among advertising consultants, design consultants, PR and brand consultants, etc. What emerged was the emphasis on a holistic and emotional brand-building which the informants thought could be achieved by stressing storytelling, humour and irony in ads and, overall, a designed and coordinated construction of a unique extended product (*meta-product*) that could appeal to teenagers. Not surprisingly, these were not just one-off efforts. It was necessary to work continuously to keep up the brand, keeping it alive and in the *forefront.* In fact, a few consultants questioned whether Diesel were still at the forefront, since their ironic ad style had perhaps been better – more daring – at the beginning (Figure 21.2).

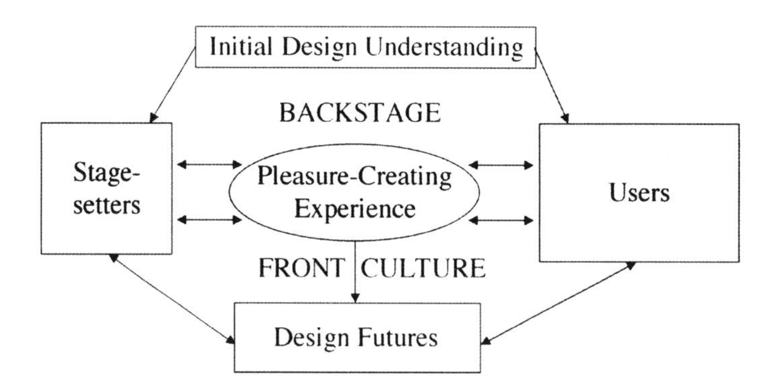

Figure 21.2 Exploring pleasure-creating experience as organised by stage-setters and users

21.9 EXERCISE 5: ANALYTICAL INTERPRETATION

Students were encouraged to analyse their field-data, looking for similarities as well as differences, in artifacts, salient practices, values, norms or other cultural issues among the groups investigated. In order to capture new practices, it may help to constantly ask: What is happening here?

Interestingly, the students were able to come up with varied interpretations. Let's consider the Extreme Sport group again: The media portrayal of base jumpers is 'Jumpers Playing with Death' (headline of an attached article). By contrast, the students found that base jumpers do not engage in this sport because it is dangerous. They engage in it to live and experience the next jump, rather than to think of it as close to death. However, there is a nuance to this. The group found that the jumpers actually belonged to two different part-culture. One mainly focused on the joy and the fulfilling experience of jumping; these members often preferred to jump outdoors and in real nature. Another group, the one that tended to be most often profiled in the media, sought out risky jumping opportunities in cities and urbanised settings; these people would break laws by, for example, jumping from the Eiffel tour or from a skyscraper.

Part-cultures that were contrasted could also present common features. There are many common characteristics between climbers and jumpers, argued the design students, one of which is the personal dedication of resources to the sport. One of the students, herself an active climber and jumper, had eaten cheap bread without any spread for weeks to finance her own parachute. It is suggested that young base jumpers and climbers plan their education, job, etc. after the sport. The time spent on the sport is substantial for both groups.

The field study also explores the differences, such as climbers complying with a set of established norms as regards safety and regulation of the sport, while jumpers are improvising individual actions some of which may be quite wild. As mentioned, media statements may enhance this portrayal, though most jumpers also tend to apply safety norms by, for example, packing their own parachute extremely carefully (20 minutes packaging). This illustrates that there are essential practices unfolding on a *backstage* that is hidden from the surface spectator (cf. Goffman 1974).

21.10 EXERCISE 6: PREPARATION OF EXCITING PRESENTATION

Staging a particular class presentation at the very end of the course invoked exciting live presentations. For example, one group investigating the recent brandbuilding of Diesel performed their presentation wearing Diesel clothes, borrowed from a local outlet in Oslo. Another group investigating the current and future Elderly, staged a play based on one day from the diary of a modern old woman (an imaginary character based on their research).

Students have also pulled together documented papers very quickly, albeit of variable quality. The best papers are the ones that *combine* a richly documented set of interviews and observations with an inquiry mode and grounded analytical reflections. Merely reporting descriptive data easily falls into the trap of overload of information without bringing new insights. The opposite pole is not recommended either — presenting analytical categories without live data is like bones without flesh. Interestingly, the combined approach often used both visual and verbal presentations and could be quite dramatic.

21.11 RESULTS

Searching the students' papers and my notes for valuable experiences, three recurring themes emerged; first, the students' tendency to focus on particular salient actors. Second, noticing the pleasure-based creation, and, third, the pleasure of searching for the pleasurable. This is elaborated below.

Focusing-in on particular salient actors

In the Beauty-focused group, the stage-setters dominated the fieldwork while jumpers and climbers were in focus in the Extreme-Sports group. This might perhaps be interpreted as a human tendency to focus on the group that is the most dominant in creating the cultural codes. For example, in the toy projects students often focused in on the multimedia-fueled toy industry, such as Disney is presenting with its package of film productions, toys and clothes. It seemed to be more difficult to capture the particular shop design (instructor needed to raise attention to this issue) or how kids actually play with toys. Besides, students were often self-oriented when inspecting toys. No doubt it evoked memories of their own childhood and the pleasures of toys.

The overall picture was, however, that even though particular groups and themes were investigated, the students were able to come up with a nuance-rich description. For example, the group studying the elderly managed to get over their initial communication difficulties and took advantage of the curiosity among the old people regarding their visits etc.

Capturing pleasure-based creation

Perhaps affected by the introductory talk of the Dream Society, the most significant common practice by students was to focus in on pleasure-based human factors. These were often not of a hedonistic or comfort type. Rather, what was pleasurable was often closely linked to a fulfilling experience, whether it was going to a café, meeting other elderly people, or going jumping.

A special case of seeking pleasure is perhaps devotion to the adrenalin kick among the Extreme Sports enthusiasts (most are in fact young men). However, what was sought among climbers, and even among many jumpers, seemed to be not merely the jump but the whole compelling experience.

Moreover, it is worth noting that the pleasurable experience was *created* rather than *ready-made*. For example, a group of elderly (pensioners) was immersed in planning a holiday. Young base jumpers looked forward to the enjoyment of the next jump. It is worth noting that these jumpers also paid great attention to their safety through the rigorous packing of their own parachute. One jumper said he did not jump to die, he jumped to live, which impressed the other extreme-sports enthusiasts.

The urban beauties were likewise carefully preparing the next party. Their kick seemed a blend of being seen, accepted and liked by others and doing it for yourself. Being on the 'guest list', meaning that you are being let in without having to pay, was one of the decisive norms valued among the urban beauties going to various clubs.

Pleasure of the futuristic field expedition

Not surprisingly, design students found it especially stimulating to take part in a blend of future-oriented and contemporary field study. The intensity and multiple approaches of the field study also helped to raise awareness and create pleasure in the process. The deadline for the common presentation of all groups seemed to mobilise some extra creative energy. The Androgyny society group for example, performed a play in which they used a time machine in order to get in touch with *Gyda Andro* of year 2217, hence vividly exploring drama methods, too.

21.12 DISCUSSION

Despite a short time period and little research training a priori, a number of exploration practices were implemented by the student groups. Methods adopted ranged from data-search through the Internet to observation and interviewing people face-to-face. Not surprisingly, this led to new and richer ways of understanding market-cultures. Various part-cultures and even the elderly were seen as something worth exploring, beyond the statistical and therefore more dry market data. Students got a rich opportunity to see and appreciate people in context. Behind the various, perhaps previously stereotyped, categories such as extreme base jumpers were fascinating experiences to understand better. Four points may be summarised for further discussion.

Empathic and critical-constructive inquiry

The task assignments involved adoption of a rapid fieldwork method and it did bring forward new material and live memory of people's behaviour and life-context. Moreover, for most groups it led to increased 'critical empathy' with users, that is the students were able to get in touch with users in new ways within their natural cultural context. Getting closer seemed to increase their understanding. However, design students also appreciated getting out of the classroom, which suggests that a test effect might also be present. Anyway, the field study practice led to a new and memorable experience that evoked multiple reflections, including both critical and constructive interpretations.

Expanding domains for designed innovation

The focus on part-cultures can uncover new behavioural and value patterns such as groups of elderly engaged in learning Internet activities. This may not merely reduce stereotyping. Rather, it can expand the innovation domain for designing. Albeit brief, the rapid field studies may bring about live impressions and material that produce novel insight for the design students. This is valuable since it may help to close the generation gap and, in particular, allow for new, not merely usable but also pleasurable, designs for the many particular needs among old people. (We are born as copies and die as originals!)

Capturing cultural flows

Since culture is fragmented and differentiated according to particular life contexts and lifestyles, it is beneficial to capture more of the differentiating dynamics. Due to rapid urbanisation and globalisation and other change factors, traditional cultural values are perhaps not solid anchors anymore. People tend to mix and match according to their various contemporary projects and moods, etc. (see e.g. Nordström and Ridderstraale 1999). The urban beauties can, no doubt, represent this continuously changeful mix-and-match tendency.

Although culture, by definition, is transmitting values from one generation or member to the next, typically somewhat changed (Klausen 1970), it is no doubt better to be seen as something dynamic rather than as a static set of values, norms and practices. Yet there may be common processes to be found among quite varied part-cultures. Based on the students' material, we may identify a search for *self-realising experiences on a personal and interpersonal level.* This process was something that was created for particular events among the 'urban beauties' as well as among the climbers and jumpers, which accord to Frönes and Brusdal's (2000) reflective essay on contemporary cultures. It is worth noting that the design students interpreted even the future elderly within a framework of self-realisation.

New tools for developing design antennas

Although industrial design is use-focused, the ability to capture actual user experiences in their natural life context with a rich description including the users' own perspectives may be challenging even for design students. My experience through eight to nine years of running this market and culture course is the enhanced understanding and vivid impressions gained through field studies. Not all students have been as eager as the last classes to conduct systematic interviews or observations. However, those that do seem to really enjoy it and one strong indication is that they want to continue to use the approach in new projects.

Studying everyday life, and I suggest *not-everyday's* more extreme experiences as well, can generate new insight and knowledge as seen from the bottom-up (Gullestad 1999). Perhaps the insight into the extreme pleasures may be thought of as insight from the front-end or (sometimes literally) the *edge*? This can help designers to imagine and construct more fulfilling experiences for, and sometimes with, the user in order to create superior value.

The definition of what is unique or superior is from the customer's perspective, Cooper argues (1993: 77) 'it must be based on an in-depth understanding of customer needs, wants, problems, likes, and dislikes.' Up to now, this seems too often a cliché in practice. Rather than making the customer an icon speaking as if the 'customer is king', it could be turned into an opportunity for capturing more of how real people interact and create meanings. Yet the soft stuff is the hard one: culture-explorative techniques are hard to implement as a rich interpretive approach handled with wisdom.

21.13 CONCLUSION

Exploring the multiplicity of live experiences is hard work for any researcher, including design students. But it can be both enjoyable and useful. This is promising since sense and respond experiences within the digitised society are more and more constructed in rapid interaction, which raises a major challenge for design to keep up with, and grasp, the many ways contemporary part-cultures and diverse communities of interest cultivate and express their values. Within a society affluent with entertainment, perhaps it needs more than a DreamWorks 'gladiator' to create fulfilling experiences continuously. Indeed, we are perhaps no longer fruitfully seen as 'a mob' to be simply entertained, even though there seems to be a plethora of simple entertainment. What is pleasurable may indeed change back to basics, albeit in new arenas such as when Scandinavian men make food together, no doubt for pleasure! To expand our perspective on pleasure, perhaps we should revisit Aristotle's emphasis on the virtues in which pleasure is some magic wholeness that follows the various actions. Based on the students' reports, we can see that this pleasure is cultivated in multiple ways that are only visible when visiting both backstage and frontstage.

21.14 ACKNOWLEDGEMENTS

The author gratefully acknowledges the funding to NIDAR's R&D-project Dynamic innovation capability and NTNU's Integrated product development project that was provided from the Research Council of Norway (Nutrition and Productivity 2005-programs respectively). Special thanks also to the students of Industrial Design, Oslo School of Architecture.

21.15 REFERENCES

Argyris, C., 1982. *Reasoning, Learning, and Action*. Jossey-Bass, San-Francisco.

Barabba, V.P. and Zaltman, G., 1991. *Hearing the Voice of the Market: Competitive advantage through creative use of market information*. Harvard Business School Press, MA, Boston.

Bowen, H.K., Clark, K.B., Holloway, C.A. and Wheelright, S.T., 1994. *The Perpetual Enterprise Machine. Seven Keys to Corporate Renewal Through Successful Product and Process Development*. Bowen, H.K., Clark, K.B., Holloway, C.A. and Wheelright, S.T. (Eds), Oxford University Press, New York.

Czarniawska-Joerges, B., 1992. *Exploring Complex Organizations. A Cultural Perspective*. Newbury Park, Sage.

Cooper, R.G., 1993. Winning at New Products: Accelerating the Process from Idea to Launch. Addison-Wesley, MA, Reading.

Frønes, I. and Brusdal, R., 2000. *Paa sporet av den nye tid: Kulturelle varsler for en naer fremtid* (On the Track of the New Time: Cultural Warnings of a Near Future). Fagbokforlaget (in Norwegian), Bergen.

Ghoshal, S. and Bartlett, C.A., 1997. *The Individualized Corporation: A Fundamentally New Approach to Management*. Harper Collins, New York.

Giorgi, A., 1975. An Application of Phenomenological Method in Psychology. In *Duquesne Studies in Phenomenological Psychology*. Giorgi, A., Fischer, C. and Murray, E. (Eds), **2**, Duquesne University Press, Pittsburgh.

Glaser, B. G. and Strauss, A.L., 1967. *The Discovery of Grounded Theory: Strategies for Qualitative Research*. Aldine de Gruyter, New York.

Goffman, E., 1974/1986. *Frame Analysis*. Northeastern University Press, MA (reprint ed.), Boston.

Gullestad, M., 1989. *Kultur og hverdagsliv* (Culture and Everyday Life). Universitetsforlaget (in Norwegian), Oslo.

Gullestad, M., 1999. *Kunnskap sett nedenfra* (Knowledge as seen from bottom-up). Nytt Norsk Tidsskrift, **4**, pp. 336-354.

Hatch, M.J., 1997. *Organization Theory: Modern Symbolic and Postmodern Perspectives*. Oxford University Press, New York.

Jevnaker, B.H., 1993. Inaugurative Learning: Adapting a New Design Approach. In *Design Studies*, **14**, 4, pp. 379-401.

Jensen, R., 1999. *The Dream Society: How the Coming Shift from Information to Imagination will Transform your Business*. McGraw-Hill, New York.

Johnson, B. and Ireland, C., 1995. Exploring the Future in the Present. *In Design Management Journal*, **6**, 2.

Klausen, A.M., 1970. *Kultur – variasjon og sammenhenger*. Gyldendal (in Norwegian), Oslo.

Leonard-Barton, D., 1995. *The Wellsprings of Knowledge*. Harvard Business School Press, MA, Boston.

Nordström, K.A. and Ridderstraale, J., 1999. *Funky Business*. BookHouse, Stockholm.

North, D.B., 1995. How Consumers Consume. In *Journal of Consumer Research*, **22** (translated version reprinted in Magma, **3**, 2000).

Pine II, B.J. and Gilmore, J.H., 1999. *The Experience Economy*. Harvard Business School Press, MA, Boston.

Sparke, P., 1987. *Design in Context*. Chartwell, London.

Stigen, A., 1996. *Aristoteles Etikk*. Gyldendal (in Norwegian, 2nd ed.), Oslo.

Trompenaar, A. Hampden-Turner, C. and Trompenaars, F., 1998. *Riding the Waves of Culture: Understanding Cultural Diversity in Global Business*. Irwin (2nd edition).

CHAPTER TWENTY-TWO

Mapping the User-Product Relationship (in Product Design)

ROEL KAHMANN and LILIAN HENZE

P5, Quality and Product Management Consultants,

PO Box 3688, 1001 AL Amsterdam, The Netherlands

22.1 INTRODUCTION

The Usability Professional is more and more consulted in the design process. This process traditionally is focused on creating reliable and technical high-quality products. Due to developments in production quality these aspects are no longer a distinguishing characteristic. Nowadays nearly all products are of reasonable quality. Thus functionality and usability become more important. Another trend is the growing application of microchips in a wide range of products. This results in an almost unlimited amount of features. The counterpart of this trend is the fact that products become more difficult to use. Therefore usability can distinguish a product from its competitors. But why map the user-product relationship? Designers are taking up the challenge of designing products that can be experienced by their owners (Overbeeke and Hekkert 1999). Product managers and marketers are confronted with the 'heart and mind conflict' in the decision making of the buyers of their products (Shiv and Fedorikhin 1999). Mass customisation is a trend in business (Kotler 1999; Pine and Gilmore 1999) This means products should be adapted to the customers needs. Our company, P5, as usability professionals, translate these needs in order to inform both designers and product managers. Usability is – according to its definition – mainly focused on how people use the product. It concerns the interaction between user and product. But mapping the user-product interaction is no longer enough to understand all the consumers needs and to inform the designers and product managers satisfactorily. Mapping the user-product relationship will become an important additional tool in P5's toolbox.

In this paper we describe our quest to find the right methods to get more insight in this relationship.

22.2 USABILITY IN DESIGN

In our practice we do a lot of usability testing – or usability evaluation if you like. Most of the time the research is focused on consumer and professional products. In the field of activity of designers and product managers a pragmatic approach is needed. Restricted by budget and time limits we are forced to search for an optimal test design.

We developed an approach called the Usability Intervention Model. This approach is described in earlier publications (Kahmann and Henze 1999) and in the mean time has been further developed. In short it can be described in form of a model. The Usability Intervention Model is based on three elements: 'Object', 'Intervention' and 'Outcome' (Figure 22.1). These ideas are described briefly below.

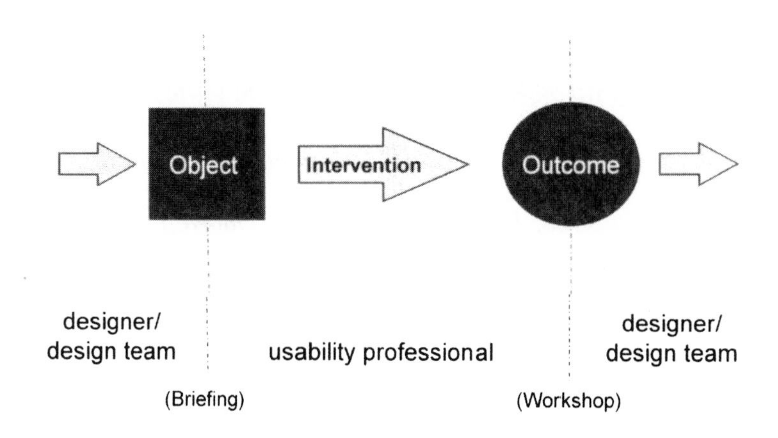

Figure 22.1 The Usability Intervention Model

The object is the subject of the usability study. Depending on the phase of the production development process, the qualities of the object can vary considerably. The object under scrutiny could be anything from a written product specification to a fully functioning prototype, or even the final product. The model can also apply to non-material things, such as services, but this is a marginal point.

The intervention is the action which P5 or the usability professional carries out on the basis of the object. The type of object determines the aim of the intervention. The aim, for instance, can be to let the requirements come to the surface on the basis of a product manager's idea, or it can be to test concepts to determine whether they meet specific requirements. The aim furthermore determines which methods are to be used, and what type of data are to be obtained. In the case of generating requirements, panel discussions and brainstorming techniques can be used, while for researching a concept, the tendency would be to use a test configuration.

Outcome is the data which emerge from the intervention in the form of information. How this information looks, and its form, can vary substantially. In the case of requirements, the information would be a programme of requirements in the form of a checklist or report, while for the concept test, the information can be a video or CD-ROM as an illustration of factors which determine use.

The concept 'result' is deliberately not used here because a result can only be realised when the outcome is incorporated into the product development process and accepted. Therefore, communication with all persons involved in the development, the design team, is very important.

A line is drawn in the model which defines when the object moves from the designer to the usability professional and where the data move from usability professional to designer (Figure 22.1). The line at the left represents the briefing from the designer or design team. It is not only handing over the object in a physical way but also communicating the goals of the intervention. The line at the right represents the workshop as a moment for communicating the outcome to the designer or design team.

The process can be repeated several times in succession. For example, when the designer offers a concept, or an adjustment has been made after the intervention, or a more finely elaborated concept or model has to be tested again. This process can be repeated as often as necessary or desirable. Schedule and budget limitations, however, severely restrict such repetition in practice.

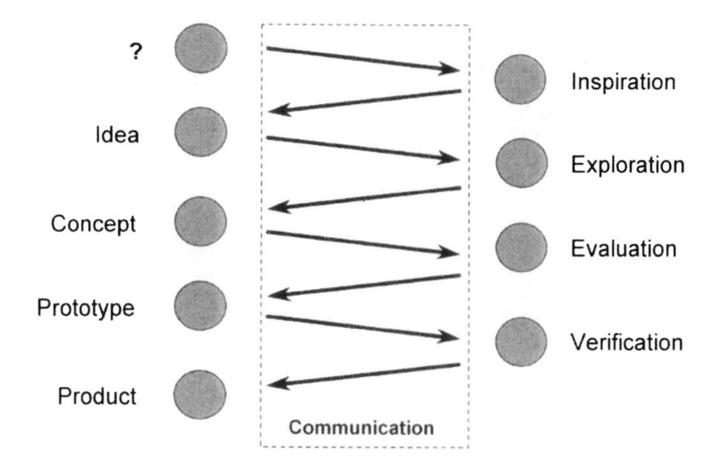

Figure 22.2 Ping-Pong Model

In Figure 22.2 – representing the process as a whole – at the left side the different stages of materialisation of the product are mentioned while at the right side the four basic phases of the usability research are positioned. This is presentation of an ideal situation. In practice some interventions will take place more than once and often the verification will be skipped.

22.3 P5 USESCAN®

In addition to the Usability Intervention model, in an earlier publication we described (Kahmann and Henze 1999) the UseScan approach. This approach consists of three phases as mentioned in Figure 22.2, i.e. exploration, evaluation and verification*.

Exploration is searching for what interaction can or should take place in using the product, and what aspects are relevant for this interaction. The following phase is the evaluation of the concept(s) of the product. This is testing the way the different interactions work out in a qualitative way. In the third and last phase, a verification will take place in a quantitative research to verify if the expected interactions work in the right way.

The information which is used during the intervention should be divided into

* Some use here validation. Verification concerns truth whereas validation concerns strength (Fuld 1997).

objective and subjective information. The objective information is what we call professional knowledge (Figure 22.3) and consists of facts; it concerns information from for example product ergonomics. This is information which is primarily obtained through scientific research, such as anthropometrics, bio-mechanics, cognitive ergonomics and standardisation, and can be found in textbooks and databases. Such data concern dimensions, permissible forces, letter sizes and colour combinations. In short, aspects of a physical and cognitive nature. In addition to the hard facts available in the literature, there is also the input from the user, consumer participation in Figure 22.3. In usability testing qualitative methods are particularly used. These data are less absolute and can be considered subjective. Through keen observation and a structured analysis, however, these data too can be translated into more or less hard facts.

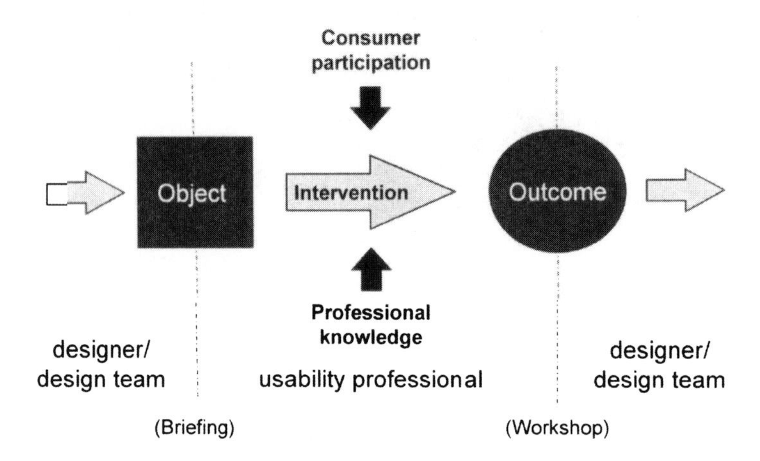

Figure 22.3 P5 UseScan & AffectionScan

The central idea in this approach is the traditional way of testing physical aspects of product use. It is focused on interaction between the user and the product and mainly based on the ISO-DIS 9241-11 terms; effectiveness, efficiency and satisfaction. Human factors is becoming established as a fundamental part of the product development process. Quantification of usability takes the issues into the product development process, and usability considerations will be part of the programme of requirements.

22.4 USER-PRODUCT RELATIONSHIP

While usability traditionally – as far as one can speak about tradition in this rather young discipline – focuses on the interaction between user and product there are more aspects playing a role. The consumer has different roles in relation to a product. In Figure 22.4 there are three of them. There is the person who buys the product, the person who uses the product and the one who owns the product. In consumer products all these roles are often filled by the same person. So besides using the product there are other aspects which are important. This scheme is distilled from more general schemes used in consumer research.

In order to look at the relationship between product and person comprehensively, it is necessary to look both at and beyond usability, touching on wider aspects of what it is like to buy, own and use a product.

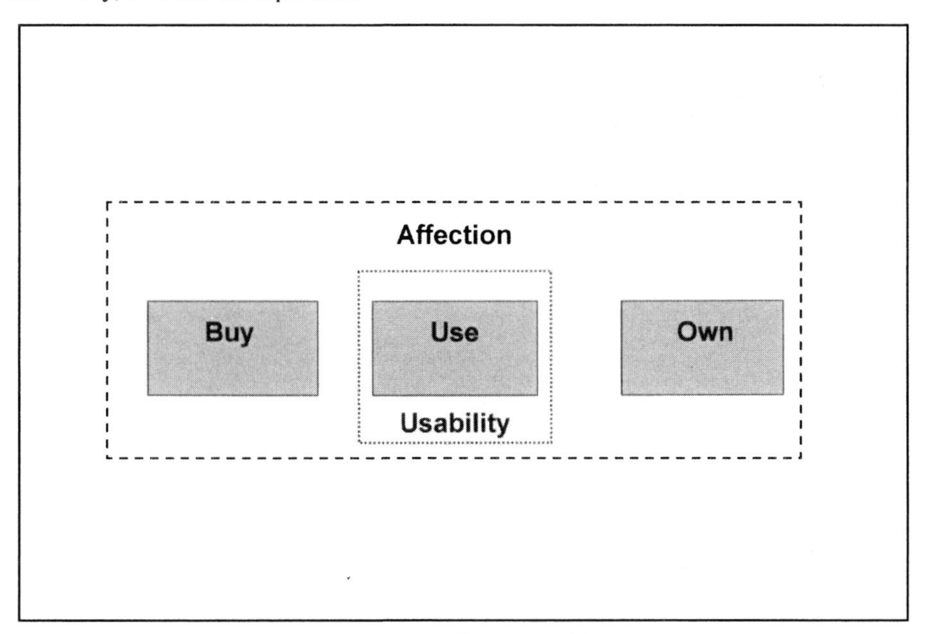

Figure 22.4 The different roles of the consumer

Jordan (1997) argues that a more holistic approach is needed. The human factors specialist should look at people, products and relationships. Besides usability, issues such as aesthetics, reliability, technical performance, emotions and values that are associated with the product are affecting the user product relationship.

This understanding has already led to research focused on issues beyond usability. Jordan (1999) focuses on the four pleasures defined by Lionel Tiger as a challenge for human factors to provide better products. Bonapace (1999) uses the SEQUAM to define sensorial qualities of products.

Pine and Gilmore (1999) state that using products is part of a wider experience. So it is important to get a grip on the experiences evoked by or associated with products.

It is therefore important to map the relationship between users and products. P5 struggles with the question: how? Traditionally market research is focused on factors that influence the way people react to a product, or why they should or will buy it. The methods used are well known and validated, but are these methods enough to understand all consumers needs? And will these methods lead to relevant information for the design-team?

22.5 P5 APPROACH

We are used to evaluating consumer products and professional products. The attitude of the respondents towards the product is important for the evaluation. This attitude influences the answers to the questions asked during the usability research. It is essential information to be related to the findings in the analysis of the test results. Therefore, in the traditional usability research we develop a questionnaire to capture the degree of experience of the respondent with the product group or type.

In doing so the degree of experience of the respondents and the role the product has in their lives becomes clear.

Being involved in earlier stages in the product development process means that the requirements of the product are not yet so clear. Some research needs to be done to find out what requirements the product should meet. Traditionally this is done by market research and not by usability testers. However, in our experience, designers don't read those vast reports they generate. The problem with market research is their distance from the designers. One of the comments is that they tell what kind of people will buy the product but not why, i.e. what do the people want or what do they expect from the product?

As a matter of fact the usability professional is specialised in this field. When these professionals can develop research methods focused on this stage of the development process, the designers will be helped.

On the other hand there is a tendency for our clients to ask for more. Not only for the strictly usability aspects, they also ask for insight into the consumer's attitude towards design, colour, texture, etc. Concluding, it can be stated that there is a need for information that goes beyond usability in a strict sense. According to the holistic approach – as mentioned above – methods should be developed to give answers to these kind of questions.

At P5 we are developing a method to solve this problem, the P5 AffectionScan. This method is meant to explore all the consumer needs not covered in the P5 UseScan.

In the next section this first step in our approach will be explained.

22.6 P5 AFFECTIONSCAN

Phase 1: Exploration

In this phase the way people look towards the specific product or product group is analysed. The goal of the exploration phase is to map what kind of attributes are important for the different users. Attributes are aspects of a product perceived by the user. In other words, in what terms does a user talk about the product. What functions does the product have to meet and what does the user expect from the product. Additional data generated in this phase are the associations users have with these types of products.

It is meant to find out information which is needed in the next phases of the research. Techniques used are in-company brainstorming, observation and, most of the time, short interviews. This qualitative research contributes to a better understanding of why particular attributes are important to the users.

Phase 2: Research

Based on the information of phase 1 the design for the research is further developed. This phase basically is split up into four different parts.

Questionnaire

At the start of a research there is a point of reference (POR) analysis. This analysis is meant to list several data of the respondents. For example:

- what kind of experience does the respondent have with this type of product;
- in what kind of circumstances does he/she use it, what is the frequency of use;
- is it an amateur or (semi)professional user.

From this information the respondent's profile is constructed. This information is used in the analysis phase (phase 3) later on in the process.

Conjoint analysis

This technique is used in market research to settle the trade-offs of a product. It lets marketers predict choice behaviour. They use it for deciding how to market a product. In this case it is used to define the relative importance of different attributes of the concept or product (Curry). De Bont (1992) concludes in his research on consumer evaluations of early product-concepts that this technique is the one most favoured in this type of research.

The technique is based on the assumption that people weight the different attributes in combination with each other. It asks the respondents to make choices in the same fashion as the consumer presumably does by trading off features, one against another. One of the approaches is the full-concept method. In the full-concept method, the respondent is asked to rank or order a set of profiles or cards according to preference. The cards are sorted by the respondents from most favourite to least favourite.

On each card a concept is printed which is fully profiled on all of the attributes. As you increase the number of attributes included in your study, both the amount of information on a card and the number of cards per interview increase. So there is a limit. With more than six to eight attributes, respondents become overloaded with information and the reliability of their answers may diminish. If there is a need for more attributes to be researched other methods are available.

Semantic differentials

In market research there are different scaling techniques available. One of them, Likert scales, is often used in questionnaires.

Nowadays different scales are used and the discussion about the amount of choices is still ongoing. Has it to be even or uneven, five- or seven-point scales? It is a technique to measure attitudes of people towards a product. This technique is used in the questionnaires as mentioned in a previous section.

Another technique is a scaling method developed by C.E. Osgood and it is called the semantic differentials or Osgood scales. In this technique, in most cases, a seven-point scale is used (1 up to 7 or -3 up to +3 inclusive). Respondents get a certain amount of scales presented with, on either end, a property with its reverse. Examples are very cheap – very expensive, modern – outdated or hard – soft (Schoormans and de Bont 1995; Hofmeester, Kemp and Blankendaal 1996).

By using semantic differentials the kind of associations people have with the object of the research can be generated. It gives insight into the position of the concept or product in the field. In contrast to the conjoint technique the attributes are not judged as a whole but apart. Not only functional but also emotional aspects can be judged.

There are different ways of presenting the concepts or products to the respondent. The complete product can be shown or only a photograph or picture of it.

Interview

In the last part of the research, before the debriefing, the different models or concepts used in the next part are shown to the participants and they are asked to choose their favourite. In addition they are interviewed about why they made this choice.

22.7 PHASE 3: ANALYSIS

Questionnaire

The questionnaires can be prepared in programs like SPSS's Data Entry. This means that there is a form designed on the computer which can be printed and at the same time used as a form to enter the data. In case the questionnaire is composed in a text editor the data can be typed in a spreadsheet of text file which can be imported in a statistical programme for further analysis.

We are used to putting every questionnaire into a database, even when there is only a small amount of them and there is hardly use for statistics. The major reason for that is efficiency in the process. When a client asks in the course of the process for more detailed information or information about other selections it is easily generated.

Conjoint

For the conjoint analysis there are statistical programmes available. These programmes not only calculate the relative importance (utilities) of the different attributes but can also generate the cards for the research.

Semantic differentials

In this technique the mean are calculated. In a strict sense it is not permitted to calculate the mean because these scales are ordinal, but in order to visualise the results it can be done. In most cases they are plotted in a graph. If there are too many different concepts these plots become quite difficult to analyse. In that case they can be plotted in a X-Y diagram. The disadvantage of such diagrams is the fact that only two attributes can be compared.

The outcome of the analysis is – finally – communicated to the designers and product managers in a workshop.

Conclusions

Measuring affection and emotion is becoming more and more important. Although this information, is traditionally generated by market research, the human factors specialist can combine it with more product focused issues.

The P5 AffectionScan is presented here as a developing tool and needs further development in near future. It generates answers for questions from designers and gives them insight into the most relevant aspects in product use. As a first step we used the more traditional market research techniques but we already knew that this would not be enough. These techniques are still based on the respondents awareness and give too little insight into the unconcious processes that are relevant aspects in the user-product relationship.

We hope that research and new techniques deriving from the creative minds of design researchers will lead to methods that make it possible to gain insight into both the mind and the heart of the consumers.

A big challenge is to work out these methods in a way that can fit into the design process with its tight budget and time limits. And, even more important, that the results can be translated into information that can easily and clearly be communicated to design teams. One of our next steps is to set up an experiment using the Sensorial Quality Assessment Method (SEQUAM described in Bonapace 1999) in co-operation with ErgoSolutions in Milan. Later on, we will dig deeper into the possibility of using methods where the respondents participate instead of being the subject of interviews and observations, like the methods used by SonicRim (Sanders and Dandavate 1999).

22.8 REFERENCES

Bonapace, L., 1999. The Ergonomics of Pleasure. In *Human Factors in Product Design; Current Practice and Future Trends.* Green, W.S. and Jordan, P.W. (Eds), Taylor & Francis, pp. 234-248.

Bont, C.J.P.M. de, 1992. *Consumer Evaluations of Early Product-Concepts.* Delft University Press (Thesis), Delft.

Curry, J., 2000. *Understanding Conjoint Analysis in 15 Minutes* and *Conjoint Analysis: After the Basics* at http://www.sawtooth.com/news/library/articles.

Fuld, Robert B., 1997. V&V What's the Difference? In *Ergonomics in Design*, **5**, 3, pp. 28-33.

Hofmeester, G.H., Kemp, J.A.M. and Blankendaal, A.C.M., 1996. Sensuality in Product Design: a Structured Approach. In *Proceedings CHI 96.* Available at http://www.acm.org/sigs/sigchi/chi96/proceedings/desbrief/Hofmeester/ghh_txt.htm.

Jordan, P.W., 1997. Usability Evaluation in Industry: Gaining the Competitive Advantage. In *Proceedings of 13th Triennial Congress of the International Ergonomics Association*, **2**, Tampere, Finland, pp. 150-152.

Jordan, P.W., 1999. Pleasure with Products: Human Factors for Body, Mind and Soul. In *Human Factors in Product Design; Current Practice and Future Trends.* Green, W.S. and Jordan, P.W. (Eds), Taylor & Francis, pp. 206-217.

Kahmann, R., and Henze, L., 1999. Usability Testing under Time-Pressure in Design Practice. In *Human Factors in Product Design; Current Practice and Future Trends.* Green, W.S. and Jordan, P.W. (Eds), Taylor & Francis, pp. 113-123.

Kotler, 1999. *Kotler on Marketing, The Free Press.* Simon & Schuster Inc., New York.

Overbeeke, C.J. and Hekkert, P., 1999. *Proceedings of the First International Conference on Design & Emotion.* Overbeeke, C.J. and Hekkert, P. (Eds), Delft University of Technology.

Pine, B.J. and Gilmore, J.H. 1999. *The Experience Economy.* Harvard Business School Press, Boston.

Sanders, E.B.-N. and Dandavate, U., 1999. Design for Experiencing: New Tools. In *Proceedings of the First International Conference on Design & Emotion.* Overbeeke, C.J. and Hekkert, P. (Eds), Delft University of Technology.

Schoormans, J. and Bont, C.J.P.M. de, 1995. *Consumentenonderzoek in de Productontwikkeling (Consumer Research in Product Development)* (in Dutch). Lemma, Utrecht.

Shiv, B. and Fedorikhin, A., 1999. Heart and Mind in Conflict: The Interplay of Affect and Cognition in Consumer Decision Making. In *Consumer Research*, **26**, 3, pp. 287-292.

CHAPTER TWENTY-THREE

Cooking up Pleasurable Products: Understanding Designers

STEVE RUTHERFORD

School of Art and Design, Nottingham Trent University, Burton Street, Nottingham NG1 4BU, UK

23.1 INTRODUCTION

We are striving to understand the design process. What kind of process leads to products which consumers 'connect' with emotionally? As products become more reliable consumers are examining more refined aspects of a product's design before deciding whether to purchase, and this puts ever more pressure on the quality of design. We need to understand more of what goes on inside the designer's head so that ergonomists can play a more integrated part in the process.

The approaches taken by researchers to analyse the design process, the product and its success vary. Some have looked at the structure of the design process (Davis 1995), some at the nature of the innovation process (Jevnaker 1999). Others have looked at the pleasures evoked by products (Taylor 1999) and one can arguably be said to have looked at the personality of the designer (Candy and Edmonds 1996). Whichever method is used there is a certain amount of dependency on the viewpoint taken by the researcher and the type of design process being examined, and this categorisation of processes is problematic – there are as many design process models as there are designers.

It is the author's experience that even under tight design guidelines and across a wide range of different companies, the influence of individuals over the process and the final design is a huge variable. The approach taken in this chapter is to observe in another way, to examine a design's personality – as represented by the forms, colours and textures which he or she creates – and examine:

- how the *Gestalt* view is used by designers in the early stages of a design project, and;
- how the context and practice of the designers contributes to the process.

Looking at the designers' environment and position in the company:

- how the human make-up of design teams might produce very different effects.

There are opportunities and warnings for ergonomists along the way. However, this analysis does not pretend to be scientific. It offers an insight into the design process and into the way a designer thinks, and it hopefully paints a picture of a possible future for ergonomics' involvement in the design of pleasurable products.

23.2 THE *GESTALT* VIEW: THE COOK'S RECIPE FOR SUCCESS

'In many cases design properties ... could not be isolated as selected variables since products are perceived as entities (the Gestalt) in which the various ... elements interact...' (Taylor 1999).

'The flavouring of cakes is a delicate branch of the baker's art, requiring the exercise of his nicest faculties... A little too much flavour is often nauseating, whereas not enough is very unsatisfactory' (Thompson Gill 1881).

We certainly get excited about individual properties of a particular design – for example the transparent casing of the Apple Imac – but the success of a product cannot rely on a few sensual surprises for market success. By the same token, we can forgive a product some fairly serious shortcomings – for example the seriously too small mouse of the Apple Imac – when overall the relationship we have with the product is still satisfying. We as consumers are taking the overall view in deciding whether we want this relationship in the first place, and whether we want to continue it after we discover the realities of living together.

This view of a product's '*Gestalt*' is why we have some relationships that are often seen as questionable by others (Figure 23.1 and 23.2):

- Q: Why would anyone own a Citroen 2CV?
- A: Outsiders see a fraction of the *Gestalt* that owners perceive, and it is the less obvious aspects of minimalist design and frugality which delight.

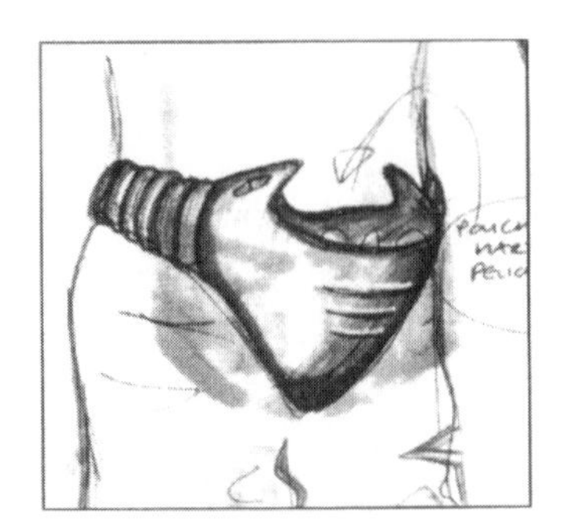

Figure 23.1 Citroen 2CV

Figure 23.2 Mobile communicator sketch, Rebecca Capper

So those of us in the know love our 2CVs and Apple Imacs.

If a product is to be judged by the sum of its parts rather than its individual qualities, how does a designer contend with this? Concept sketching at the start of a project tries to give a flavour of possible directions to a project, in order to elicit feedback from the design team, management and potential users. At this point the designer is designing from the outside in and sometimes detail is not a part of what is communicated. What is the context of the design's personality at this stage? Is it a set of technical details – function, form, colour and texture? Or is it an almost subconscious view of the design's '*Gestalt*' inside the designer's head? Based on the author's own experience and work with furniture and product design students, the *Gestalt* view of a product is often first to emerge, an image of form and colour, occasionally with a certain material in mind. Once this picture has emerged it is possible to develop it in terms of form and colour, and to zoom in on it to examine material and textural possibilities.

This results in information which could be of immediate use to ergonomists in terms of identifying aspects which users might connect with. It is possible within minutes of starting a project to have hand drawn sketches showing very life-like products, including form, materials, textures and colours. The immediacy of this communication and the usefulness will be reinforced throughout this chapter.

23.3 FORM, COLOUR AND TEXTURE: THE IMPORTANCE OF INGREDIENTS

'He chose to work with steel, which he knows how to imbue with warmth by working on its form, surfaces and coupling it with other essential materials, thus creating objects that are unique in style, function and aesthetic' (Alfieri 1999).

'It is well to remember that a single musty egg will spoil a ton of cake or anything else it is put into' (Vine 1898)

The *Gestalt* view emerges as an important conductor of the product's personality. We will now see how the detail of that view – form, colour and texture – fit into the designer's control of the design process.

Despite the individualism brought to a product by its designer, fashion plays a large part in the use of form, colour and texture in three-dimensional design. We are all very adept at guessing the age of an old product by its visual detail and references (Figure 23.3). This has been so since the birth of industrial design in post-WW2 western civilisation. Form, colour and texture are still manipulated by designers, but often within limits set by the times in which the designer operates. As Kalviainen (1999) states: 'Even fashion or general changes over time should be considered as providing the larger framework in the network of influences.'

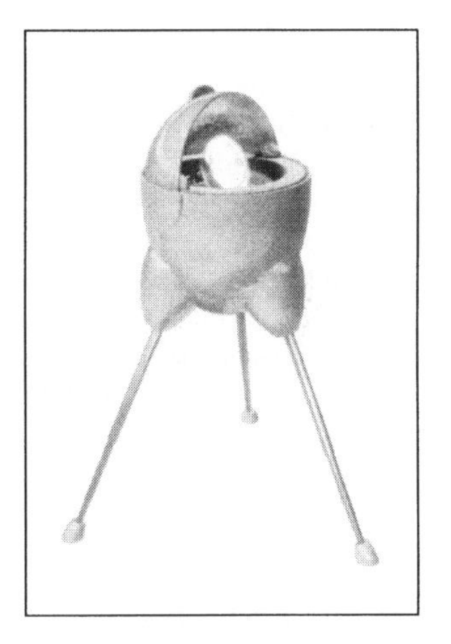

Figure 23.3 Domestic lamp, Chris Bainbridge 1999

This is especially true of the design of the largest, most expensive of products – the car. The wild tail fins of the 50's were calmed down through the 60's. The 70's oil crisis curbed the exuberances of the industry, but the 'me' generation of the 80's demanded their angular 'folded paper' designs, culminating in the aggressive Audi Quattro. This may well prove to be the end of the original period of car design as the industry has recently been gripped by retro-fever and is constantly looking over its shoulder into the history books for inspiration. The 90's have seen a new Beetle, Mini, Rover, Jaguar and BMW designed which wouldn't look out of place on Carnaby Street 30 years ago. The influence of car design over product design during this period has been evident in the High Street. Sit a Roberts radio on the leatherette seats of a Ford Anglia and you'd swear they were made for each other. Ditto a Sony MiniDisc player and an Audi TT.

Where does this leave the designers? They operate within some quite tight parameters and, especially in a market as large as cars, the designs must be commercially viable. As we know, reducing risk is important in new product development. How do they do this? Is it possible that they operate within such tight constraints, and communicate with each other informally to such an extent, that they are reasonably sure of the viability of what they produce? Even a cursory glance at the consumer product market place shows a repetition of certain forms, colours, textures and materials. Is creativity a closed shop with tight rules?

Most of the world's car designers attended the same college. Most product designers are produced by a small handful of colleges. It's a small world. Do they play it safe most of the time? And if they do go over the top now and again, can they rely on customer focus groups to weed out anything slightly wild that may put off the buyer? The reality is that there is quite a closed shop of acceptable creativity in the design world, due to its close community, its following of fashion and its need to be commercially viable. Also, designers and companies are not averse to virtually bypassing the design process altogether and jumping on someone else's bandwagon. The translucent plastic fad attributed to the Apple Imac has spread to the extent that every steam iron in John Lewis Partnership stores in the UK has a translucent blue/green water compartment in it. Also, nearly every vacuum cleaner manufacturer now has a model with a clear, plastic, Dysonesque cyclone unit on the front.

Within all of this the ergonomist looks for a part to play. As a designer *and* ergonomist, the author looks especially hard and as yet the picture is not easy to decipher. On the one hand we can interrogate potential users, breaking down a product into various readable qualities, or assess it through a useful construct, such as its perceived gender (Mcdonagh-Philp and Denton 1999). On the other hand, the set of possible solutions is already constrained by the need for commercial viability and the fashion-oriented outlook of the designers. This leaves a dauntingly refined set of possibilities for an ergonomist to interrogate (Figure 23.4).

At this point it is worth remembering our ergonomists' war cry – 'Involve us from the start, please'. There are already aspects of the design process which require early input of ergonomics, even into the pre-designing 'design specification' stage. From a functional, traditional ergonomics viewpoint, we want to be there at the start. We should add to this early involvement the analysis of the conceptual possibilities of the project, as mentioned earlier. This would mean sitting down with designers in a studio on day one, discussing first ideas over lunch and throwing the ideas into a focus group type activity at 2:00 pm.

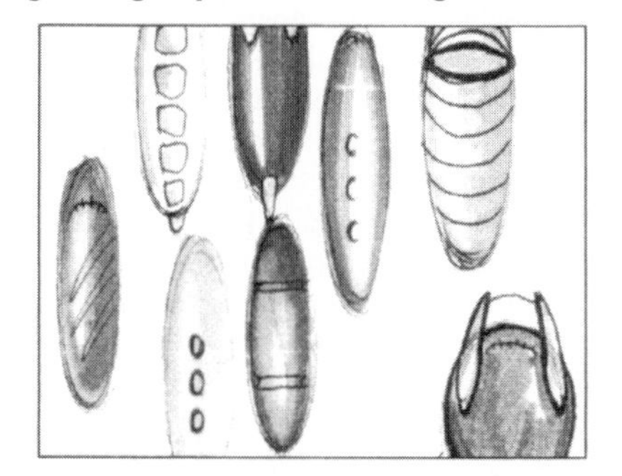

Figure 23.4 Concept sketch, Rebecca

The author's experience of large companies suggests this does not happen very frequently, if at all. In smaller companies, it can happen to some degree, if not quite as quickly as the example above. In student design projects at Nottingham Trent University it has been encouraged at a very early stage in the design process and, although it's done very informally by the student designers themselves, there has been a healthy exchange of information in some instances. The beauty of the principle is that, with very little investment in time, feedback on directions the project might take, can be built into the design process, both from the start and at frequent intervals throughout the project.

What has been described above is not ground-breaking. It is probably a part of many healthy design courses' teaching. As a principle it's sound, however its implementation is another matter. The third part of this paper explains why this common sense approach is not very widespread. It relates to the importance, perceptions, creativity and personalities of the client, the manufacturer, the designers and their ergonomists. It hopefully demonstrates the variety of ways design works within different company structures, and suggests where new ergonomics methods might fit into these patterns.

23.4 THE MANUFACTURER, THE DESIGNER AND THE ERGONOMIST: THE COOKS

'Taste is not so much about what things look like, as about the ideas that give rise to them. Intention is the key to understanding taste' (Bayley 1991).

'The variety and names of buns are determined by the skill and ingenuity of the bakers' (Kirkland 1927).

The parts that the many cooks play in new product development differ from company to company and evolve over time. The design process not only evolves over time but is subject to standardisation via ISO guidelines and seismic shifts due to technology. Even in the complex area of vehicle design each company has its own recipe for success. The following examples show different approaches leading to different results and give an insight into how pleasure-based ergonomics might fit into the system. In the first a safe, 'design by committee' process leads to questionable results. In the second a recently enlightened company gives a non-vehicle design guru his head with surprising results. In the final example the vision comes from within the company and at a high level, and is the result of design becoming a powerful force within the company management structure.

The Company:	**Toyota**
The Committee:	The Genesis Group
The Car:	Toyota Echo

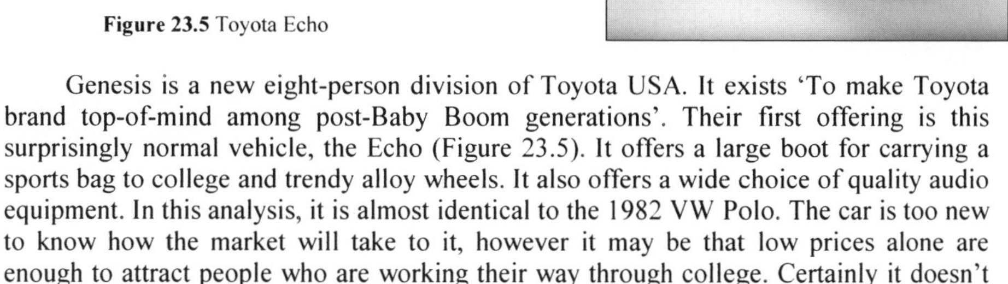

Figure 23.5 Toyota Echo

Genesis is a new eight-person division of Toyota USA. It exists 'To make Toyota brand top-of-mind among post-Baby Boom generations'. Their first offering is this surprisingly normal vehicle, the Echo (Figure 23.5). It offers a large boot for carrying a sports bag to college and trendy alloy wheels. It also offers a wide choice of quality audio equipment. In this analysis, it is almost identical to the 1982 VW Polo. The car is too new to know how the market will take to it, however it may be that low prices alone are enough to attract people who are working their way through college. Certainly it doesn't have the contemporary fashionable image that many small cars are cultivating at the moment.

It is easy to protest that the formation of the Genesis Group has precluded the need for 'design' at all. The work of the Group – assessing needs; laying down priorities for the design specification – seems to mirror in some ways the input that ergonomic methods might have had early in a design project. The car is, in fact, a development of an existing, ironically quite trendy, hatchback, the Yaris. The results are, from a designer's perspective, very disappointing. There is a reminder here that, at times, ergonomics should take precedence over design (function; safety) or work with design (usability; fit) but not get in the way of creativity, as can be argued in this case.

The Company: **Ford**
The Guru: Marc Newson
The Car: 021C

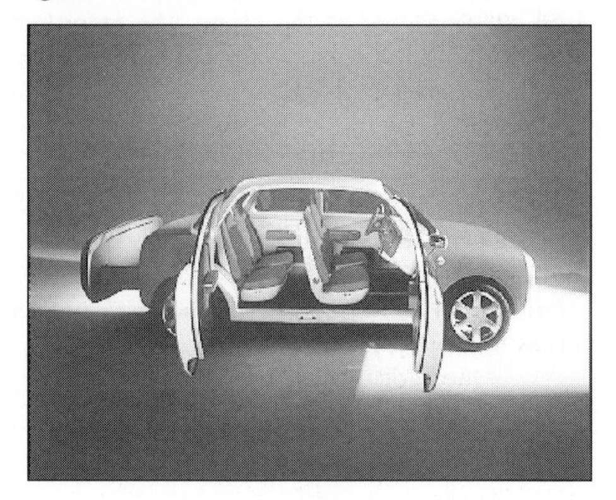

Figure 23.6 Ford 021C

Marc Newson designs lighting, furniture, interiors and products and is famous for it. Ford design chief J Mays wants his company to communicate with Marc Newson's world. The collaboration is refreshing as it gives one person responsibility for a car's design, from the concept to the details (Figure 23.6). It has been an educational process for all concerned. The product is a complex one for Newson, and his method of working must necessarily be very different to the teams of designers and levels of management involved in a normal project.

The experience has been interesting for both sides. Is there room for an ergonomist within this design process? The principle was to give the designer responsibility. The level of creativity and personal ownership of the design by Newson is very high. If his personal stake is so high what are the chances of an outsider having the nerve to challenge his authority? Probably low. It is the author's opinion that the personal aspect and the risks involved in proposing something very innovative preclude any useful analysis of the pleasurable aspects of the design. Only by taking what seem at the time to be risks can truly new ways of looking at objects be discovered. Ergonomists involved at this stage could well discover a mismatch between the product image and public opinion. However, given the chance, it is possible that public opinion may catch up with the designer's opinion.

The Company: **Renault**
The Visionary: Philippe Guedon
The Car: Renault Avantime

Figure 23.7 Renault Avantime

The last example perhaps gives ergonomics a more hopeful vision of the future. The structure of design management within Renault would seem to encourage the visionary approach within the framework of more traditional vehicle design methods. At the 1999 Barcelona Motor Show, Renault filled a stand with an outstanding display of a succession of concept vehicles from the last ten years. The Renault system seems to produce results.

The idea for this particular car, the Avantime (Figure 23.7), a large, 2-door, sporty people carrier, originated with Philippe Guedon, who runs Matra, the company which assembles the present Espace for Renault. The Director of Design and Quality, Patrick Le

Quement, encouraged the idea. In turn, the exterior is the work of designer Thierry Metroz and the interior is by Stephan Janin. It is rare for an important vehicle to be the identifiable work of such a small number of people. The benefits, as with the Ford 021C, seem to be a chance for innovation to prosper. In an interview with Richard Bremner of Car magazine (1999), Le Quement states that: 'The car is designed with magical functions, like modern furniture, with flaps and drawers and compartments. We want a car that nourishes the soul, as well as the body.'

The Avantime could be described as 'designed by committee'. However, given the positive result, the committee would seem to have a recipe that brings out the full flavour. The differences between this design process and a typical vehicle design process are principally the personal levels on which it operates – individuals are identified – and the high level at which design is represented within the company. The situation here would seem to offer many of the advantages of the Ford experience without the drawback of investing everything in one person's vision.

If the industry develops in the direction of the Renault process model the signs for ergonomists will be very positive. Design in an enlightened environment is open to innovation in product image and design process. The Renault example has demonstrated good results from their present methodology, however results may be achieved earlier and with more certainty with the utilisation of subjective ergonomics evaluation techniques.

23.5 CONCLUSION

This chapter has highlighted something of the initial process of designing, demonstrating that '*Gestalt*' images of a product are there from the start of the design process and are given form by the designer's concept sketches. Many aspects of a product's potential portfolio of pleasures can be communicated visually, and this would give an ergonomist something to measure subjectively at an early stage in the design process. Further on in the design process, market constraints and the tight commercial framework within which many companies operate may produce a set of closely defined possibilities which are more difficult for the ergonomist to discriminate between.

The usefulness of subjective ergonomics evaluation within different design methodologies is predicted with the vehicle design examples. This points out possible problem areas for ergonomics as it widens its remit in the design world. There is a danger that it can get in the way of creativity if it is too prescriptive and detailed early in the design process. In other instances, if the vision of an individual designer is to be trusted, the data that a subjective ergonomics evaluation produced might be untrustworthy – the designer is in a position of educating the consumer and could produce a ripple in the market when consumer opinion catches up. In the right environment, however, as in the Renault example, subjective ergonomics evaluation could enhance and support creativity. If the design process is recognised at a high level within the company and if it also retains elements of designers' individuality, an ergonomist's early communication with the design team and exploitation of the designers' visualisation skills could give the project team confidence in the early vision of the designers.

23.6 ACKNOWLEDGEMENTS

Gill Chapman, School of Cultural Studies, Sheffield Hallam University.
Hilton Holloway, News Editor, Car magazine.
Chris Bainbridge, undergraduate, Furniture & Product Design, Nottingham Trent University.

Rebecca Capper, graduate, Furniture & Product Design, Nottingham Trent University.
Cooking quotes taken from:
David, Elizabeth, 1977. *English Bread and Yeast Cookery*. Penguin, London.

23.7 REFERENCES

Alfieri, F.M., 1999. *Cini and Nils Clarity and Forms.* Domus Lighting File, pp. 6-7.
Bayley, S., 1991. Taste: The Secret Meaning of Things. Faber & Faber, London.
Bremner, R., 1999. Van extra ordinaire. In *Car Magazine*. EMAP, Peterborough, pp. 130-134.
Candy, L. and Edmonds, E., 1996. Creative Design of the Lotus Bicycle: Implications for Knowledge Support Systems Research. In *Design Studies*, **17**, 1. Butterworth-Heinemann Ltd., Oxford, pp. 71-90.
Davis, G., 1995. Ergonomically Designed? In *Proceedings of the Ergonomics in Consumer Product Design and Evaluation Conference*. Chilworth Manor, Southampton.
Jevnaker, B.H., 1999. Integrated Product Innovation: Dilemmas of Design Expertise and Its Management. In *Proceedings of Design Cultures, The European Academy of Design, Sheffield*. European Academy of Design, Salford,.
Kalviainen, M, 1999. Customer Taste as a Challenge in the Design Process. In *Proceedings of Design Cultures, The European Academy of Design, Sheffield*. European Academy of Design, Salford.
Kirkland, J., 1927. The Baker's ABC. In *English Bread and Yeast Cookery*. David, E. (Ed.), 1977, Penguin, London.
Mcdonagh-Philp, D. and Denton, H., 1999. Using Focus Groups to Support the Designer in the Evaluation of Existing Products: A Case Study. In *The Design Journal*, **2**, 2. Gower Publishing, Aldershot, pp. 20-31.
Taylor, A.J., 1999. The Relationship between Ergonomics and Industrial Design in New Product Development. In *Contemporary Ergonomics*. Hanson, M., Lovesey, E.J. and Robertson, S.A. (Eds), Taylor & Francis, London.
Thompson Gill, J., 1881. *The Complete Bread, Cake and Cracker Baker, Chicago*. In *English Bread and Yeast Cookery*. David, E. (Ed.), 1977, Penguin, London.
Vine, F.T., 1898. *Saleable Shop Goods*, 6th edition. In *English Bread and Yeast Cookery*. David, E. (Ed.), 1977, Penguin, London.

CHAPTER TWENTY-FOUR

Prolonging the Pleasure

GARY DAVIS

Davis Associates Ltd., Wyllyotts Place, Potters Bar, Herts EN6 2HN, UK

24.1 INTRODUCTION

The reality of our consumer society is that the pleasure of owning or using a product is a transitory state. Today's favourite, most treasured product will be found in tomorrows' car-boot sale. Perhaps the biggest challenge to product developers is, not only to enhance the pleasure of the person-product relationship (PPR), but also to ensure that the pleasure is prolonged for as long as possible.

Many people find pleasure in owning or using products. The pleasurable responses have as much to do with the character and needs of the individual, as with the design of the product. As a result, not everyone experiences the same level of pleasure from the same product. For example, a pair of in-line skates might be extremely pleasurable for a teenager, but would give no pleasure to an elderly person suffering from osteoporosis.

Although there are many differences in peoples' responses to products, there is one factor which has an overriding influence on the PPR — the passage of time. A sense of pleasure at the beginning of the PPR is no guarantee that the pleasurable experience will continue for the serviceable life of the product. Conversely, initial negative responses can sometimes turn to a positive sense pleasure after a period of use.

The ability to predict the effects of time on the PPR would be a powerful tool in the product developers' armoury, making it possible to design-in features and characteristics which help prolong the pleasure of owning and using the product.

24.2 THE LIFECYCLE OF THE PPR

A technique which has proven useful in predicting the effects of time on the PPR, is to consider its lifecycle in terms of an interpersonal relationship. It is a simple but effective technique which assists product developers to envisage how the needs of users will change over time, and what factors are likely to produce pleasurable responses at each stage of the PPR lifecycle.

The PPR lifecycle model consists of the following stages:

1. **First encounter** - first impressions and setting of expectations;
2. **Commitment** - the decision to purchase;
3. **Honeymoon period** - getting to know the product intimately;

... and then one of two possible outcomes:

4. **The relationship breakdown** - an unpleasant experience; or
5. **The enduring relationship** - a long and pleasurable experience.

24.3 FIRST ENCOUNTER

The nature of the first encounter is an important factor in shaping the user's expectations for the longer-term PPR. For example, if the product is first encountered in an advertisement, expectations can be set artificially high due to the intrinsically positive and targeted message that the marketeers have projected. However, if the product is first encountered in use by a friend, the perception of the product's benefits may be more realistic.

First impressions can be very influential in determining the attraction of a product. As discussed above, different individuals will find different attributes attractive. For example, any of the following may be important:

- the latest, most up-to-date model;
- feature/function list;
- perceived ease of learning and ease of use;
- aesthetics, style and fashion;
- sociological factors (e.g. association with certain cultural groups);
- novelty, rarity;
- reliability and durability;
- cost and value for money.

The features which are found attractive initially, will have a bearing on the life-cycle of the PPR. For example, the pleasure of owning the latest, most up-to-date model is likely to evaporate when the next, new model is launched. Contrary to that an attraction based on the product's good reliability record or its established ease of use is more likely to lead to a longer-term PPR.

User-focused research is regularly used to identify which factors are the most attractive to the target market sector. Sadly, this effort is usually focused on achieving sales and not at prolonging the pleasure of the PPR. However, consumer groups, such as the UK's Consumers' Association, specialist magazines, and consumer-based television programmes, have all raised awareness of the longer-term attributes of products, and consumers are increasingly less tolerant of sub-standard products. Although most of this publicity focuses on functional, rather hedonistic attributes, many consumers do look beyond the 'showroom appeal' when making significant purchases.

24.4 COMMITMENT

Rather like committing to a marriage, the decision to purchase a product, particularly if it is of high cost, can be very significant for the individual(s) concerned. For some people, the act of shopping, and perhaps the freedom and means to do so, is exciting and a pleasure in itself. However, the anticipation of owning a new product is sometimes accompanied by anxiety that the wrong purchase decision will be made.

Research before the point of commitment often includes consideration of a wide range of factors, including a comparison of alternative products. In these cases, the probability of a long-term PPR is far higher than with an 'impulse purchase' — the sort of decision which might be made at an airport for example. Purchases made through mail order or through the Internet — without direct contact with the product — can also run a higher risk of PPR breakdown.

24.5 THE HONEYMOON PERIOD

The honeymoon period follows the acquisition of a new product. It is a period of exploration, discovery and playfullness, as the new user becomes more intimately familiar with the product. This process can be a pleasure in itself, learning the product's operating characteristics and discovering surprising features. Innovative and unnusual features of a new product can provide great pleasure at this stage.

It is often with great pride that the new product is shown-off to friends, and the user seeks reassurance of the wisdom of his/her purchase decision.

24.6 RELATIONSHIP BREAKDOWN

A breakdown of the PPR can happen immediately after the honeymoon period or after a period of pleasurable use.

An early breakdown in the PPR can result from a mismatch between the user's expectations and the reality of the product. This can be due to misleading advertising or point of sale information or a lack of pre-purchase research on behalf of the user. The user may have formed misconceptions about many of the very same factors which appeared attractive during the first encounters. For example, the product may prove difficult to use, some of the features may prove to be of little practical use, it may prove unreliable or inefficient, or the use of the product simply fails to be as pleasurable as expected.

A later breakdown of the PPR can occur for various reasons, for example: when a newer, more attractive product is launched, when what seemed to be novel product features become annoying traits, or if there is a failure of the product.

The PPR might also be doomed to fail if the user was unable to contribute to the purchase decision, for example, if they were issued with a less than desirable mobile phone by their employer. The right to be a 'user-chooser' is in itself a pleasurable experience for most people.

When the PPR breaks down, it usually means that the product is either disposed of, or simply no longer used. But some users are forced to continue using products which they no longer like — perhaps because they cannot afford to replace them, or for other reasons.

24.7 THE ENDURING RELATIONSHIP

In contrast to a breakdown of the PPR, a prolonged relationship occurs when the needs of the user(s) are more fully satisfied by the product, and the passage of time does not reduce the pleasure obtained from the product. In some cases, the sense of pleasure can actually increase over time as the user gains confidence in the product and it becomes a familiar, trusted friend. Many people develop sentimental attachments to products which have given pleasure or have served them well.

The passage of time can also distort some negative effects into sources of pride and pleasure. For example, users of early versions of Microsoft Word learnt to work with a less than ideal user interface, yet when improved versions became available, many people were reluctant to upgrade because of the time and effort they had invested in learning the early version.

For many consumers, a product which is durable and which remains functional for longer than expected, provides the greatest pleasure. But products also remain satisfying if they can maintain a sense of engagement and fun, without being gimicky. As the user's needs, skills and abilities change over time, products which can be adapted or can be used in a flexible way, will also provide greater satisfaction and pleasure.

24.8 WHY SEEK TO PROLONG THE PPR?

A cynical observer might conclude that manufacturers are highly motivated to shorten the PPR. If their products remain in use too long, they will make fewer sales and less profit. It is true that marketing specialists put a great deal of thought into strategies which encourage purchases of new products earlier than is really necessary.
However, there are distinct advantages of prolonging the PPR:

- direct advantages to the user;
- indirect advantage to the manufacturer due to brand strengthening.

An owner or user of a product which has given long and trouble-free service is far more likely to be loyal to the brand (or even the same product) when the time comes to purchase a replacement. That person is also likely to be far more vocal in recommending the same product or brand to friends. As was mentioned above, metrics of customer satisfaction are increasingly reported in consumer magazines and other media, and the bad reviews can easily damage the brand image.

24.9 METHODOLOGIES FOR PROLONGING THE PPR

New methods are being developed to investigate the generation of positive feelings of pleasure from products, but a great deal can also be gained from adapting more established methods. For example, semantic differential scales, such as advocated by Osgood (1967), are successfully used to investigate pleasurable responses to products.

Essential to the process is the adoption of a user-focused approach to design, and the use of inter-disciplinary project teams. It is essential that the team works closely together and shares the goal of achieving a prolonged PPR, which delivers pleasurable responses well beyond the honeymoon period.

With this in mind, established methods must also be adapted to consider the various stages of the product life-cycle. Of particular importance is the ability to learn as much as possible from existing users of similar products and to understand how their PPR has evolved. The use of exploration techniques enables users to contribute directly to the innovative process by drawing upon their own experiences.

When testing new concepts it is essential to allow users to use the products in the appropriate scenarios and environment, and importantly, to arrange for prolonged trials whenever possible. Enabling testers to take a prototype home and use it for a week or more, provides far high quality and more reliable feedback than a one-hour test in a laboratory.

24.10 CONCLUSIONS

The design of products which deliver a positive sense of pleasure has become a goal for many product developers. Human factors specialists, who traditionally have focused on product usability, are now developing new techniques and adapting established techniques to provide scientific rigour to the process.

Perhaps the biggest challenge to product developers is, not only to enhance the pleasure of the PPR, but also to ensure that the pleasure is prolonged for as long as possible. There are direct advantages for the user, and indirect advantages — derived from brand strengthening — for the manufacturer.

One simple technique which has proven effective in predicting the effects of time, is to consider the PPR lifecycle in terms of an interpersonal relationship. This technique assists product developers to envisage how the needs of users will change over time, and which design features are likely to produce pleasurable responses at each stage of the PPR lifecycle.

Methodologies which are being used to produce pleasurable responses, must therefore be extended to consider the various stages of the PPR life-cycle. This involves capturing the PPR lifecycle experiences of existing users and enabling them to contribute to the innovation process. It also requires prolonged user testing within realistic scenarios and environments.

24.11 REFERENCES

Osgood, C.E., 1967. Semantic differential technique in the comparative study of cultures. In *Readings in the psychology of language*. Jakobovitis, L.A. and Miron, M.S. (Eds), Prentice-Hall, Englewood Cliffs, N.J., pp. 371-397.

CHAPTER TWENTY-FIVE

Comfort and Pleasure

DENIS A. COELHO

Department of Electromechanical Engineering,

University of Beira Interior, Cal ada do Lameiro, 6201-001 Portugal

and

SVEN DAHLMAN

Department of Human Factors Engineering, Chalmers University of

Technology, 41296 Göteborg, Sweden

ABSTRACT

The ambition of this paper is to position the concepts of comfort, pleasure and usability in relation to each other, using as a basis the theoretical framework on pleasure with product use. The theme for this chapter is the relationship between comfort and pleasure in product use. The starting point considered is that pleasure with products goes beyond usability. In this context, comfort can be considered as an aspect of both usability and pleasure. The distinction between comfort and pleasure, what differentiates them and also how and where these two concepts intersect, is one focus. The empirical background for this study comes mainly from the context of car seat comfort, where the authors have performed two experimental studies. A wider interpretation of the concept of comfort is also brought into the discussion, based on a study of office seating and work on human comfort. Furthermore, a small questionnaire study about pleasure and comfort related to cars and car seats was conducted. This provided data to illustrate the relationship of comfort with pleasure in the context of the evaluation of the design of automobiles and their seats. The studies performed allow us to suggest hypotheses about the interrelation between comfort, pleasure and usability. Implications for industry include the findings that interpretations of these concepts are overlapping and would benefit from more distinct understanding. Practitioners are working with and need these concepts in order to create good, worthwhile and successful products. A final section argues the role of human factors in product development, considering comfort and pleasure in the light of environmental sustainability.

25.1 INTRODUCTION

Comfort and pleasure are both concepts that are receiving growing attention as a possible means of adding value to products. Sawaki and Price (1991) reported on the Human Technology Project in Japan, showing that factors such as comfort, enjoyment (a synonym for pleasure) and usability have increased in relative importance as part of product quality. Usability-comfort enjoyment constitutes a new paradigm in the goal of product development, adding to the functionality-reliability-cost paradigm.

In historical terms, ergonomics' role in product development has shifted and widened over the last half of the 20th century. The history of human factors and ergonomics (Meister 1999) shows that in its early stages ergonomic intervention was meant to assure safety, health and performance for the users of products (e.g. the design of World War II aircraft cockpits, radar systems, etc.). This was followed by a stage where functionality was a goal of ergonomic studies in product development, initially through enabling increased performance and later by enhancing the usability of products. Comfort came as the next stage, although in some domains it might be regarded as an aspect of usability (e.g. software development). However, comfort may also be seen as an independent goal in itself, as is the case in car seat development. Finally, the emergence of pleasure as a goal in ergonomic product development completes the progression so far.

The boundaries between the different goals for human factors interventions are not clearly established, and a single intervention may of course accommodate several of these goals. Moreover, these boundaries may be subjective, or domain specific, creating variants of the distinction between these goals and concepts. As an example of this, usability may in some cases be seen as an aspect separate from performance. For other domains, usability may instead be seen as an aspect of performance.

25.2 AIMS, RATIONALE AND METHOD OF APPROACH

'This study aims at positioning the concepts of comfort, pleasure and usability in relation to each other, using as a basis the theoretical framework on pleasure with product use' (Tiger 1992; Jordan 1997).

The study also aims at positioning comfort in the hierarchy of user needs (Jordan 1997), which relates functionality with usability and pleasure (Figure 25.1).

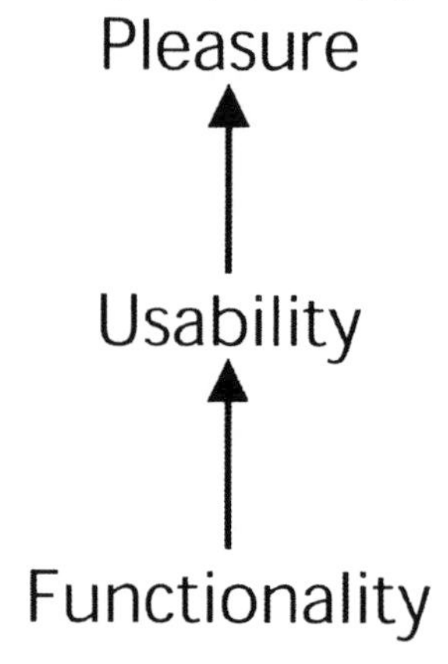

Figure 25.1 Hierarchy of user needs (fr. Jordan 1997)

A revision is proposed to this hierarchy, under the light of environmental sustainability. This aspect should not be forgotten in a review of the present role of human factors in product development, given the urgency of placing environmental concerns among the highest corporate priorities (ISO/DIS 14004). The focus of this study is on the distinction between comfort and pleasure and their intersection.

This being a first venture into a new area, it is our ambition to generate hypotheses about the interrelations of comfort, pleasure and usability, and later to test the hypotheses suggested with empirical data in studies to come.

This analysis is based on referenced definitions of comfort and pleasure. The case of pleasure and comfort in car use and car seats illustrates the relationships between these concepts, backed up by the results of a questionnaire.

25.3 LINKS BETWEEN COMFORT AND PLEASURE

Comfort and pleasure are linked and intersect each other as concepts, as can be extrapolated from the views and definitions presented below. Slater (1985) defined comfort as 'a pleasant state of physiological, psychological and physical harmony between a human being and the environment'. From an evolutionary point-of-view, comfort and pleasure, as well as pain and discomfort, have been receiving great attention by human beings since the birth of the species. Tiger (1992) defends this idea with the following argumentation: 'Our ancestors found comforting pleasure in nothing more complex than sitting by a fire and watching its ever varied motion. (...) our pleasures are as much related to our history as a species and products of it as they are products of our invention. (...) Ancient parts of the brain constantly monitor the comfort of the body and obviously seek to reject pain and seek pleasure'. It comes as no surprise, therefore, that contemporary human factors approaches to product development are giving greater emphasis to comfort and pleasure, adding up to previously established goals, i.e. safety and health, performance and usability.

Tiger (1992) distinguishes among four categories of pleasure: physiological, psychological, sociological and ideological pleasures. Jordan (1997) reinstated these and illustrated them with case studies of product use, defining pleasure with products as the 'emotional, hedonic and practical benefits of product use'. In a similar manner, comfort can be considered to have three different categories: physiological, physical and psychological (Slater 1985). While physiological pleasure is clearly linked to physiological and physical comfort, psychological comfort may be linked to psychological pleasure. However, sociological and ideological pleasures cannot be directly linked to comfort, except in the case where comfort is considered as an aspect of the quality of life (Maldonado 1991).

Zhang, Helander and Drury (1996) identified factors of comfort and discomfort in office sitting. The following descriptors of sitting comfort were brought forward in the study, based on a survey of 42 office workers:

agreeable,	at ease,	calm,	content,
cozy,	happy,	luxurious,	not think about workplace,
pleasant,	pleased,	plush,	refreshed,
relaxed,	relief,	restful,	safe,
softer,	spacious,	supported,	warm and well-being.

These descriptors of comfort reinforce the idea that comfort and pleasure are intersecting concepts. The hedonic benefit from comfortable sitting is conveyed by 'well-being, safe, pleased, pleasant, content'. Physiological pleasure can be linked to the descriptors that are related to physiological or physical comfort, such as 'cozy, plush, refreshed, relaxed, relief, softer, spacious, supported, warm'. Psychological comfort may be reflected in the terms 'agreeable, at ease, calm, happy, not think about workplace, restful'. Sociological and ideological pleasure are the most unlikely to be linked to the above descriptors, although the term "luxurious" might be thought of in such a context.

Table 25.1 Links between comfort and pleasure found from the definitions of Slater (1985), Tiger (1992), Jordan (1997) and from the study of Zhang, Helander and Drury (1996)

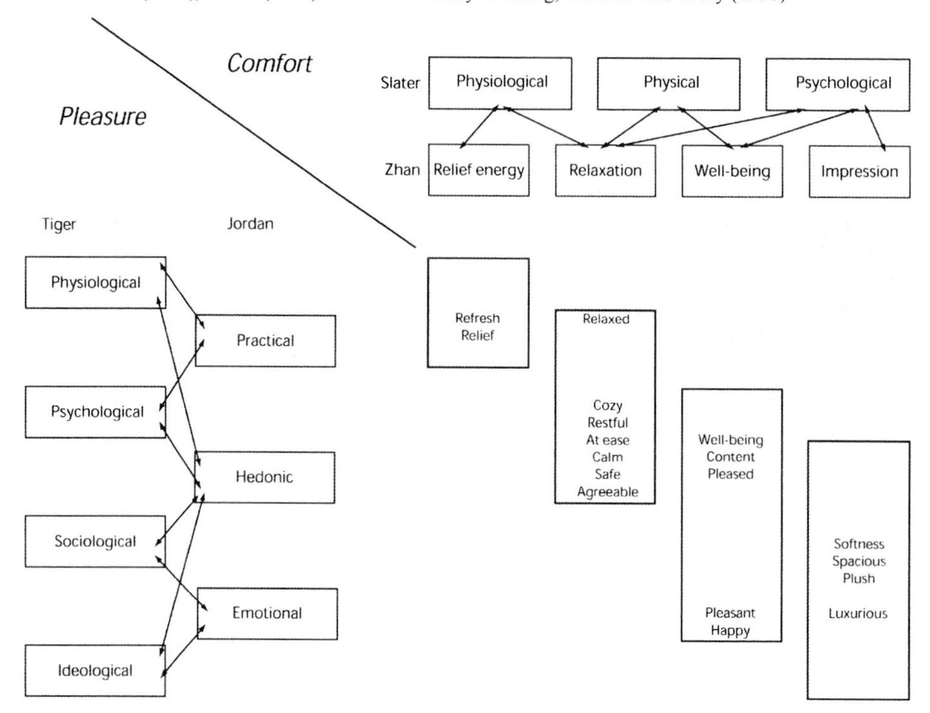

Table 25.1 describes a subjective interpretation of the authors, trying to bring together models of pleasure (Tiger 1992; Jordan 1997) and relating them to models of comfort (Slater 1985; Zhang, Helander and Drury 1996). Tiger and Jordan suggest ways of breaking down 'pleasure' into separate dimensions. Similarly, so do Slater and Zhang with 'comfort'. In Table 25.1, we have interpreted the definitions of the dimensions and tried to relate them within comfort and pleasure respectively, but also between comfort and pleasure. A frame indicates a suggested relation. The comfort descriptors described by Zhang have then been positioned in the comfort dimensions according to the results of Zhang, and by us subjectively in the pleasure dimensions. This exercise suggests that the descriptors of comfort derived by Zhang, Helander and Drury (1996) all seem to be relevant from a pleasure point-of-view. We further suggest that comfort is a constituting part of pleasure and it seems that pleasure holds dimensions not included in comfort: performance pleasures, skill pleasures, aesthetical pleasures, etc. The exercise also triggers the idea of performing an analogy to the Zhang study, but focused on pleasure, to more stringently grasp the dimensions of pleasure.

The exercise suggests that the boundaries between comfort and pleasure in product use are blurred and would benefit from more strict consideration in the light of a spectrum of empirical cases. Table 25.1 summarises the links found between comfort and pleasure. One may as well consider a similar exercise relating discomfort and displeasure in sitting, which is the discussion presented in the following paragraph.

We believe it is not a straightforward cause-effect relationship that discomfort in sitting leads to displeasure. An uncomfortable seat may not allow a pleasurable sitting experience, since sitting in that seat is an unpleasant experience, but the seat might be pleasurable in other aspects. The seat may have a nice looking design or a soft touch that would enable other pleasures (sociological), but not physiological pleasure, since physiological or physical comfort are absent. A distinction between pleasurable products and pleasure derived from activities can be made in this example. The seat (product), although pleasurable to look at and to touch, does not enable pleasurable sitting (activity). A negative emotional response to the physiological or physical discomfort occurring during the sitting activity may be present as physiological displeasure.

One can analyse displeasurable sitting and relate it to sitting discomfort, but can pleasurable sitting be related to sitting comfort? To distinguish between seats that are comfortable one must look at the pleasurable aspects of each seat. This idea is supported by Zhang, Helander and Drury (1996) who suggest that comfort-discomfort are two interrelated variables, although not a continuum. This has a bearing on the hypothesis that pleasure is something more than comfort. Seats can be comfortable in many different ways, as with different assumed postures, or with different seat cushion stiffness, or different seat fabric properties, or different seat contours, or different seat styling, aesthetics, and so on. Different seats (although comfortable) will enable distinctly different pleasurable sitting experiences, since pleasure can take the form of any or all of its four categories: physiological, sociological, psychological, and ideological.

25.4 INTEGRATING FUNCTIONALITY, USABILITY, COMFORT AND PLEASURE - THE CASE OF CAR SEATS

This section applies Jordan's (1997) hierarchy of user needs to an example of a product, i.e. car seats. Car seats are commonly evaluated in terms of comfort. In this section the argument is brought forward claiming that car seat comfort is built upon basic seat functionality and usability of the seat and its controls. The section presents arguments supporting the intersection of comfort with functionality, usability and pleasure in the case of car seats.

The comfort of a car seat depends on the characteristics of the seat and, in general terms, for a car seat to be comfortable it must provide functionality. This can be seen as adjustment features, such as height of the seat, reclining the backrest, adjusting the distance from the pedals, and so on. However, the inherent functionality of a car seat is also to support the occupant at ease in a driving posture. One can judge if it is easy to adjust the seat settings, and this would be part of evaluating the usability of the seat. But having easily adjustable settings is not the only way in which the seat can be easy to use. Being easy to use (e.g. its usability) also means it is easy to get in and out of the seat (egress/ingress characteristics), and that it is easy to use the seat in all that it is used for. Focusing the attention on the driver seat of the car, as there are more demands on its functionality than in the other seats in the car, the seat is used primarily for sitting while driving. Reynolds (1993) considers the car seat as an interface between the car and the driver. Therefore, besides the usability of the seat adjustments, being easy to use (the seat) means it is easy to drive, to see the road and to reach the controls while sitting in that seat. Ultimately, usability of the seat means it is easy to sit in or to stay seated in the seat while performing the task of driving. In other words, part of the usability of the driver's seat is that it is comfortable. We have also seen how the functionality of the seat is linked to comfort, since basic functionality, such as the adjustment possibilities available, has an impact on comfort in the seat.

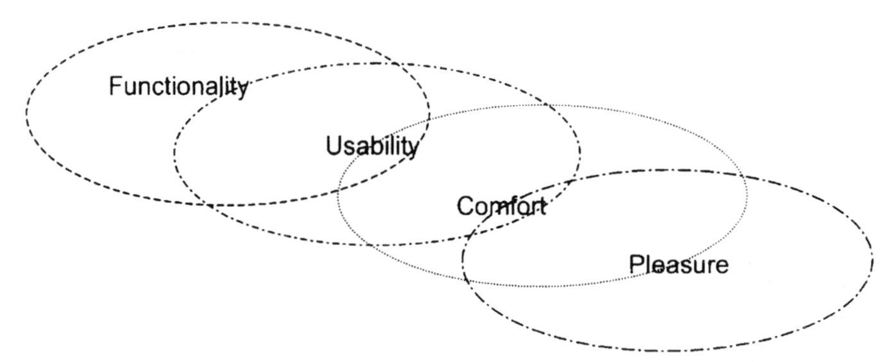

Figure 25.2 The blurred boundaries and intersections of functionality, usability, comfort and pleasure - the case of the driver's car seat

In the previous section, a study of comfort in office seating (Zhang, Helander and Drury 1996) led us to suggest that comfort could be seen as an aspect of pleasure. This analysis of the usability of the driver's car seat suggests that comfort may also be seen as an aspect of usability, defined as the effectiveness, efficiency and satisfaction with which users can achieve tasks with a product (ISO DIS 9241-11). Assuming that Zhang, Helander and Drury's (1996) descriptors of comfort can be applied to the driver's car seat, one concludes that functionality, usability, comfort and pleasure in the seat are concepts that intersect each other, and do not have strict boundaries (as shown in Figure 25.2). Furthermore, the element of satisfaction in the definition of usability may be linked to pleasure.

25.5 EMPIRICAL OBSERVATIONS

Comfort and pleasure have been discussed and links have been emphasised between these two concepts in the preceding sections, crossing the boundaries of functionality, usability, comfort and pleasure. A structured questionnaire was designed to test the applicability of Tiger's (1992) 'Four pleasure' classification and Jordan's (1997) 'Hierarchy of user needs' for driving, cars and car seats. The results of the questionnaire reveal evidence to support the links exposed between comfort and pleasure and the inter-relationships in the levels of user needs. The methods used and the results of the questionnaire are presented in the following paragraphs.

Thirteen Swedish automobile drivers answered the structured questionnaire, which had both open-ended and multiple-choice questions. The subjects were seven men and six women, aged from 30 to 65 years old, who had driven regularly for at least five years, driving between 5 and 20 thousand kilometers per year. The questionnaire administration was integrated with a larger research study on car seat comfort (Coelho and Dahlman 1999). A further selection criterion of the subjects was their sitting height. There were four subjects in each of the following sitting height intervals: 82-84cm and 92-94cm. There were five subjects in the 87-89cm sitting height interval. Each subject responded to the questionnaire during his or her leisure time, after performing two two-hour-long sitting trials in the laboratory, separated by approximately a week. A telephone number was indicated in the questionnaire header for support, however, none of the subjects called to clarify questions in the questionnaire.

Selected results of the questionnaire are presented in the following paragraphs. The question 'Do you find driving pleasurable?' yielded the following results:

'Not at all'	**8%**
'Somewhat'	**46%**
'Rather a lot'	**38%**
'A lot'	**8%**

These results show that the vast majority of the subjects considered driving as a pleasurable activity. This does not directly imply that the product that supports the activity (the car, the seat) is pleasurable. A suggested hypothesis (see above) is that the pleasure derived from the interaction with a product can either originate in the activity, or from the product qualities. In the case of car driving a combination of both seems to apply, as can be seen from the results of this question and those following.

To the question 'Do you think driving is hard?' subjects answered:

54%	**'Never'**
31%	**'In adverse conditions'**
15%	**'Driving is hard in adverse conditions, but it is pleasurable to succeed in doing it'**

The 15% of the total of 13 subjects who gave the last answer seem to derive a sort of psychological pleasure from completing a hard driving task in adverse conditions. This may lead one to suggest the hypothesis that pleasure with an activity may exist even though psychological discomfort is present.

The 13 subjects participating in the study were probed about the pleasurability of the interiors of their cars. This yielded the following results:

54%	**considered that the interior of their car was 'somewhat pleasurable'**
23%	**thought it was 'pleasurable'**
23%	**thought that it was 'not pleasurable at all'**
0%	**considered it to be 'very pleasurable'**

Although the subjects had cars of different makes, size and age, these answers show that pleasure seems to be a relevant attribute for the design of car interiors.

When asked to describe pleasurable attributes of their car, the subjects mentioned (number of subjects mentioning the attribute is indicated in parentheses):

- comfortable (5)
- performance (3)
- lovely road handling (2)
- nice looking design (2)
- safety (2)
- automatic transmission (1)
- gives me freedom (1)
- strong construction (1)
- well built (1)
- well equipped (1)
- quick (1)
- sportiness (1)
- big car (1)
- comfortable seats (1)
- good seats (1)
- cozy (1)

This question had open-ended possibilities for the subjects' answers, which might explain the spread in the results. Still, the answers show that 5 of the 13 subjects answering the questionnaire considered that their car was comfortable and that this contributed to make it pleasurable. This result supports formulating the hypothesis that comfort enables pleasure. It also shows that people understand the separation between a pleasurable activity (driving or socialising in the car) and pleasurable attributes.

Subjects were asked if it was a pleasurable experience to socialise with others in their car, yielding the following results:

54% **'Yes, always'**
38% **'Yes, on relaxed occasions'**
8% **'No, never'**

These results show that, for the vast majority of the subjects, pleasure can be derived from the company of others in the car, thus falling into the category of sociological pleasure.

The subjects were also requested to indicate the persons whose company they appreciated in the car, with the following answers:

77% **enjoyed the company of their spouses while driving**
69% **enjoyed the company of their children while driving**
62% **enjoyed having their friends while driving**
31% **enjoyed having their colleagues while driving**

(Not all of the 13 subjects were married or had children living with them and there were two retired persons in the group.) By enabling the presence of others in the car (a functionality attribute) sociological pleasure can be derived from using the vehicle (product). This is an example of how pleasurable interaction may be built directly on basic functionality.

The subjects were asked if their car reflected their ideological values, yielding the results presented below:

38% **Thought that their car did 'not' reflect their ideological values**
23% **Answered that their car did 'somewhat' reflect their ideological values**
38% **Replied that their car reflected their ideological values 'rather a lot'**
0% **Answered that their car reflected their ideological values 'a lot'**

One of the subjects who answered that her car reflected her ideological values rather a lot added a comment to the answer, saying that her car was meant to carry her and her spouse and their things and food, in an economic, quick, comfortable and safe manner. The wording of the question and its results support suggesting that identification with the product derives from a match between users' requirements on the product and its affordances.

The subjects were asked to rank seven attributes of car seats in order of importance to them. The results are shown below (most important - top of the list):

- The seat is comfortable.
- The seat has the right adjustment possibilities.
- The seat cover has a soft touch.
- The seat looks good.

- The seat cover is beautiful.
- The seat cover does not need to be washed.
- I look good when I sit in the seat.

The combined answers of the 13 subjects have a Kendall coefficient of concordance of 0.83, significant for $p<0.001$. Comfort of the seat ranks highest in the list, followed by the adjustments available (which can be considered an aspect of functionality or usability of the seat). Physiological (tactile) pleasure as well as physiological comfort may be derived from the soft touch of the seat cover, which comes next. The aesthetic qualities of the seat rank in fourth and fifth place, these are necessarily pleasure-linked attributes that may lead to sociological pleasure once others recognise the beauty of the seat. Attribute six may be connected to functionality and usability of the seat. Finally, the lowest ranking attribute may also be considered an aesthetic quality, leading to sociological pleasure.

The results of the questionnaire show the relevance of the 'four pleasure' framework to the specific products considered (cars and car seats). Furthermore, subjects' answers show that pleasure can be derived from the activity of driving and the products that support it. Comfort is seen as one of many possible attributes that contribute to a pleasurable car. Functionality and usability attributes are also present in the results of the questionnaire. These results, and the conclusions of the first part of this section, support revising the model of the hierarchy of user needs (Jordan 1997) to encompass the aspect of comfort. Figure 25.3 depicts the suggested adaptations to the model.

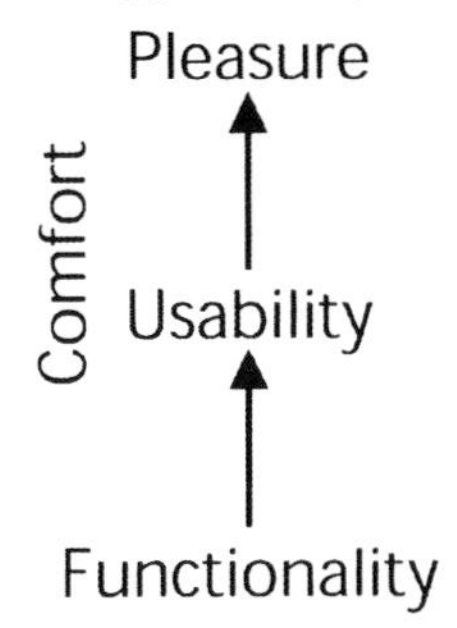

Figure 25.3 Hierarchy of user needs, adapted from Jordan (1997) to encompass the Human Factor goal of comfort, in its relationship with pleasure, usability and functionality (based on the analysis of the case of car seats)

25.6 SUSTAINABILITY AND USERS' NEEDS

The goal of human factors intervention and its role in product development is considered to be supporting the development of products, which are functional, user-friendly and pleasurable (Jordan 1997). Although for many products there is still a long path ahead towards achieving pleasurable interaction, a reflection and argumentation can be made concerning the importance of the users and developers' role also in attaining environmental sustainability. Environmental planning is encouraged throughout the product life cycle (ISO 14004), and it is generally accepted that moderate and wise use of the planet's resources is a pre-condition to assure a sustainable environment. As well as the need to give users functional, user-friendly, and pleasurable products, products might also have an embedded appeal which would encourage suitable and sustainable use and promote discarding in an environmentally friendly way, e.g. recycling.

Above the pleasurable level in Jordan's (1997) model of user needs (shown in Figure 25.1) one should consider the ceiling of sustainability which helps to set designers and users on the right track. That means putting users of products in a perspective that stresses global awareness and what their role is in preserving the environment and assuring a sustainable life style for the coming generations.

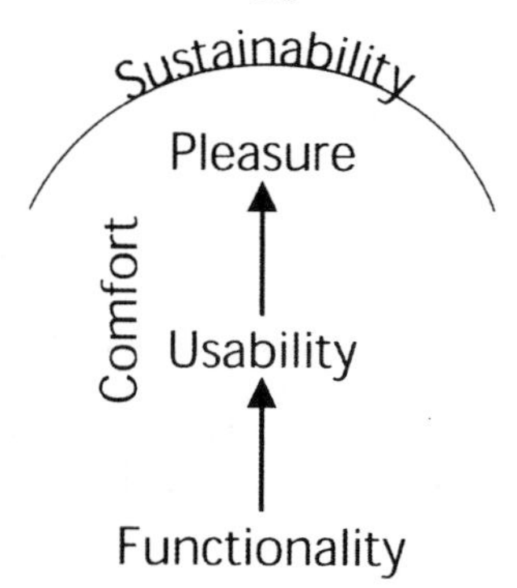

Figure 25.4 Model of the hierarchy of user needs, adapted from Jordan (1997), considering comfort and limited by sustainability

The role of human factor specialists in this enterprise is to assure that the emotional responses users attain through interacting with products are ones that will be good for us as individuals, for society, for humanity, and for the planet. The immediate benefit is environmental sustainability. Jordan's (1997) model can thus be further adapted to encompass this sustainability ceiling in the hierarchy of user needs, as shown in Figure 25.4. Further studies are needed to understand and explore the impact of this limitation on product development and users' interaction with products.

25.7 CONCLUSION

The exercise of linking comfort and pleasure suggests that comfort fits nicely into being an aspect of pleasure, and that the boundaries between comfort and pleasure in product use are overlapping and specific to the context and situation. It seems that pleasure holds dimensions not included in comfort: performance pleasures, skill pleasures, aesthetic pleasures, to mention a few. The ambition to clarify and define the overlapping and intersection of these two concepts should be pursued as a more strict consideration in the light of a spectrum of empirical cases. Table 25.1 summarises the links suggested between comfort and pleasure, while Figure 25.2 gives a graphical interpretation of the overlapping boundaries of these concepts and functionality and usability.

Taking as a basis Zhang, Helander and Drury's (1996) suggestion that comfort-discomfort are two interrelated variables, although not a continuum, the authors suggest that different seats (although comfortable) may enable distinctly different pleasurable sitting experiences, since pleasurable can take the form of any or all of the categories considered by Tiger (1992): physiological, psychological, sociological and ideological. In this context one could consider the relation between discomfort and displeasure in sitting, since a seat that is pleasurable to look at and to touch may not enable pleasurable sitting.

This study also suggests that part of the usability of the car driver's seat is that it is comfortable. In the same line of thought, seat functionality can be linked to comfort, since basic functionality has an impact on comfort in the seat. Furthermore, basic functionality of the automobile (accommodating other occupants besides the driver) is a pre-condition for sociological pleasure, derived from the company of others in the car.

A suggested hypothesis resulting from the analyses presented here is that pleasure may either derive from the interaction with a product (given the product's qualities), or it may originate in the activity supported by the product. Other hypotheses were formulated, based on empirical data and the analyses conducted. These included the hypothesis that identification with the product derives from a match between users' requirements and ideological values and the product's affordances.

Finally, this contribution revises Jordan's (1997) hierarchy of user needs, considering the addition of comfort and in the light of environmental sustainability. The result is shown in Figure 25.4. Further studies are needed in order to understand and explore the impact of the limitation of sustainability on the development of pleasurable products and on users' interaction with products.

Implications of the study for industry include the findings that the boundaries between functionality, usability, comfort and pleasure may be indistinct for most products. This urges practitioners to look both at and beyond each of these concepts in order to create good, worthwhile and successful products.

25.8 ACKNOWLEDGEMENTS

This study was conducted under the auspices of the Department of Human Centred Technology (its name has since changed to Dept. of Human Factors Engineering), Chalmers University of Technology, Göteborg, Sweden. The lead author's role was funded in part by Fundação para a Ciência e a Tecnologia, Lisboa, Portugal, with grant no. PRAXIS XXI / BD / 16059 / 98.

25.9 REFERENCES

Coelho, D.A. and Dahlman, S., 1999. Articulation at Shoulder Level - A Way to Improve Car Seat Comfort? Technical report - University of Beira Interior, Covilhã, Portugal and Chalmers University of Technology, Göteborg, Sweden, 159 pp.

ISO/DIS 14004, 1995. Environmental Management Systems – General Guidelines on Principles, Systems and Supporting Techniques.

ISO/DIS 9241-11, Ergonomic Requirements for Office Work with Visual Display Terminals (VDTs) - Part 11: *Guidance on Usability.*

Jordan, P.W., 1997. A Vision for the Future of Human Factors. In *Proceedings of the HFES Europe Chapter Annual Meeting 1996.* Brookhuis K. *et al.* (Eds), (University of Groningen Centre for Environmental and Traffic Psychology), The Netherlands, pp.179-194.

Maldonado, T., 1991. The Idea of Comfort. In *Design Issues*, **8**, 1, pp. 35-43.

Meister, D., 1999. "The History of Human Factors and Ergonomics", Lawrence Erlbaum Associates, Mahwah, New Jersey, USA, 382 pp.

Reynolds, H.M., 1993. Automotive Seat Design for Seating Comfort. In *Automotive Ergonomics.* Peacock, B. and Karkowski, W. (Eds), Taylor & Francis, London, UK.

Sawaki, Y. and Price, H., 1991. The Human Technology Project in Japan. In *Proceedings of the Human Factors Society 35th Annual Meeting - 1991*, Santa Monica, California, USA, pp. 1194-1198.

Slater, K., 1985. *Human Comfort.* Springfield - Illinois: Thomas Books, USA, pp. 3-11.

Tiger, L., 1992. *The Pursuit of Pleasure.* Little Brown and Company, Boston, USA, pp. 52-60.

Zhang, L., Helander, M.G. and Drury, C.G., 1996. Identifying Factors of Comfort and Discomfort in Sitting. In *Human Factors*, **38**, 3, pp. 377-389.

CHAPTER TWENTY-SIX

Collecting Stories on User Experiences to Inspire Design – a Pilot

ANU MÄKELÄ

Human Factors Researcher, Helsinki University of Technology,

PO Box 5900, FIN - 02015 HUT, Finland

and

TUULI MATTELMÄKI

Industrial Designer, University of Art and Design Helsinki, c/o Helsinki

University of Technology, PO Box 5900, FIN - 02015 HUT, Finland

26.1 INTRODUCTION

In this paper a pilot of a user study method is presented. The method is a combination of two field inquiry techniques, collecting stories and self photographing.

The focus of the user study was on users' experiences with mobile phones. It was assumed that by studying the users' experiences on current mobile devices, new and interesting phenomena could be found to inspire future design of mobile communication applications. This assumption was tested by presenting the essential findings of the user study in the form of storyboards to a group of designers working for a company developing mobile phones.

26.2 USER STUDY

In recent years designers (Gaver, Dunne and Pacenti 1999; Erickson 1996) have developed their own user study methods, to be used at the beginning of a design process. They have felt that methods adopted from social sciences have not succeeded in serving their need for an explorative and playful design process.

For example, Cultural Probes (Gaver, Dunne and Pacenti 1999) was developed by designers to collect inspirational data on the user's beliefs, desires, aesthetic preferences and cultural concerns. The designers did not want an objective view of the users' task-based needs and thought that methods adopted from social sciences would provide only that kind of view.

The piloted user study method does have a background in social sciences, but the aim has been to modify it to make it suitable for explorative design work that does not focus only on the cognitive or task-based aspects of user experience.

26.3 COLLECTING STORIES

Background

The story collection technique used in the pilot is mainly based on Flanagan's (1954) 'critical-incident technique'. Flanagan's technique is assumed to allow an interviewer to get many stories of real situations and behaviours from users in a very short time (Hackos and Redish 1998). Each interviewee is asked to tell about a critical incident they recall related to a phenomenon under study. After getting enough information about the incidents, the interviewees were asked to recall another incident, and so on.

Erickson (1996) thinks that stories have not been respected as a design tool because they are subjective, ambiguous and particular. In other words, they do not provide objective, generalisable and repeatable findings. In Erickson's opinion, in part because of this conflict, stories are good design tools. They are memorable and their informality is suitable for design-related knowledge that often is uncertain.

Social scientists have also noted the value of stories in doing research in their field. Some of them have collected people's stories in order to understand, for example, how adults see their lives (McAdams 1993), or how they have recovered from addictive behaviours (Hänninen and Koski-Jännes 1999). People's stories can give meaning to a phenomenon that a researcher is studying. The stories are not only a review of an event, but include also causal, emotional, moral and ethical concerns related to the event (Hänninen and Koski-Jännes 1999).

Pilot

Altogether 21 mobile phone users were interviewed in groups. The interviews were done in a commercial exhibition of computers and home computing, Kotimikro messut, in Helsinki, Finland. Those interviewed were exhibition visitors. Their ages ranged from 17 to 68 years.

One group interview lasted around 30 minutes, and all interviews were finished in three hours. The interviews were video-recorded to avoid too much writing down during the discussion. At the end of the interview the participants were given tickets for the cinema as a reward for their time.

In an interview session a group of people (from two to four persons) who were friends or members of the same family were asked to tell stories about their experiences in using their mobile phones. The interviewees were given a pile of pictures the size of postcards to view freely during the interview. The pictures were selected by the authors from magazines, comics and postcards, and were both abstract and representative.

Although no projective tools have been recommended for use with Flanagan's (1954) technique, it was thought that the pictures would prompt the participants' storytelling. Indeed, the interviewees made associations from the pictures, and started to tell about their experiences. The same picture raised different associations, depending on the person. (see Figures 26.1 and 26.2).

Figure 26.1 A picture used for prompting storytelling

This Christmas postcard (originally red) prompted several different stories, like: 'I can call relatives abroad at Christmas with my mobile', 'I have a red cover on my phone, because I like the colour red', 'At Christmas everyone places their phones in a row on the table'.

Figure 26.2 Another picture used for prompting storytelling

This postcard (orginally black and white) prompted several different stories, like: 'I use text messages, because it is the easiest and cheapest way, if something just pops into your mind and you don't want to call. It's like email or a post card but gets there faster', 'Coloured covers and ringing melodies are personal'.

26.4 SELF PHOTOGRAPHING

Background

The designers (Gaver, Dunne and Pacenti 1999) who developed Cultural Probes, and Maypole human factors researchers (Maypole project team 1999), have used self photographing in their user studies. Using a disposable camera, users are asked to take photographs of certain activities, places, things or people in their lives.

Social scientists have used self photographing in a similar way. They have asked subjects to take photographs that represent their attitudes and behaviours, or in other words personal orientations, in certain contexts. The motivation for using self photographing has been to avoid the problem of influencing subjects while observing them (Ziller 1990).

Pilot

Self photographs that had been taken in previous user studies done by the authors were used in the pilot for illustrating user environments. The photographs were suitable for the purposes of the pilot since they were about people, places and things related to communication among family members and friends. The photographs had been taken by children, teenagers and adults with disposable cameras in the Finnish capital Helsinki in 1998-1999.

26.5 SELECTING IMPORTANT AND INTERESTING STORIES

Background

The user data gathered with the Cultural Probes was not analysed in a systematic way, or summarised and communicated formally. The designers reflected the ideas they got by sorting through the raw material directly in their design proposals (Gaver, Dunne and Pacenti 1999).

However, documentation of essential findings is often needed because today's products are developed by multidisciplinary teams, whose members change and are located in different places.

Pilot

After the interviews in the exhibition the users' stories were written down on the basis of the notes and video recordings. The authors selected and combined the most important and interesting stories to be presented to the designers in the workshop. The selection was based on intuition.

26.6 COMBINING THE USER'S STORIES AND PHOTOGRAPHS INTO STORYBOARDS

Background

Storyboards are one way of documenting and communicating the essential findings of a user study. They are visualised scenarios that include a story of user activity in certain context. The main purpose of developing scenarios is to stimulate thinking (Jarke, Bui and Carroll 1998; Black 1998), and communicate user research results (Erickson 1995; 1996) and new designs (Moggridge 1993; Beyer and Holtzblatt 1998) to design team members, other stakeholders, and users.

Storyboards are good tools in design since designers can transfer abstract requirements into scenarios, and use scenarios to create solutions. In other words, scenarios help people to think about abstract things with concrete examples. Moreover, they are good for starting the design from a concrete point of view (Gedenryd 1998).

Pilot

25 specific stories or comments on mobile phone use experiences were visualised. The storyboards were made in a sketchy manner using the self photographs in the background (see examples in Figures 26.3 to 26.6).

Figure 26.3 'Without a mobile phone it is really hard to set up a date or meeting in town. You can't call if you are late. Meeting times or places are often agreed or changed ad-hoc, which can be both a good and bad thing.'

Figure 26.4 'On Christmas Eve everybody put their mobile phones on the same table in a row. Everybody has chosen their own ringing melody. Also the colour covers were personal. The same thing happens at parties with friends.'

Figure 26.5 'The mobile phone is a really clever invention. You don't have to carry separate games with you, especially when you are bored with waiting for a bus or train to come. It would be cool to combine a mobile phone with TV, radio or a mini disc player...'

Figure 26.6 'Often the phone rings, and you can't find it in your bag. My family calls my handbag 'the black hole.' Sometimes when rushing in a noisy public place you can't hear your phone ringing. It is irritating, but sometimes it is good depending on who had tried to call.'

26.7 WORKSHOP WITH DESIGNERS

Presenting the user study findings

Five mobile phone designers from Nokia Corporation were invited to a half-day workshop. They were shown the storyboards with an over-head projector. The story related to the storyboard was read aloud with the spoken language the users had used in the interviews.

Discussion about the findings

After that the content of the stories was discussed. During the discussion eight user experience categories were mentioned:

- **Feeling safe.** Having a mobile phone creates a feeling of safety
- **Playing**. People use mobile phones to kill time by sending short text messages or by playing games.
- **Personalising**. People choose the colour and/or melody of their own phone.
- **Being in contact**. It is important to maintain contact with family and friends from a distance. Short text messages are being used in places where speaking is not possible.
- **Being together**. Mobile phones are being kept visible by placing them on the same table. This happens when a family or group of friends gathers together to celebrate a special occasion or to party.
- **Annoying.** Other people's mobile phone use can be annoying. The phones are ringing everywhere, and people talk – often too loud – everywhere on their phones.
- **Learning to use.** For the elderly starting to use a mobile phone depends on learnability. Also, being able to see a small screen and press small buttons is crucial for them.
- **Carrying.** Women have problems finding their mobile phone in their handbags. Moreover, carrying the mobile phone when doing sports is not so easy or convenient.

Brainstorming and sketching

The next task was to brainstorm new design ideas on the basis of the stories presented. The designers were given papers and pens, and time to create ideas. They worked individually for a while by sketching and writing down, and after that the ideas were discussed in the group.

The designers were familiar with most of the presented user study results since their company has done mobile phone user research. However, there was one totally new phenomenon, 'placing mobile phones on the same table in a row when gathering together with family or friends' (see Figure 26.4) that really inspired them.

The ideas that were generated from this finding were mostly about designing a stand for a group of phones (see Figures 26.7 and 26.8) but there were also other explorations, such as competition among phones, phones socialising (Figure 26.9), and a phone band (Figure 26.10). However, the designers would have liked to know more about the discovered phenomenon. They were interested in why people leave their phones on the same table – because of comparison or as a sign of being together?

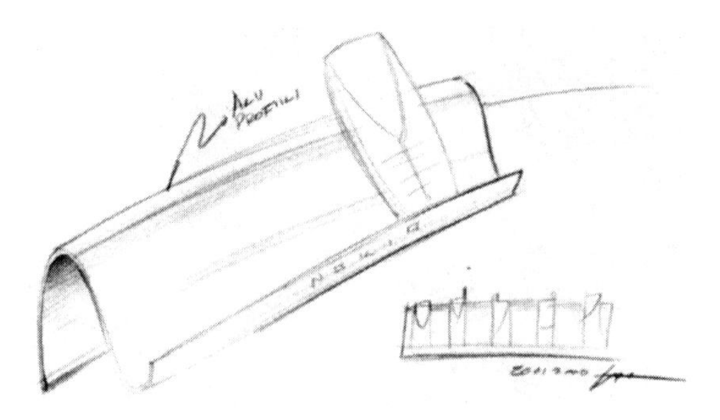

Figure 26.7 A group stand for mobile phones

Figure 26.8 A group stand for mobile phones. A phone represents its owner

Figure 26.9 Mobile phones socialising

Figure 26.10 A mobile phone band (= playing their ringing melodies in synchronisation)

Two other findings inspired the designers. First, a story about a woman who always had difficulty hearing her phone ringing, and finding it in her handbag. The woman's family called the handbag a 'black hole'. This inspired the designers to create a mechanism to clip or attach the phone to the inside of the handbag, in order to keep it from dropping into the 'black hole' (see Figure 26.11). Second, two stories about how lovers send text messages gave ideas for creating new solutions for romantic communication.

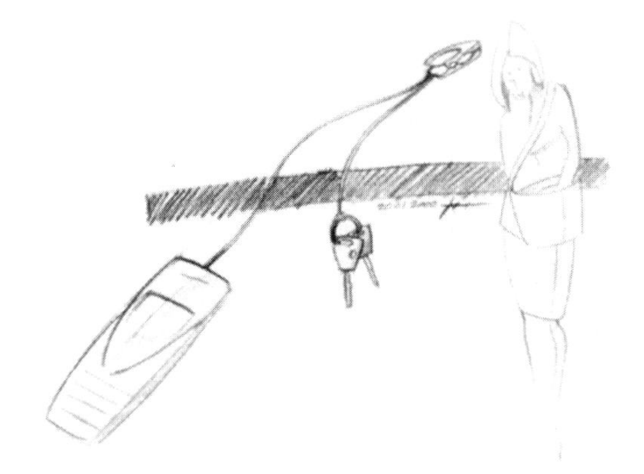

Figure 26.11 A mechanism to clip or attach the phone to the inside of a handbag in order to keep it from dropping into the 'black hole'

Conclusions

A promising method to be used in parallel with other methods

The user study method described as used in the pilot provided inspirational user data from many perspectives of user experience, such as play, safety, learning, being together, and annoying things. Nevertheless, it took only three hours to conduct the interviews plus collecting the self photographs.

The workshop indicated that the results gathered with the method were similar to those that the designers were familiar with from studies done in their organisation. There were also findings that really inspired them to think about new concepts.

The method did not provide the insights needed to do more detailed design than was done in the workshop. Instead, it helped to discover interesting phenomena for starting the design, and identified issues that could be investigated further with formal user study methods, such as Contextual Inquiry (Beyer and Holtzblatt 1998).

The method also mainly provided findings on experiences that were critical or significant for the users. Therefore, other user research methods that are used for gathering temporal, everyday data, such as diaries (Nielsen 1995; Sanders and Dandavate 1999) or shadowing (Sachs 1993), are recommended to be used in parallel with it.

Using the self photographs as a background for the storyboards helped to illustrate user environments. However, it is important to note that the method described in this paper is only one way of using self photographs in user research. For example, users could be asked to make collages that would include their self photographs and other material. Sanders and Dandavate (1999) call such collages 'Make Tools' that show or tell users' stories and dreams.

Difficulties in communicating the user needs to the designers

Presentation technique

One of the drawbacks in the pilot was the way in which the user's stories were presented to the designers. According to the feedback from the designers, the users in the visualisations should have had stronger characters, and there should have been less users in less contexts. As it was, it was difficult to identify with the users. The visualisations should have included two or three more pictures, according to the designers.

Moreover, the presentation technique with overhead projector and reading aloud was not on the same level of quality that the designers were used to in their own organisation. As well as multimedia presentations, modelling user environment and role playing could also have been helpful (see e.g. Iacucci *et al.* 2001).

Different approach

In order to gain a common understanding of the user needs it would be useful to discuss them with designers. Finding a common language requires intensive collaboration and an understanding of different approaches.

In the workshop, the focus of the designer' discussion, brainstorming and sketches was mainly on solutions (see Figures 26.7 and 26.11). The atmosphere of the workshop was similar to Cross's (1995: p. 107) view of the design work:

- produce novel, unexpected solutions;
- tolerate uncertainty, working with incomplete information;
- apply imagination and constructive forethought to practical problems;
- use drawing and other modelling media as means of problem solving.

Only one of the designers was able to visualise and discuss user needs (see Figures 26.8 to 26.10) instead of solutions, but as he said: 'I have been working more than others with psychologists and sociologists'.

26.8 ACKNOWLEDGEMENTS

We would like to thank the users who participated in the pilot. We are also grateful to the five designers Tapani Jokinen, Turkka Keinonen, Totti Helin, Anna Valtonen, and Juha Hemanus from Nokia Corporation. Turkka Keinonen helped us also in writing the paper. Many thanks also to our dear colleague Katja Battarbee for helping us in collecting the stories, and in writing the paper in english. Jarmo Parkkinen, Sari Kujala, Professor Kari Kuutti and participants of his post-graduate student seminar also gave constructive feedback on the paper. The pilot and the paper were done as a part of eDesign research project funded by the Academy of Finland.

26.9 REFERENCES

Beyer, H. and Holtzblatt, K., 1998. *Contextual design.* Defining customer-centred systems. Morgan Kaufmann Publishers, Inc.

Black, A., 1998. Emphatic design. User focused startegies for innovation. In *Proceedings of New Product Development.* IBC Conferences.

Cross, N., 1995. Discovering design ability. In *Discovering design - explorations in design studies.* Buchanan, R. and Margolin, V. (Eds), The University of Chicago Press, p. 107.

Erickson, T., 1995. Notes on design practice: Stories and prototypes as catalyst for communication. In *Scenario-based design. Envisioning work and technology in system development.* Carroll, J.M. (Ed.), John Wiley & Sons Inc, NY, pp. 37-58.

Erickson, T., 1996. Design as Storytelling. In *Interactions*, July and August, pp. 31-35.

Flanagan, J.C., 1954. The critical incident technique. In *Psychological Bulletin*, **51**, 4, pp. 327-358.

Gedenryd, H., 1998. *How designers work – making sense of authentic cognitive activities.* Jabe Offset AB.

Gaver, W., Dunne, T. and Pacenti, E., 1999. Cultural probes. In *Interactions*, **4**, 1, January and February, pp. 21-29.

Hackos, J.A.T. and Redish, J.C., 1998. *User and task analysis for interface design.* Wiley Computer Publishing.

Hänninen, V. and Koski-Jännes, A., 1999. Narratives of recovery from addictive behaviors. In *Addiction*, **94**, December.

Iacucci, G., Mäkelä, A., Ranta, M. and Mäntylä, M., 2001. Visualizing context, mobility and group interaction: Role games to design product concepts for mobile communication. In *Proceedings of Fourth International Conference on the Design of Cooperative Systems*, (COOP'2000).

Jarke, M., Bui, T. and Carroll, J.M., 1998. Scenario management: An interdisciplinary approach. In *Requirements Engineering*, **3**, pp. 155-173.

McAdams, D.P., 1993. *The stories we live by.* Personal myths and the making of the self. The Guilford Press.

Maypole project team, 1999. What makes kids tick? In *Interactions*, **6**, 6, November and December.

Moggridge, B., 1993. Design by story-telling. In *Applied Ergonomics*, **24**, 1, pp. 15-18.

Nielsen, J., 1995. Scenarios on discount usability engineering. In *Scenario-based design.* Carroll, J.M. (Ed.) Envisioning work and technology in system development, John Wiley & Sons Inc., NY, pp. 59-83.

Sanders, E. B.-N. and Dandavate, U., 1999. Design for experience: New tools. In *Proceedings of the first international conference on Design and Emotion*, pp. 87-92.

Sachs, P., 1993. Shadows in the soup: Conceptions of work and the nature of evidence. In

The Quartely Newsletter of the Laboratory of compartive Human Cognition, **15**, 4, pp. 125-133.

Ziller, R.C., 1990. *Photographing the self.* Methods for observing personal orientations. SAGE Publications Inc., London.

CHAPTER TWENTY-SEVEN

Usability Perception

RICCARDO EMMANUELE and BARBARA SIMIONATO

NCB Design, Via Leopardi 10, 21047 Saronno – VA, Italy

PREMISE

The Italian industrial context presents some peculiarities, well known to all the international economic culture. It is made up of a large number of small- and medium-sized companies, highly specialised in their marketing fields. This specific characteristic of Italian manufacturing has promoted a very special approach to industrial design, strongly based on continuous research and experimentation, both in design and in technology. In addition to extreme attention to detail and finishing in product manufacturing, (derived from an historical handicraft tradition) what really determines the success of Italian production abroad are the skills and talents of Italian designers who, since the end of the second world war, have been able to establish a prolific relationship with the manufacturing industry.

This explains, to some extent, the important role designers play in Italian industry today. They are not simply art directors but are sometimes also directly involved in the management and strategic definition of a company. Designers are considered as demiurgic figures, able to perceive and read new social and economic trends, to suggest new ideas and solutions, without giving up the promotion of original and individual research. The confidence of the companies in the designer's capabilities has prevented a broad diffusion of design theories and methodologies based on a user-centred approach. In fact, apart from the big companies operating in the global market with conspicuous human and economical resources, Italian manufacturing companies ignore the important implications of user-centred design techniques and methods in order to innovate their products. That is why in Italy it is very difficult to offer and sell services such as product testing or explorative and comparative usability research founded on user-centred design techniques, although all the marketing operators are aware of the increasingly high quality performance required by the market. Given these preliminary remarks, it is clear that the most sustainable way to promote usability and user-centred design techniques inside the Italian manufacturing context is as follows:

- to emphasise the competitive advantages obtainable from usability-based approaches;
- to supply the industry with effective, fast and low-price consulting services;
- to collaborate synergically with designers or design staff.

Based on awareness of the real operative conditions existing in Italian industry, the study conducted by NCB design represents a concrete contribution towards fostering the diffusion and applicability of user-centred design techniques.

27.1 A THEORETICAL APPROACH TO USABILITY PERCEPTION

Usability has become a well-known concept in design and manufacturing contexts, and in practice it is often seen as a method of examining and testing new products before they are launched on the market. Although ISO 9241-11 has established a theoretical and practical frame for usability, by codifying key-words and suggesting methods and procedures by which to measure it, it must be said that usability is not a readily standardisible concept. Its meaning can be construed differently according to the context in which is it placed. Scrutiny of the literature makes it possible to realise how many different approaches actually characterise the research in this field. A more important issue in recent usability studies and research is the distance which exists between the theoretical approach and the practical implications. That is, the difficulty in translating effective theoretical methods based on usability into very practical and operative tools with which to develop, and test products. In these terms the success of quick-and-dirty methodologies in evaluating the usability criteria for a product can be appreciated. Starting from these considerations, the research on usability perception carried out by NCB design represents an attempt to consider carefully the Italian operative (cultural and economic) context in order to give practical and effective answers to the issues of usability tests and evaluations.

Usability perception: a definition

The present research starts from the assumption that it is possible to make reference to usability as a perceived value; that is, it is possible to think about usability as an implicit and ingrained quality of a product. According to this premise, the purpose of this study was to define an effective tool for evaluating usability as a perceived value, without referring to a structured usability test or evaluation in which the user is asked to test and manipulate the product in a defined context, with precise tasks but simulating the more real situation of a potential user/customer in front of an object/product. This point-of-view is closer to the market condition in which a potential customer has to make a choice between similar products without testing them all.

Is it possible to believe that, even before actually using a product, a potential user can predict if he/she will have trouble with it, or will get benefit or pleasure from it?

What features will motivate a potential user to address his/her attention towards a product, making him/her presume it would be functional and easy-to-use?

It is evident that design features play a prominent role in motivating a subject to estimate the functional and usability values of a product; shapes, colours, formal features suggest to a subject the object's functional and usability features, even without conducting a deeper examination of the object itself. In a subject/object relationship, first the subject elaborates an unconscious perceptive strategy to verify if the object, by its design and formal features, might bring him some emotional benefits (a certain pleasure in use); afterwards he focuses his attention on the functional and usability characteristics. Therefore, because of the importance of the emotional aspects which motivate a subject to express a judgement on a product, it is worth considering usability as a value resulting from a perceptual approach, a perceptual experience with the product. Perceptual in this context means the subject's capability of interpreting data and information extracted from reality by his/her sensorial apparatus; every subject, once they have gathered all the possible data and information related to an experience, elaborates the perceived stimuli, gives a meaning to that situation and finally expresses a value about the experience he lived (Pedon 1987).

Every perceptive experience with an object could be divided into three different phases:

- *Visual phase*: the subject looks at the object and pauses in front of those elements which have attracted his/her attention either for the shape or for the colour;
- *Tactile phase*: the subject touches the parts of the object which have attracted his/her curiosity due to materials, shape and features;
- *Test or simulated use phase*: the subject tries the object in order to understand how it works, whether it is functional, handy, easy to use; he tries to verify if the expectations he rapidly formulated in his mind can be satisfied.

These aspects, drawn from research on cognitive psychology, constitute the basis of the present work.

27.2 METHODOLOGY FOR PERCEIVED USABILITY TESTING

The three psychological assumptions we began with define the individual based on how he/she relates to the world, i.e. how he/she interacts with the external impulses.

Those three assumptions are:

- The individual elaborates the information.
- The interpretation of an impulse by an individual depends on
  (i) the characteristics of the object.
  (ii) the previous expectations of the individual.
  (iii) the terms of comparison.
- The individual arranges his/her own experiences.

At the time of purchasing an object, an individual may not effect an analysis of the functionality of the object, of its use or of the pleasure deriving from using it. Instead it may happen that the individual builds his/her own expectations on the functionality, the performance of the object itself, based on the first impression he/she had on the object.

Our goal is to verify those expectations, i.e. to determine how the individual comes up with such expectations, how they come to think about the functionality of the object, and finally how they get the impression that the object will be usable in its entirety. We therefore intend to measure the usability perception.

Operative definition of perceived usability

In order to provide a definition of 'perceived usability', we need to first clarify the meaning of the terms 'usability' and 'perception'.

As to usability, reference is made to the definition in the ISO 9241 text, where ***usability*** is defined as follows:

The extent to which a product can be used by specified users to achieve specific goals with effectiveness, efficiency and satisfaction in a specific context of use.

As to the meaning of ***perception***, it is defined as follows, according to Darley *et al.*, 1986:

The interpretation of the information perceived by the senses (...) almost all the perceptions are a series of suppositions on the world, worked out and based on the previous experiences.

By merging these two theoretical assumptions, we can define **perceived usability** as follows:

The usability value perceived and attributed by a specific user to a specific product, further to (as a consequence of) his/her first appraisal of the product itself.

The perceived usability indicators

After thoroughly analysing the existing literature on the subject, we have realised that the various theories on usability, in terms of the relation between an individual and an object in a defined context, are very similar in definition and meaning. It is also possible to find such definitions and meanings in the theories of other authors.

Thanks to an analysis of the concepts, we have been able to classify all the indicators, dividing them into five macro groups which define the usability:

- Likeability: enjoyment, pleasure, comfort.
- Attitude (acceptance): trust, lack of stress and frustration, control, understanding (self-understanding).
- Perceived utility: use frequency, adaptability and compatibility.
- Perceived effectiveness: relation functions/use, flexibility.
- Perceived efficiency: learning time schedule, need of support, margin of errors.

Perceived usability, effectiveness and efficiency are to be seen in terms of the expectations of the individuals before they try the object. That is an evaluation of the object at first sight, at first approach, which is fundamental as it represents the time when the individuals develop their willingness to verify their expectations and their attraction towards an object.

Perceived usability evaluation

A tool has been developed in order to evaluate the perceived usability, in relation to what has been described so far. This tool had to have the following characteristics in order to meet the requirements described in the introduction:

- Short time schedule.
- Easily answered.
- Flexibility towards different contexts and different objects.

The tool was therefore structured in the form of a questionnaire divided into the following three parts:

- Data collection relative to the people interviewed (age, sex, studies, profession, family status, etc.).
- Structured questionnaire formed by a variable number of items (max 21) relative to the analysed object, and developed through a series of statements with which the interviewed people had to agree or disagree (max disagreement (1) and max agreement (5)).
- Semantic Differential, consisting of a variable number of pairs of antithetical

adjectives that could describe the feeling of an object from an emotional point-of-view. It is requested that a degree of correspondence of the adjective or its opposite to the object be expressed. The scores are marked on a scale of 1 to 5, i.e. 5 is the maximum score for an adjective while 1 is the maximum score for its opposite.

The adjectives were selected on the basis of previous experiences in the ergonomic field. They were obtained from interviews in which the interviewed people had defined the objectives using their own adjectives. Thanks to a long and thorough analysis and evaluation, 22 pairs of adjectives have been selected.

Both the statements of the survey and the pairs of adjectives from the Semantic Differential have been inserted, as they identify the five indicators of the perceived usability as defined earlier, i.e. likeability, attitude, perceived utility, perceived effectiveness, perceived efficiency.

The items of the survey and the number of pairs of adjectives are quite flexible and may vary in the order in which they are applied to the various objects, which differ in terms of shape, size and display. However, the variance of items and number of pairs of adjectives may not alter the nature of the tool or change its goal, which is the identification of the degree of perceived usability through a quick and easy tool.

Finally, a debriefing phase follows the survey, in order to analyse the reasons behind the answers given during the survey itself.

27.3 APPLICATION EXAMPLES

This methodology has been applied in the evaluation of different kinds of objects.

Following are two examples of its application. The first example relates to the applicability of the methodology in a comparative context. The other was conducted on behalf of a major Italian automobile manufacturing company interested in perceived usability issues. The company was interested in the possible impact on the public of a newly conceived object, an object with particular implications from a cultural and communications point-of-view. The survey was therefore conducted using a non-functioning, three-dimensional prototype.

All processing was done through the software programme ‘Statistica’ by StatSoft, version 5.

The interval data were processed through the descriptive, parametric statistics, the ordinal data were processed through the non-parametric statistics, while the nominal data were processed through the calculation of the frequencies and relative percentages.

The debriefing data were processed through a qualitative analysis of the answers provided by the survey participants.

Steam irons

The first example refers to the analysis of three different steam irons. The objective of this first test was to evaluate the intelligibility of the survey items, to verify the time schedule needed to complete the survey and the concepts meant to be measured, rather than to evaluate the objects themselves.

The objects chosen for this first test – steam irons - are commonly used objects, of medium size, probably present in all Italian households, and potentially usable by both males and females of any age, culture, social standing, etc.

The reasons for a multiple object situation are mainly twofold. First, multiple choice represents a situation similar to that in a selecting and purchasing environment. Secondly, this situation makes it possible to evaluate the actual degree of measurability of the proposed items, through the analysis of the variance effected through the Friedman test.

Ten subjects represented the test market, 40% male and 60% female (since the steam iron is mainly used by females), aged between 22 and 59 years old, the age of potential purchasers having been reduced.

The comparison of different objects enabled us to verify whether or not the survey items could define different elements, thus obtaining different evaluations of different objects. Since 13 out of the 22 pairs of adjectives and 14 out of the 21 items included in the survey represented significant differences according to the evaluating situation, we can assert that they realistically indicate the parameters they were meant to measure.

This test was not aimed at finding out which of the three objects was considered better than the other two by the test market, from a perceived usability point-of-view. However, one of the three irons did, in fact, score higher than the other two.

Furthermore, when asked which of the three irons they would choose, 80% of the answers reflected the survey results. Therefore this result seems to confirm that usability represents a significant parameter when choosing a product.

A three-dimensional prototype

In this second test, the analysis of the perceived usability was only a phase of a broader project of evaluation of a multi-tasking information car system, on behalf of one of the most famous Italian companies in the automotive sector.

The main goal of this project was to study, from both a conceptual and formal point-of-view, the equipment, which would have an innovative design and would be easily used.

The goal of the survey was to seek opinions from potential car purchasers on the possible installation in the car of a multi-tasking information system. The information sought through the survey referred both to the conceptual aspects of the innovation and to the practical impact that a new means of communication could have.

For this purpose 30 subjects aged between 19 and 59 years were interviewed. All were drivers but with different frequencies of use.

The first phase of the survey consisted of checking the possible services that a system like the one-under-examination could provide; the second phase consisted of the examination of a three-dimensional prototype and its evaluation from a perceived usability point-of-view.

A real-size prototype was installed in a non-functioning car, so that the perception would be near to reality when a subject was seated in the driver seat. The interviewer conducted the interview from the passenger seat.

The survey was divided into four parts:

- The first part was relative to the subjects' personal data and their driving attitudes.
- The second part requested the subjects to express agreement/disagreement with 17 statements.
- The third part, the semantic differential, asked the subjects to give a score from 1 to 5 on a few pairs of opposed adjectives in relation to the object.
- The last part was a debriefing in order to define the reasons for the individual choices of answers and evaluations.

It took an average of 15 minutes for all the interviewed people to complete the survey.

An analysis of the survey highlighted the positive impact of the object on the interviewees. The negatively conceived items had a low score, in disagreement with the assertions, while the positively conceived items had a high score, in accordance with the assertions.

The subjects interviewed expressed a willingness to try the object; they acknowledged its utility from a general and a personal point-of-view; they thought the learning time schedule would be a maximum of one week, although they would require a manual in order to use it. The accessibility to the commands scored quite highly, indicating a good level of accessibility.

The second part required subjects to assign scores from 1 to 5 to 20 pairs of opposite adjectives in relation to the object. The analysis of this second part confirmed the results of the previous analysis.

The data obtained from the survey were confirmed by the debriefing sessions. The object was considered easily accessible, good to look at and not too difficult to use. The data were processed through the software programme 'Statistica' by StatSoft version 5. The interval data were processed through the descriptive, parametric statistics. The ordinal data were processed through the non-parametric statistics, while the nominal data were processed through the calculation of the frequencies and relative percentages.

Another interesting application of the method was carried out for a well-known espresso-machine manufacturer, requesting the evaluation of a domestic espresso machine already launched on the market, in order to innovate the product by acquiring new information on the expectations of the public in terms of usability.

27.4 CONCLUSIONS

The outlined process presents some advantages:

- Although it does not presume to be a substitute for an analysis of the objective usability as specified in ISO 9241-11, it provides an opportunity to collect qualitative and quantitative data on the degree of perceived usability, by simulating a relation between the product and the purchaser.
- It provides an opportunity to effect comparative tests in short time schedules and at limited cost, thus enabling the evaluation of products form the competition.
- It is adaptable to different product and trade contexts. By analysing the products from a perceptive point-of-view, it may be applied to simple mock-up or by-dimensional prototypes, thus enabling the monitoring of the products in the different phases of planning.
- Finally, it is useful for checking the usability requirements that a product needs in order to meet the users' expectations from a purely conceptual point-of-view.

27.5 ACKNOWLEDGEMENT

Special thanks to La Cimbali Spa and Fiat Auto-Department of Ergonomics for their collaboration.

27.6 REFERENCES

Bandini Buti, L., 1996. *Ergonomia e progetto*. Maggioli Editore, Rimini.

Bevan, N., 1995. Measuring usability as quality of use. In *Journal of Software quality*, **4**, pp. 115-130.

Bonapace, L. and Bandini Buti, L., 1997. The pleasant object: the SEQUAM. In *Ergonomia*, **10**. Moretti & Vitali, pp. 11-14.

Darley, J.M., et al., 1986. *Psychology*. Il Mulino, Bologna.
Emanuele, R. and Simionato, B., 1998. Usabilità percepita: sviluppo di un metodo sperimentale per la valutazione dei prodotti con utenti. Tesi di Master, DITEC, Politecnico di Milano, Milano.
ISO FDIS 9241-11, 1997. Ergonomic requirements for office work with visual display terminals (VDTs). In *Guidance on Usability*, **11**. International Organisation for Standardisation.
Jordan, P.W., Thomas, B., McClelland, I. and Weerdmeester, B.A., 1996. *Usability Evaluation in Industry*. Taylor & Francis, London.
Jordan, P.W., 1998. Human Factors for pleasure seekers. In *Ergonomia*, **11**. Moretti&Vitali, pp. 6-20.
Pedon, A., 1987. *Introduzione alla psicofisica sociale: analisi epistemologica dello scaling psicofisico per la misura degli atteggiamenti.* Libreria Universitaria Editrice, Verona.
Rubin, J., 1994. *Handbook of Usability Testing*. John Wiley & Sons Inc., New York.

CHAPTER TWENTY-EIGHT

Applying Evaluation Methods to Future Digital TV Services

MARTIN MAGUIRE

HUSAT Research Institute, The Elms, Elms Grove

Loughborough, Leics LE11 1RG, UK

ABSTRACT

This paper discusses the evaluation of leisure-based electronic services such as digital television (DTV). Usage of such services is typified by: a diverse user population (the consumer public), single or shared use in the home, users being in a relaxed state of mind, as well as having the discretion to use the service or not. These characteristics can have implications for the way in which such systems or services are evaluated. The paper draws on the author's experience of performing evaluations of digital TV services with members of the public to highlight some of the main issues associated with such studies. For user trials these include choice of user sample, task setting, practice and introduction and user attitudes to their experience with the service. For discussion groups, issues include choice of group structure, factors affecting user opinions and the role of demonstrations in the discussion sessions. The paper also refers to a study based on discussions with families in their own homes, about their management of domestic finances. It considers whether people would like financial services to be delivered via entertainment and other domestic devices such as TVs, or whether they prefer the two functions to be separated. It also comments on the process of interviewing family groups in their own homes.

28.1 INTRODUCTION

The emergence of digital television has produced the means for people to receive interactive services in their homes without needing to own a computer (Hunt 1999). These include the following (Bains 1999):

- Receiving programme information and TV listings.
- Manipulating the TV picture.
- Sending and receiving email.
- Browsing the World Wide Web.
- Receiving commercial services (e.g. home shopping and banking).
- Playing games.

Watching television is essentially a passive leisure activity. The development of interactive TV demands greater mental effort from the viewer who has learn about operating these new services. Speck (1999) discusses his first use of Sky Digital's interactive service for football matches. This allows the viewer to select camera angles, replays, highlights of the match and to tap into statistics. He comments that using this interactive TV service can now require a lot of hard work and is not for 'couch potatoes'.

An interesting issue is whether such services will enhance people's pleasure in watching television, with on-screen guides providing greater scope for them to plan their viewing and interactive TV allowing them to manipulate the programmes themselves.

The success of digital TV will also depend on its ability to deliver additional services, such as allowing people to check their bank accounts while sitting in their living rooms, and whether people wish to work in this way.

Evaluation of digital TV services will be an important focus for human factors practitioners in the future, and also how to carry out such studies in an effective way. This paper discusses some of the work carried out in evaluating a developing digital TV service and in assessing user attitudes to the delivery of new services through a television set. It comments on the evaluation activities performed, including planning and running users trials and holding discussion group sessions, both at the developer's site and in a domestic setting. The issues raised should have relevance for the evaluation of leisure related products in general.

28.2 DIGITAL TV DEVELOPMENT

A new stage in Internet access is via digital TV and it is predicted that this will be the basis for an ecommerce boom (Clarke 1999). This is based on survey findings such as those of Gallup, who found that nearly twice as many people would prefer to buy goods using their TV rather than their PC (Bains 1999). TV-based interactive services may be attractive to people less familiar with computers, although many older or retired people may still be reluctant to take up such services, seeing them as unnecessary and requiring extra payment, possibly for a phone line connection and purchase of a set-top box. However, currently in the UK competition between digital TV service providers is strong and set top boxes are available free from Sky and ON-digital. To access the Web, the user interacts via the TV handset or a laptop keypad, requiring them to become familiar with indirect selection of items on screen, rather than direct pointing with a mouse. Other user interface issues that arise are whether on-screen text will be large enough to be read at a typical viewing distance (10 to 15 feet), and whether the restructuring of a website so that scrolling is not required will demand a greater depth of structure and more interface complexity. Current webpages, which may employ complex layout formats and integrated controls, will need to be modified if they are to be easy to use via digital TVs (Clarke 1999).

However, it is argued by Norris (1999) that the digital television revolution is failing to live up to the hype. In a report by the Consumers' Association, a number of negative aspects were found of which potential consumers may not be aware. It was reported by the researchers that picture and sound were no better than on traditional sets, unless the consumer lives in a poor analogue area. Digital broadcasting does not suffer from the effects of a poor analogue signal (resulting in ghosting or a snowy picture) but it occasionally produces a 'jerky' motion with the image freezing momentarily, or a 'blocky' picture with the image breaking up. There may be less flexibility in recording and viewing programmes than at present with analogue TVs and video recorders. In order to use on-screen listings, a dedicated channel may be needed, so users cannot watch a programme at the same time. Also, for one digital TV service it was stated that the cost of using a decoder could add £9-14 to the annual electricity bill. In the UK there was also the possibility that subscribers to digital TV would have to pay an extra £24 for their TV licence to help fund the development of BBC digital services. This has recently been ruled out by the Government as it was thought that this would be a disincentive for customers to take up digital television (Schaeffer 2000; Gibson 2000).

28.3 CHARACTERISTICS OF DIGITAL TV USE

Digital television will provide the basis for many households to receive interactive services in the future. Yet there are several ways in which it is used that contrast with use of traditional office-based computer systems:

- Broad user community: Currently purchasers of digital TV and DTV interactive services are likely to be people who are knowledgeable about, and experienced with, information technology. However, as a wider range of members of the public acquire digital TV it will be necessary to cater for this diversity.
- Discretion to use: Users of digital TV services (as well as electronic consumer products in general) have the discretion to use them or not. Thus if the service or features of it are not easy and enjoyable to use, they will fall into disuse. Users also have discretion to learn to use features in a partial way and at their own pace.
- Group watching: In a domestic situation, digital TV will be watched by both individuals and groups. Thus there will be different and competing interests in what programmes to watch. In the family home some parents will wish to control what their children see on TV.
- TV as a leisure product: TV is generally watched as a leisure activity and many people view TV in the evenings for relaxation (as opposed to, for instance, watching a programme for home study purposes). Arguably therefore many viewers will not be inclined to work hard to learn or use interactive services. Conversely, users may be prepared to spend more effort in using an intermediate service if they are in a relaxing environment.

The evaluation of digital TV services with users is an important activity for human factors practitioners. The following sections report on the author's experience in running user trials and discussion groups in relation to digital TV services.

28.4 ISSUES IN RUNNING USER TRIALS OF DIGITAL TV SERVICES

The user trials principally involved evaluation of prototype versions of an on-screen digital TV guide. The approach adopted was to run these as user trials in a traditional way, devising and setting tasks for users to perform. As a result of running the trials a number of issues were raised about the procedures for evaluation. This section discusses some of these issues which may have an influence on the way in which such trials may be performed in future.

Specifying subjects and recruitment

In carrying out user trials it is necessary to specify the characteristics of the users who will act as subjects within the study. The approach adopted was to include users from different age groups, with an even mix of males and females, and with different levels of experience (of cable or satellite TV, or with computers). While age, gender and experience are characteristics often used to select users for human factors studies, there may be other more relevant criteria for user recruitment, such as their acceptance of or motivation to use new technology, as well as their background experience with it. It would then be possible, for instance, to compare users with a relatively positive attitude towards technology to those with a relatively negative attitude. Such a questionnaire for measuring the level of technology acceptance has been developed and applied by the author.

The kinds of response that might be expected from different categories of user, based on their previous experience of new technology or similar services and their acceptance of new technology are presented in Table 28.1.

Table 28.1 Possible effect of categorising user subjects by experience and technology acceptance

	Experienced users	**Inexperienced users**
Technology-accepting users	Will provide informed and constructive responses. Should result in useful ideas for improving service based on user's previous experience.	Will provide idea of how inexperienced consumers will react when they first acquire the product.
Technology-resistant users	Will highlight concerns from a technical perspective and which features are of value, even to technology resistant consumers.	Will identify concerns of the wider consumer population. Useful if aim is to launch a simple service of interest to mass market.

Market researchers split the market for a product into groups such as 'early adopters', 'mainstream followers' etc. It may be helpful to use such categories for subject selection. One question is whether people who adopt technology easily should be the focus of user studies. It may be argued that this group will have a better understanding of technology generally, and will be able to comment in more depth than less experienced users. The products and services will also need to match the needs of early adopters if they are to take them up and lead consumer demand. However, for consumer products, it is always important to consider the needs of less experienced users, older users and people with disabilities in order to understand what problems they will face when the product or service does get taken up by mainstream consumers.

Recruitment during the digital TV trials was carried out through an external agency who recruited users in the vicinity of the developer's site. This process worked well but occasionally produced subjects not matching the specified criteria, e.g. one person was included who stated that he rarely watched TV. It is, of course, essential that third-party recruiters have some understanding of the application area, e.g. the difference between a standard satellite TV service and a digital satellite service.

On one occasion double booking of subjects occurred, which meant that the two subjects had to work together to carry out the evaluation tasks. Despite this, it was found that using two subjects together had certain advantages. They worked together to solve problems, vocalised their thought processes in a natural fashion, and produced useful feedback for the evaluator to record.

Task setting

The digital TV service that was tested was an early prototype of the on-screen guide and did not include the TV broadcast programmes. The user trials with the service were based on the traditional approach of setting each user a series of tasks such as:

- Finding out what programmes are on at a certain time of day.
- Finding programmes of a certain type.
- Booking events to be paid for.
- Searching for programmes by subject.
- Setting up a favourite list of channels.
- Setting up a video recording or a reminder to watch a particular TV programme.

This structured approach to the trials was necessary in order to test different aspects of the system with all users. Of course most users would not naturally carry out such a range of tasks in a short period of time. In practice, they will act differently in their own home, where they have discretion about their use of the system, how they explore it and whether they seek help from others in the family. Following the trials, users were asked to give satisfaction ratings for the simulated service (without seeing the TV programmes). While this can be argued to be a reasonable measure to take, clearly the users' real level of satisfaction will depend on seeing real channels appear when they complete tasks. In other words, user satisfaction will depend on the viewing goals that users set for themselves, their ability to perform the tasks to achieve those goals (in a relaxed way in a domestic setting), and seeing some positive output, e.g. a TV programme they enjoy. The measurement of user satisfaction for a simulated service is a general issue that may apply to trials of domestic services in the future.

How much practice and introduction to give?

During the evaluation trials, users were given a short introduction to the on-screen programme guide before being asked to perform specified tasks. This was thought to be similar to the kind of introduction they might receive from a dealer or salesperson. They were shown how to use the main menu for the on-screen guide, and how to make selections. Users then had to use the system or service themselves to perform the tasks. This raises the question of how much introduction to give users for leisure-based products and services in general.

Providing no help, or a minimal level of help, does test the system stringently and allows the evaluators to tell how easily users will be able to learn to use the system. However for services such as digital TV, it may be argued that users will often seek help from several sources (the installer, friend, other family members, the handbook, etc.) and will explore the service gradually when they are learning about the system and to overcome problems. Thus it may be considered that requiring users to operate a service without allowing them to seek assistance in some way is unrealistic.

Considering these different sources of help may also suggest new ways of providing online assistance for leisure-based interactive services. These may include an online synthetic agent, natural language interaction, or video contact with a human technical support agent. One method in current use is to broadcast a video-based introduction or tutorial for the user to watch. Such an approach allows the user to be shown visually how to operate the service. However, for many systems and services online help or an online tutorial is not developed at the prototype stage but when the development is near completion. Thus there is less scope to develop online help iteratively and for this process to influence the system or service itself.

User attitudes

For practical reasons, the prototype on-screen TV guide was tested within the offices of development company. Three test areas were set up with a TV, sofa, rugs, plants and coffee tables to provide a domestic living room setting. The aim was to create a relaxed atmosphere for the user subjects (members of the public) to work in. Users experienced problems in using the system. This stemmed partly from the use of a Windows-like style of user interface which included drop down menus which made some facilities hard to find and to use with a handset.

User attitudes to the on-screen guide were generally positive, even though many needed assistance from the evaluator for the different tasks they had to perform. Figure 28.1 shows for each task the percentage of user subjects needing some form of help and their ratings of the ease of performing each task.

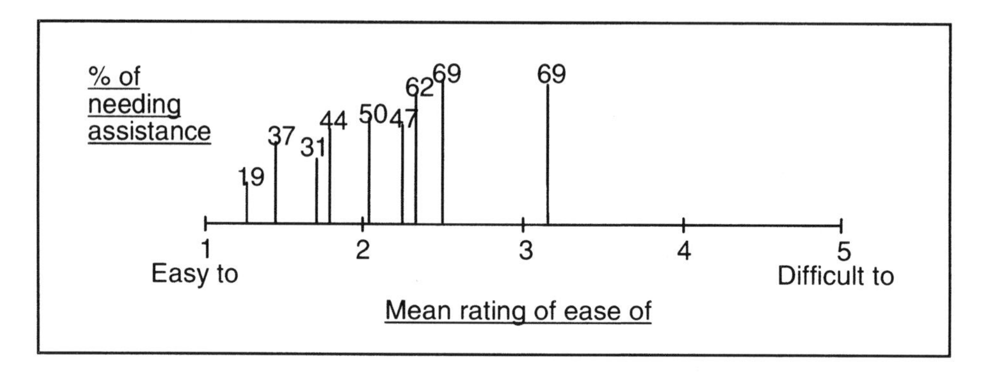

Figure 28.1 Relationship between need for assistance and ratings of ease of use

This seems to show a relationship between the ease/difficulty of use rating and the amount of help needed. However it is also noticeable that the ratings are strongly oriented towards the 'easy to use' end of the scale, even for tasks where a high proportion of users needed help from the evaluator.

It has been found in previous studies that there is a general tendency for users to struggle with interactive tasks and yet rate the user interface as easy to use. This may be because they are relieved to complete tasks and do not want to appear to be struggling to use the service. It could be helpful to consider ways in which users attitudes can be measured which reflect accurately the problems they have experienced.

28.5 ISSUES IN HOLDING DISCUSSION GROUPS CONSIDERING DIGITAL TV SERVICES

A series of three discussion group sessions were held as part of the DTV trials involving members of the public. Each group was presented with demonstrations of two digital TV services and invited to comment on them. The benefits of holding discussion group sessions are:

- They are a good way of bringing stakeholders into the design process to explore new ideas, design options, costs and benefits, screen layouts, etc.
- They are an efficient way of obtaining opinions from a range of people and **contrasting them**.
- The discussion process may stimulate new opinions and ideas that individual interviews may not elicit.

The following sections discuss some other implications of the discussion group format.

Group membership

It was intended that the discussion or focus groups would include people who were users of cable or satellite TV. It was also specified that they should be parents, and so would be able to comment on issues relating to parental control of their children's viewing, and control of payments made to purchase films. Two groups (aged between 45 and 64) were parents of teenage children, while another group (aged between 25 and 44) were parents of children of 12 and under.

The recruitment process (via a market research agency) succeeded well. It illustrated the compromises that may have to be made. The original idea was to base the study on digital TV users. As the number of digital subscribers in the UK was small at the time of the study, this proved difficult to achieve in a short time and so the groups were based on standard cable and satellite users. However, one digital TV user was identified who was included in Group 1 as it was thought that his experience would be valuable. Group 2 was recruited at short notice during the daytime (as the prototype was available for a longer period than expected). Here an all female group was recruited. In Group 3, one participant did not match recruitment criteria and was found to have only terrestrial TV (not satellite or cable).

It was found that having groups recruited according to different criteria did provide different perspectives. Group 1 (aged between 45 and 64), for example, appreciated the increased convenience of being able to use the on-screen guide to set up recordings and reminders easily. They also liked the ability to filter and reduce the wide range of channels and programmes available by choosing them by subject category. However several participants thought the whole on-screen guide too complex, and stated that they would prefer to leave primary use of it to their children. Comments were:

- 'Elderly people will have difficulty.'
- 'It's getting too technical – you want to relax. It's getting too much like computers.'
- 'My kids would use it. I'd get fed up.'
- 'I might want to see a programme about animals, say, but don't want to get too much information about animal programmes though.'
- 'With all this choice, you would get confused.'
- 'It takes too long to browse. I will miss the programme.'
- 'I don't want more decisions. I just want to flick through.'

Also the inclusion of a younger, digital user, while differing from the characteristics of the others in the group, promoted some interesting discussion between him as a strong acceptor of digital TV services and the others who were more sceptical of them.

Group 2 (aged between 45 and 64) were all members of families that owned satellite TV but were not the main users of it. They were concerned about the cost of purchasing films and events and the ability to control children's viewing.

Group 3 (aged between 25 and 44) were not daunted by use of the on-screen guide but seemed to think it had too many functions that could be reduced. Several participants felt that the on-screen guide and programme planner was more for 'telly addicts' and would be of more interest to their parents.

Factors affecting user opinions

In one of the discussion groups, there was a difference of opinion over certain features such as the 'Now and Next' window. This facility presented the name of the programme currently showing on each channel, and what was coming up later. The window appeared for a few moments when the user changed channel and then disappeared when the user stopped interacting with it. While some users found this feature annoying, others liked it. It transpired in the discussions that several of those preferring the feature had already had experience of it on their own or on their friends' TV service. This highlighted the importance of understanding the background experience of users in order to be able to interpret their opinions. However such user experience and preferred ways of working with their current system (e.g. the ability to type in channel numbers as well as selecting them) can help to suggest facilities that the current demonstrations may not have.

One facility discussed was a programme planner that allowed viewers to select programmes an place them into an on-screen diary, which would then remind them to watch the programmes. This concept of a programme planner was thought to be a good idea generally. However several group participants stated that they would not need them personally although 'their parents would love them', or they would be of 'more interest to teenagers'. This reaction against planning TV viewing may have been due partly to participants not wishing it to appear that they watched a lot of TV or depended on it.

Using demonstrations in discussion groups

Providing a demonstration of the digital TV service allowed participants to see the system in totality, to give their first impressions and to comment on the overall style of the user interface. The group setting also mirrored group watching of TV programmes so that people were able to comment on whether they could read screen information on the TV screen at a distance.

It is important to address the general issue of text size on TVs if interactive services are to be used successfully and enjoyed. For the focus group discussions, the TV services were displayed on large screen TVs, which made viewing easier. It is, however, recommended by Gill (1999) that for accessibility to interactive television by older users and people with visual impairments, 'character size should be under user control with clear typefaces on non-patterned backgrounds'.

The demonstrations of the TV services did not highlight many of the interaction problems that were apparent during the individual user trials. The groups commented generally on whether they liked the layout of a screen, whether the functions looked useful or otherwise, whether the interface looked too complex. However, participants were more accepting of interaction features (e.g. setting up a list of favourite channels) which individuals found quite complex when they tried to use them in the trials. Thus system demonstrations to groups should be supplemented with hands-on user trials.

Table 28.2 summarises the issues raised:

Table 28.2 Showing demonstrations to support group discussions

Advantages and disadvantages of using demonstrations in group discussions	
Advantages	***Disadvantages***
• Allows users to understand the design concept being discussed. • Satisfies possible concerns about safety and security for some types of system.	• May raise unrealistic expectations of the capability of the future system. • Users may take explanations of the usability of the system for granted. • Prototype may lack features (e.g. broadcast sound) which may influence usage

28.6 DISCUSSIONS WITH FAMILIES ABOUT MIXING TV WATCHING WITH ACCESS TO FINANCIAL SERVICES

Digital TV may be used as a means of offering services to help consumers with household finances, e.g. banking, investing, etc. A study was carried out to discuss how people manage money at present, and how they might like financial services to be delivered to them in the future. One aim was to assess participants' reactions to receiving services via digital TV or other household devices. Interviews were carried out with thirteen families (including single and two-parent families and retired couples) in their homes.

The family discussion format worked quite well and people were happy to discuss how they managed money. However, it was realised that the process had to be handled sensitively, and it had to be stressed that the questions were of a general nature and did not ask for personal financial details (e.g. how much is held in a particular account). A video camera was used to record the discussions and to supplement the researchers' first-hand note taking. Only two of the thirteen families objected to this but were prepared to be audio-recorded. The participants were also happy for the researcher to video locations of devices such as TVs, radios, telephones and PCs, etc. to see how they might be adapted to receive financial services. Good ideas arose from people who were already familiar with the Internet and Information Technology, e.g. receiving bank statements by email and having a portable device (like a pager) for checking an account balance. Participants also discussed the possible benefits of receiving financial services electronically. The main benefits for electronic access were seen to be:

- Ability to receive banking transactions in electronic form and to only keep paper copies of recent material, thus saving storage space.
- Ability to retrieve from the archive quickly to perform financial transactions.
- Ability to scan financial offers more quickly and so deal with them rather than just considering them as junk mail to be discarded.
- As a step towards the e-cash society where there will be no need to handle cash.

There were seen to be some significant drawbacks:

- The cost of the equipment.
- The perceived lack of security with electronic services, particularly Internet-based.
- The lengthy procedures that will be involved in identifying oneself for telephone banking.

Locations for performing financial tasks in the home

The most popular location for performing financial tasks is the dining room table (8 families out of 13) since it allows paperwork to be spread out and lighting is normally good. Other locations used are: the bedroom (2 families), on the kitchen table (2), the study (2) and on the sofa (1). The main reason given as to why few people preferred to work on the sofa was that they like to keep financial and relaxation activities separate, while the TV would cause a distraction.

The teenagers who were involved in the discussions tended to keep bank account books, statements, etc. with their parents financial items or in their own rooms. They did not perform extensive financial tasks but were generally happy to carry them out anywhere in the house, including when watching TV. They seemed to be less concerned to keep financial and relaxation areas separate.

Receiving financial services via TV and other devices

Table 28.3 shows the percentage in favour of accessing financial services via specific device or location, as a total and split up between children and adults. The sample consisted of 19 adults and 8 children:

Table 28.3 Preferences for access to financial service via specific devices or locations

Percentage in favour of access to financial service via specific devices or locations			
	% Adults	*% Children*	*% Total*
TV	74	75	74
Arm of armchair	44	36	42
Telephone	79	50	70
Bedside device	53	25	44
Kitchen device	50	27	44
In the car	36	60	44

It can be seen that access to services via the TV and telephone are much more popular than the other methods (selected by 74% and 70% of the whole sample). This is perhaps because people are familiar with the telephone as a device for service access, and because the TV is seen as potentially being able to offer similar interaction to that of a computer. TVs and telephones are also devices that the public normally have and so will be less expensive to employ.

These results indicate that domestic users are willing to consider receiving financial services via their TV. However, in practice, this may cause friction between different family members wanting to use the TV for both entertainment and financial purposes at the same time.

When asked to suggest further ways of accessing financial services, these included access from work or school, as these are places where a lot of time is spent and where technology is often available. The lounge was also seen as a good place for browsing through an on-line shopping catalogue (as opposed to drawing up a shopping list, where the kitchen was preferred). One teenager felt that the most flexible solution was to have one screen per person, wherever they are located in the house, and a mobile device that could be used anywhere required.

28.7 SUMMARY AND CONCLUSIONS

The growth in the use of a wide range of electronic services by the general public, often as a leisure-based activity, provides new issues for human factors specialists to consider in obtaining user involvement in the design process. A number of different methods can be used for evaluating digital TV services. Issues relating to their use are summarised in the Table 28.4 below:

Table 28.4 Application of different evaluation methods to leisure-based services

Issues relating to different evaluation methods for leisure-based services	
User trials	• Should emphasis be placed on experienced and motivated purchasers or less experienced personnel? (Ideally a mix of the two categories should be achieved.) • Should tasks be fixed or should users be allowed to use the service as freely as they wish? • How much practice or introduction should be given? • How much assistance should users be given during the trials? Should they be allowed to seek limited help, as they might in a real domestic setting, or forced to complete tasks? • Can improvements be made to the collection of attitude statements to reflect problems users experience? Could this possibly be achieved by recording user's attitude after each sub-task rather than just at the end each main task or at the end of the trial session.

Issues relating to different evaluation methods for leisure-based services (continued)	
Focus groups	• Should homogeneous or mixed discussion groups be arranged? • Need to present concept and how it is likely to be implemented technically if users are to avoid making judgements based on false assumptions. • Focus groups can take an overview of whole system or service. If demonstrated, participants can comment on visual aspects (the 'look') of user interface but user trials are needed for feedback on interactive aspects (the 'feel'). • One aim of group discussion is to gain consensus. However users should be allowed to write down personal views in questionnaire at end of session, enabling them to express views they were not able to make or which they thought others would disagree too strongly with in discussion.
Interviews with families in their homes	• Not all family members may be able to be present. The interviewer should not try to force everyone to be included. • Videoing of the discussion is possible if family agrees. Do not expect to be able to video whole house, e.g. bedrooms. • Include all family members in the discussion and make an effort to involve those family members who are quiet during the session. • Remain neutral rather than taking sides with one family member.

During discussion sessions with family groups at home, there was an acceptance of many of the ideas proposed for accessing financial services in different ways, particularly through the telephone, the TV, and the computer.

Regarding location for access, there were differences between those who wanted to keep the location for performing financial tasks separate from the relaxation areas in the house, and those who did not mind the two areas being the same. Some people, particularly children, seemed less concerned about performing financial tasks while sitting in the lounge watching TV. Similarly some parents were happy with the idea of a facility in the bedroom that would enable them to carry out financial tasks away from the children, and where they could discuss financial matters in private. The majority of adults, however, liked to work on a desk in an office, or on a table in the dining room or the kitchen, where work could be spread out.

Formal laboratory trials will often lead to users trying hard to perform well (the 'Hawthorne effect'). Yet it should be remembered that many leisure-based services such as digital TV will be used on a casual basis as a means of relaxation. Evaluation studies for leisure services should therefore be carried out in the same spirit — in an informal and friendly way, rather than placing users under stress. This will help to put users in an appropriate state of mind, encourage them to act as they might when relaxing at home, and should help to produce more realistic results.

In conclusion, this paper has outlined a number of issues and concerns in performing user-based trials and running discussion groups for digital TV services. While most of the issues relate to the evaluation of systems generally, they are of particular concern when evaluating leisure related systems and services in the future.

28.8 REFERENCES

Bains, G., 1999. The golden screen. In *Computer Active*, **46**, pp.102-104.

Clarke, G., 1999. Preparing for e-commerce with the TV, not the PC. In *Computing*, 22nd July 1999, pp. 16-18.

Gibson, J., 2000. Television licence fee to increase by £3. In *The Guardian*, Tuesday 22nd February 2000, p. 1.

Gill, J., 1999. Telecommunications - guidelines for accessibility. Published by *Royal National Institute for the Blind* on behalf of Cost 219bis, 224 Great Portland Street, London W1N 6AA, UK, ISBN 1-86048-022-5, email: jgill@rnib.org.uk.

Hunt, J., 1999. How to get on the Internet without buying a computer. In *Daily Mail*, Tuesday 31st August 1999, p. 49.

Norris, D., 1999. Digital TV turn-off. In *Daily Mail*, 1st April 1999, p. 31.

Schaeffer, S., 2000. Smith adds £3 to TV licence to fund BBC digital. In *The Independent*, Tuesday 22nd February 2000, p. 8.

Speck, I., 1999. Digital means no rest for couch potatoes. In *Daily Mail*, Monday 23rd August 1999.

CHAPTER TWENTY-NINE

Activity and Designing Pleasurable Interaction with Everyday Artifacts

VESNA POPOVIC

School of Design and Built Environment, Department of Interior and Industrial Design, Queensland University of Technology, Brisbane, Qld 4001, Australia

ABSTRACT

The paper addresses some issues of pleasurable interaction with everyday artifacts. It covers artifacts and their contextual environment of interaction. It stands on the premise that dynamic interaction between human artifacts and their contextual environment is essential. Its theoretical construct is based on Popper's theory of objective knowledge. The paper suggests an approach for user–designer–artifact interaction that is able to support the design of everyday artifacts to be pleasurable to use. It emphasises that the activity and user–artifact interaction are the foci of the design.

29.1 INTRODUCTION

The interaction between people and artifacts has existed for centuries. Its complexity has increased in parallel with the development of human civilisation. The number of activities humans make influences this. This means that designers need to understand the nature of the activity and its participants. In order to do this they need to share the knowledge about the activity with people.

It is possible to link the artifact–human–designer interaction with Popper's theory of objective knowledge. In his philosophical approach to knowledge, Popper (1989; 1989b) suggested a pluralistic view of *three worlds*. These are (a) the world of physical objects or the world of physical states, (b) the mental world or the world of mental states and (c) the world of objective content of thought or the world of ideas. The correspondence of the three worlds with human expertise and knowledge engineering is outlined by Gaines (1987), who delineated three environments that correspond with Popper's worlds: (a) the physical environment, (b) the social environment and (c) the knowledge environment. These three environments, or three worlds, are linked with an artifact interface that is the main communication channel between them.

For the purposes of this paper this scheme is epitomised as follows:

- World 1, the world of physical objects (artifacts) or physical states or *the artifact environment.* World 1 is a product of Worlds 2 and 3.

- World 2, the world of states of consciousness or mental states or behavioural disposition to act. *The social environment* 'subjective experience of people and their mental experiences and feelings'. Artifact designers and human users are part of this environment. The second world represents their thought processes through which they grasp third world contents.
- World 3, the world of 'objective content of thought' which is a natural product of humans. It interacts with the first and second worlds. This is the *knowledge environment* of theories and their 'logical development'. Users' and designers' knowledge and problem situations belong to this environment.

These three environments correspond with the contextual environment of artifact interaction at the appropriate levels. Each contextual environment consists of three environments that correspond with Popper's three worlds (1989; 1989b). In his study, Gaines (1987) applied Popper's theory of three worlds to explain relationships between knowledge engineer, expert and environment. A similar analogy can be drawn to explain relationships between designers, artifact users and the contextual environment of their interactions (Figure 29.1) (Popovic 1998; Popovic 1998b).

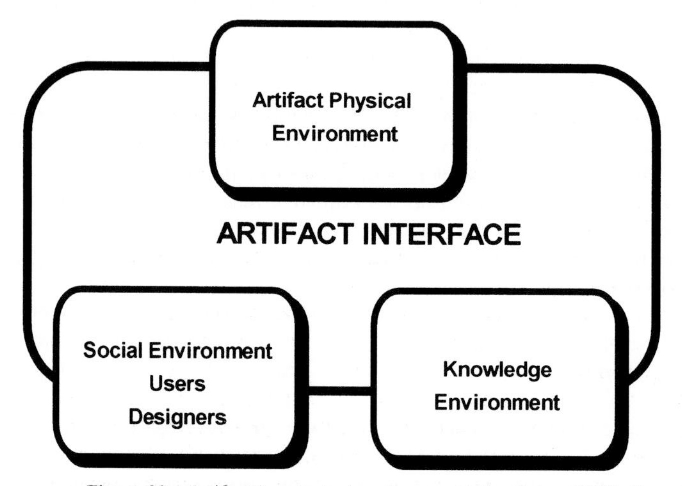

Figure 29.1 Artifacts' contextual environment (after Gaines 1987; Popovic 1998)

Figure 29.1 illustrates the contextual environment of an artifact which includes:

- *The artifact's physical environment,* which consists of the artifact's technical context, function and life cycle. In this context, they have their behaviour and function.
- *The social environment,* which consists of the artifact users who use them and the designers who design them. In this context, a user's experience and a designer's expertise need to interact. The role of the designer is to understand the *activity* and artifact users' concepts and, through design, to support them.
- *The knowledge environment,* which consists of a user's and a designer's knowledge. The designer is an expert who knows about design and knows how to respond to design constraints. She has design strategies, general design knowledge and domain-specific design knowledge relevant to the design task. Knowledge about the human users belongs to domain-specific design knowledge. Users possess the knowledge in

> the domain of their expertise and experience and have concepts about the artifacts. The user's concept and the designer's concept should be compatible. The knowledge environment exists within artifacts (interactive devices) but it depends on the user's expertise and experience within the activity. The designer attempts to represent this through her knowledge of the second and first worlds. Therefore, designers' models and users' models are part of the third world (Thimbleby 1990).

These three environments or three worlds are linked with an artifact interface that is the main communication channel between them (Popovic 1998). Consequently, this paper is going to suggest an approach for user–designer–artifact interaction that will be able to support the design of everyday artifacts to be pleasurable to use. The design of artifacts should be able to provide understandable devices and systems with interfaces that bridge the gap between people and products and are pleasurable to use during an interaction.

29.2 DESIGNING PLEASURABLE INTERACTION

Traditionally, task analysis conducted in Human Factors and Ergonomics and HCI (Human-Computer Interaction), is based on the idea that a description — containing all the necessary information to understand the use of an artifact (or device) — can be made of the sequences of steps that it takes for human users to conduct tasks within an activity. This analysis thus contains a description of each step required by an individual user for successful interaction. However, during the observation of people interacting with the artifact, it is possible to capture some of the knowledge that they require in order to perform skilled activities in the actual work process. Nevertheless, it is not possible to accurately predict their performance in future situations. People do not react until the situation occurs and the context and conditions of the environment trigger their actions. What makes a good artifact interface is often a good *conceptual model* behind the system that is reflected by the 'system image' (Norman 1986). This requires an understanding of the activity and its context (Popovic 1998).

The process of designing is an activity itself. It can be seen as the discovery of new ways to improve the quality of human life by exploring and designing new artifacts by applying knowledge and the designer's imagination. Design concerns with the ways of how things ought to be. It aims to change an existing situation into a preferred one. Designers attempt to predict the behaviour of an artifact using their knowledge and expertise (Popovic 1998). The inner system of any artifact can be predicted accurately (Alexander 1973). The most difficult part to predict is the interface of interaction, as the environments in which the artifacts are used are very often ignored. Human beings, during their interaction with artifacts, cannot envisage all obstacles that may occur. This is very significant when they interact with complex systems in order to achieve the desired goal. In this case, one has to use one's knowledge about the artifacts (systems) and adapt to the contextual environment of its operation. The adaptation is achieved through the process of learning and transmission of knowledge about the requirements of the task. To enjoy an interaction, the human has to understand it. To achieve this, designers have to understand what is the knowledge structure domain that people have about activity, artifacts and their contextual environment and how this knowledge is exchanged in order to support interactivity (Figure 29.1). This requires the inclusion of the activity into the design process.

Technologically interactive devices or cognitive artifacts (Norman 1991) that are with us everyday, impose another constraints to both – users and designers. According to Norman (1991), cognitive artifacts are 'devices designed to maintain, display, or operate upon information in order to serve representational function'. However, little is known about the 'information-processing role by artifacts' and their interaction with human information processing (Norman 1991). The whole contextual environment of artifacts affects levels of interaction and artifacts use. The way they are used and level of interaction which is to be achieved depend on users' tasks and expertise. This remoteness of an artifact and its contextual environment generates conflicts between people and artifacts.

Norman (1991) proposed two views of artifacts: the systems view and the personal view. The systems view consists of the person (user), the tasks (tasks within an activity) and the artifact. Here, the artifact is enhancing the performance of its user. The personal view consists of the task and an artifact. According to this view, the original task is changed; a new task is introduced, and it has a different cognitive capacity. The user has to face new and different cognitive requirements.

Human interaction with artifacts can be direct, remote or virtual. All these three levels of interaction occur in their contextual environment (Popovic 1998). The term *direct interaction* refers to a user's direct engagement during the activity, such as using a hammer or screwdriver, or chopping vegetables. A direct relationship exists between the user's movement of the device and the task outcome. *Remote interaction* occurs when a task within the activity is done either by instructing someone else to do it, or by controlling it remotely, such as in some surgical procedures. *Virtual interaction* occurs in a virtual environment in which an artifact is present, such as the monitoring of a plant through a computer display or a building simulation using a computer interface.

Norman (1986) pointed out that there are discrepancies between people's 'psychologically expressed goals and the physical controls and task variables'. Psychological variables are goals and intentions. Physical variables refer to the task to be performed. Here, people need to use physical means to achieve this — that is, controls, levers, or displays and controls concurrently. They need to translate psychological goals and intentions into physical actions (the variables are interpreted). Very often, physical variables are not the ones that the person is concerned about. For example, a person intends to withdraw money from an automatic teller machine (ATM). It is envisaged that the user will walk to the machine, insert the card and personal identification number, and do the banking. Thus the user's goals and intentions are translated into physical actions. This does not always fulfil the user's expectations, however, because of tasks that the user may have to do in order to achieve a goal. Sometimes users forget their cards, or the machine does not return them. Alternatively, the screen may be illegible or a card inserted the wrong way or the machine may work too slowly (Rogers and Fisk 1997). In this particular situation the needs within the activity are constantly changing and designers are facing two sets of needs: (a) bankers' needs and (b) users' needs. Both must be translated from psychological needs to physical needs. This requires an interface that is easy to use and interact with. The interface where the user can manipulate the physical variables in order to achieve the desired goal with a pleasure. The most common sequences of translation between psychological variables and physical variables are those in which the goal and intentions are translated into actions which consist of tasks as action paths. During use, users evaluate their tasks from the feedback received until the desired outcome is reached. In order to interact with an interface, users have to map the problem (task or sub-task sequences) by understanding buttons or other elements on the interface. For example, with ATM users, mapping is essential to distinguish where on the ATM interface the physical variables are to be used to translate the goal state into physical actions. This involves interpretation of an interface, which means interpretation of the perceived system. To match psychological and physical variables, the mapping should be simple and should convey the conceptual model of an artifact.

There are distinct approaches in cognitive sciences that are relevant to the understanding of human users. One is the traditional symbolic processing approach on which cognitive science was founded. The other is the so-called *situated action* approach that emphasises the context, the environment, the situation in which the activity occurs, and cultural and social settings (Suchman 1987). The third is the distributed cognition – the study of the knowledge representation on the system level and its unit of analysis is a 'cognitive system composed of individuals and the artifacts they use' (Flor and Hutchins 1991; Hutchins 1991) (in Nardi 1996). The unit of the analysis has been moved to the system and in practice is referred as 'functional system'. It is to say that distributed cognition is concerned with structure – representation inside and out side of the head. It differs from the well-established notion of an individual cognition as it is attempting to understand 'coordination between individuals and artifacts' in order to understand how task coordination is distributed (Nardi 1996).

In the field of Human Factors and Ergonomics, *situation awareness* theory has emerged with its conceptual basis ill-defined. Situation awareness theory proposes that there are aspects of situations for which awareness must be maintained — for example, cues, evolving situations and special characteristics of situations or higher-order goals (Gaba, Howard and Small 1995). However, the situation awareness approach may have an impact on an operational setting (Gilson 1995) and artifact interface design. Situation awareness is important in decision making for some tasks. It should occur simultaneously with the user's goals and intentions. This means that a user's goal directs which aspect of interface is attended to, and the perceived artifact contextual environment is brought into line with the user's goals based on that understanding (Endsley 1995) and responds to the particular situation (Popovic 1998).

In the area of human-computer interaction (HCI) activity theory has emerged as potential field that would be able to support human interaction with cognitive artifacts within the context (Nardi 1996). The field is not new and started in the former Soviet Union in the 1920s but its relevance to HCI is just recognised recently. Nardi (1996) compared activity theory, situated actions and distributed cognition in relation to studying the context. Situated actions emphasise human responsiveness to the environment and an artifact setting. The focus is on the notion that an activity will develop and grow out directly from the situation (Suchman 1987). It is said that this approach overlooked to treat an environment as an agent that shapes an activity in which people are engaged in a flexible way. Situated action approach concentrates on representation as an object of study (Nardi 1996).

Activity theory stands on the premise that the elements of the activity are not fixed. They change and transform as an activity itself evolves. The main idea is that artifacts mediate the activity as introduced by Kuutti (1991). Actions are seen here in a similar way what is referred to tasks (HCI) or Human Factors and Ergonomics. They overlap and they have operational aspects. It is understood that the operations become unconscious routines that come with practice. According to Leont'ev, 1974 (in Nardi 1996) their operational structure will change only. The activity is seen as a system consisting of objects, actions and operations which is seen as the *context*. This context is generated by people and it is 'external' or 'internal' to them. Term 'external' refers to artifacts, other people and settings; term 'internal' refers to specific object and goals. They are inseparable (Nardi 1996).

The way in which information is presented through an artifact interface will influence situation awareness by determining what kind of information is acquired and how it is compatible with users' needs. Interface design should provide needed information to the user without imposing a cognitive effort or a mental workload. During interaction, an artifact user is involved in visual search tasks through an interface. Therefore, interface knowledge is very important for directing the selection of interface cues in order to achieve the stated goals.

It has been emphasised that the most important aspect of a product's design is a design of its interface. People are confronted with two different mapping stages (Norman

1986) while using interactive artifacts. These stages are: (a) mapping from the psychological variables to the physical variables and (b) mapping from the physical variables to the psychological variables. The input mechanism (interface) is a mediator between two representations. This is a key point for achieving an interaction. Therefore, the design of the interface should focus on mapping and how it can support the user to accomplish the task without difficulties. This confronts designers with many unanswered questions, as the users' requirements are variable and each user is different in knowledge level, skills, needs and artifact concepts. This leads to compromise solutions or to interface design that incorporates visual cues to help users' interactions and assist users to understand the system.

Norman (1986; 1988) suggested that a good design model and relevant system image should be provided. Human beings can change their levels of interaction with an artifact interface. One change is by design. The other is by the user's expertise and experience level. The interface design should present the appropriate 'system image' to artifact users to help them form a users' concept. Artifact users must not feel that they are out of control. On the contrary, they should enjoy interacting with the artifact and have 'a pleasurable engagement' during its use (Laurel 1986). Therefore, a conceptual model of the system is very important. It supports the user's interaction and is essential for novice users to assist them to learn how to use artifacts and to develop a user model of an artifact that is to be more consistent with the design model.

The theory relating to user involvement has expanded in the human-computer interaction community, as it has been easier to study users and involve them in the design process. The gap between users and designers still remains, but more and more user's viewpoint is taken into consideration. Nevertheless, the human user is an 'information processor' whose behaviour is unpredictable (Sutherland 1994). This is another reason why designers need to understand their users and be able to model users' activities and tasks during the early stages of the design process (Popovic 1999). Activity and task analysis are important steps in interface design and when done, should reveal various ways of human interaction. Task analysis that relies on reason only is dangerous, as any new task is evolutionary, discovering new goals as it proceeds (Hammond *et al.* 1987). Interface design does not involve task analysis only. It requires the designer to predict what users will know or be able to learn. Designers are supposed to map psychological principles into their design decisions. It is important for the designers to understand the humans and their activities and how users' social environment, artifact physical environment and knowledge environment are shared (Figure 29.1). This share of knowledge might contribute to the formulation of an activity and artifact concepts. When human interact with artifacts they have an intention. This is what guides them as they have a concept behind the activity. It is essential to discover users' intentions and incorporate them into the user's concept about the artifact.

Pleasure is not defined by rules. However, it incorporates some aspects of human factors that contribute toward humanisation of our living and working environment (Jordan 1999). Therefore, this imposes a higher level of complexity to an artifact design. To achieve pleasurable interaction an artifact design should incorporate the following process: (a) research, (b) scenario and user's concept formulation, (c) design and application of relevant research findings and (d) design development and production (Figure 29.2).

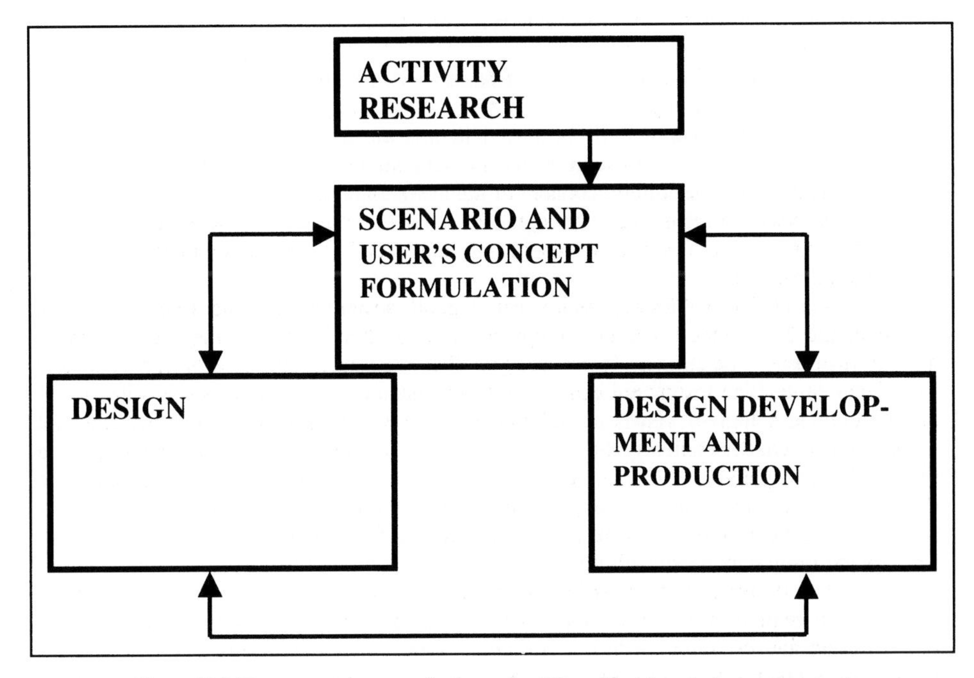

Figure 29.2 The suggested process for the study of the artifacts' contextual environment

Figure 29.2 illustrates the suggested process for the study of the artifacts' contextual environment. Its stages are explained below.

Activity research

It is envisaged that the following research, when appropriate, should be undertaken:

- Research into the nature of the activity in order to define its relation to the three environments (Figure 29.1) that constitute the artifacts' contextual environment.
- Research into the life style of the activity players in order to define their needs and culture and how pleasure is represented within that particular activity.
- Research and understanding of users' interaction within the activity and how these correspond with the artifacts' contextual environment.
- Identification of activity players' knowledge base and how it is shared in the social context of that particular activity.

Scenario and user's concept formulation

Scenario formulation should be based on the research related to the activity, its players and the knowledge they share (Figure 29.1). The scenario developed can be used as a formulation of the project brief. It should convey a user's concept on whose basis the design is to be developed. It should convey user's intentions. An artifact should be a mediator of human thoughts and behaviour (Nardi 1996). This is where the link between psychological and physical variables (Norman 1986) is to be identified.

Design and application of relevant research findings

It is envisaged that the relevant research findings will be applied throughout the design process.

Usability of the design is to be tested in relation to a user's concept developed during the scenario formulation in order to achieve the best possible compatibility between the user's and designer's concept of the artifact and its associated activity.

The following criteria should be applied:

- compatibility between the user's and designer's concept;
- representation of physical and psychological variables that supports the user's concept;
- ease of use, interaction and support of its dynamic structure;
- simple interface design and information organisation;
- simple and logical visual organisation of information on an interface that will convey visual cues clearly;
- interface mapping;
- appropriate tactile information that evokes appropriate feelings;
- appropriate colours that evokes specific feelings;
- sound that can enhance an interaction and feelings but not to destruct them;
- olfactory information to enhance an interaction ;
- artifact form / shape that corresponds to culture and life style which conforms to the appropriate aesthetics – culture to be considered;
- artifact form / shape that conveys humour or joy when perceived – culture to be considered, and;
- flexibility and adaptability of tasks during the activity.

Design development and production

It is envisaged that at this stage an artifact will not go through the radical changes regarding its interactivity with the users. However, final test of user's concept and its compatibility with the designer's concept of the artifact and the associated activity should be done.

Many aspects listed above are influencing humans' perception of artifacts by saying, 'it feels good', 'that is right', 'cool', 'cute' or 'looks different'. These are popular attributes used by people when they see or use artifacts. Nevertheless, they convey a lot of qualitative values that make some artifacts to contribute to the enhancement of the activity and its transformation (Nardi 1996). When people see artifacts they make a 'perceptual categorization' (Clancey 1999) of visual form or an interface layout. Thus, visual aspects of artifacts play an important part in developing a user's concept and how they might contribute in achieving an enjoyment during the interaction.

29.3 CONCLUSION

Activity theory, plans, mental models, situated awareness, situated actions, distributed cognition and cognitive maps are different approaches to assist in understand activities and humans' interaction. Within any activity people have social engagement everyday, be it work or pleasure. They are linked on social and individual level (Nardi 1996). The activities are in constant change that influences an artifact transformation (Kuutti 1996). Any developmental process of an activity can generate a new activity (Nardi 1996). This is done through different tasks or actions. The historical evolution of artifacts and activity associated in one generation of artifacts are reflected in artifact development in the next generation (e.g. aeroplanes, helicopters, sewing machines, computers or any artifact used everyday). In this case the design can be seen as an agent for change. It is trying to change the activity by introducing a new activity, which may lead to a new design and new challenge and enjoyment.

Social structure plays an important part in any activity (Nardi 1996). These include artifact users, organisation and its culture and environmental structure. It is very important for the designers involved in artifact design to understand the process that occurs behind the activity. The user is an agent who directs the whole interaction (Laurel 1986) especially as interactive interfaces are becoming easier to use. The new designs should challenge its users to enjoy different levels of interaction. They should experience new pleasure every time they interact or use the artifacts. The design of the dynamic structure of the artifacts should support exploration, flexibility and adaptability during interaction. The concept of *'form follows function'* is evolving to *'form follows pleasure'*.

29.4 REFERENCES

Alexander, C., 1973. *Notes on the Synthesis of Form.* Harvard University Press, Cambridge.

Clancey, W., 1999. *Conceptual Coordination.* Lawrence Erlbaum, Mahwah, NJ.

Endsley, M.R., 1995. Toward the Theory of Situation Awareness in Dynamic Systems. In *Human Factors*, **37**, 1, pp. 32-64.

Gaba, D.M., Howard, S.K. and Small, S.D., 1995. Situation Awareness in Anesthesiology. In *Human Factors*, **37**, 1, pp. 20-30.

Gaines, B., 1987. An Overview of Knowledge Acquisition and Transfer. In *J. Man-Machine Studies*, **26**, pp. 453 - 472.

Gilson, R.D., 1995. Special Issue Preface. In *Human Factors, Special Issue: Situation Awareness*, **37**, 1, pp. 3-4.

Green, W.S. and Jordan, P.W., 1999. Introduction. In *Human Factors in Product Design – Current Practice and Future Trends.* Green, W.S. and Jordan P.W., (Eds), Taylor & Francis, London, pp. 26- 36.

Hammond, N., Gardiner, M.M., Christie, B. and Marshall, C., 1987. The Role of Cognitive Psychology in User-Interface Design. In *Applying Cognitive Psychology to User Interface Design.* Gardiner, M. and Christie, B. (Eds), Wiley, Chichester, pp. 13-52.

Jordan, P.W., 1999. Pleasure with Products: Human Factors for Body, Mind and Soul. In *Human Factors in Product Design – Current Practice and Future Trends.* Jordan, P.W. and Green, W.S. (Eds), Taylor & Francis, London, pp. 206 –217.

Jørgensen, A.H., 1990. Thinking-aloud in User Interface Design: A Method Promoting Cognitive Ergonomics. In *Ergonomics*, **33**, 4, pp. 501-507, 502.

Kuutti, K., 1996. Activity Theory as a Potential Framework for Human-Computer Interaction Research. In *Context and Consciousness: Activity Theory and Human – Computer Interaction.* Nardi, B. (Ed.), The MIT Press, Cambridge, Mass., pp. 17-44.

Laurel, B., 1986. Interface as Mimesis. In *User Centred System Design.* Norman, D. and Draper, S. (Eds), Lawrence Erlbaum, Hillside, NJ., pp. 67-85.

Nardi, B., 1996. Studying Context: A Comparison of Activity Theory, Situated Action Models and Distributed Cognition. In *Context and Consciousness: Activity theory and Human – Computer Interaction.* Nardi, B. (Ed.), The MIT Press, Cambridge, Mass., pp. 69-102.

Norman, D., 1986. Cognitive Engineering. In *User Centred System Design.* Norman, D.

and Draper, S. (Eds.), Lawrence Erlbaum, Hillside, NJ., pp. 32-61, 33.

Norman, D., 1988. *The Psychology of Everyday Things.* Basic Books, New York.

Norman, D., 1991. Cognitive Artifacts. In *Designing Interaction, Psychology at the Human-Computer Interface.* Carroll, J. (Ed.), Cambridge University Press, New York, pp. 17, 26.

Popovic, V., 1998. *Novice and Expert Users Model.* Ph.D. Thesis, University of Sydney.

Popovic, V., 1998b. Lighting Design Interfaces, Lighting 98. In *Proceedings of the 43rd Annual Conference of Illuminating Engineering Society of Australia.* Brisbane, pp. 173-178.

Popovic, V., 1999. Product Evaluation Methods and Their Importance in Designing Interactive Artifacts. In *Human Factors in Product Design – Current Practice and Future Trends.* Jordan, P.W. and Green, W.S. (Eds), Taylor & Francis, London, pp. 26-36.

Popper, K.R., 1989. Epistemology without a Knowing Subject. In *Objective Knowledge: An Evolutionary Approach.* Popper, K. (Ed.), Clarendon Press, Oxford, pp. 106-152.

Popper, K.R., 1989b. On the Theory of the Objective Mind. In *Objective Knowledge: An Evolutionary Approach.* Popper, K. (Ed.), Clarendon Press, Oxford, pp. 153-190.

Rogers, W.A. and Fisk, A.D., 1997. ATM Design and Training Issues. In *Ergonomics in Design*, **5**, 1, pp. 4-9.

Suchman, L.A., 1987. *Plans and Situated Actions.* Cambridge University Press, Cambridge.

Sutherland, S., 1994, *Irrationality.* Penguin Books, London.

Thimbleby, H., 1990. *User Interface Design.* ACM Press, New York.

Visser, W., 1992. Designers' Activities Examined at Three Levels: Organisation, Strategies and Problem Solving processes. In *Knowledge-Based Systems*, **5**, 1, pp. 92-104, 103.

CONCLUSION

Conclusions

PATRICK W. JORDAN

Contemporary Trends Institute,

PO Box 31958, London W2 6YD, UK

The chapters in this book give an overview of a new approach to human factors. This approach is about fitting products to people in a holistic manner. It is based on a recognition that the quality of the relationship between people and products depends on more than simply product usability. People are more than just 'users'. They have hopes, fears, dreams, aspirations, tastes and personality. Their choice of products, and the pleasure or displeasure that products bring to them, may be influenced by these factors.

Some of the chapters in this book have dealt with the theoretical framework within which issues relating to pleasure-based human factors may be considered (e.g. Reinmoeller, Chapter 9; Dejean, Chapter 11). Whilst it is recognised that usability may be a key component of what makes using a product a pleasurable experience, it is likely that there will be a number of other factors which influence the pleasurability of a design. These will include the aesthetics elements of a product and the experiential associations that users attribute to particular aesthetic properties, such as form, colour, and tactile properties.

Other chapters describe methodologies associated with the creation and evaluation of products (e.g. Taylor, Bontoft and Galer Flyte; Chapter 14, Boess *et al.*, Chapter 17). Creating pleasurable products requires the definition of user requirements specifications that define the person-product relationship holistically. This may mean, for example, including specifications that relate to the social and aspirational qualities that a design should convey. Understanding user requirements on a holistic basis requires a rich understanding of the roles that products play in people's lives. Ethnographic methods may be particularly useful in this context as their use may give rich insights into these roles. Such methods also tend to be qualitative, requiring few advance assumptions about the issues that are likely to be important to users.

Another theme to emerge was that of how to manage the implementation of holistic requirements specifications in the design process (e.g. Fulton Suri, Chapter 13; Rutherford, Chapter 23). Broadly, two complementary approaches emerged. One of these is to specify quantitative goals against which the pleasurability of a product can be assessed. The ability to quantify usability goals is arguably one of the main contributory factors to the success to date of the human factors profession in influencing the product creation process. By defining quantitative goals the usability requirements become 'hard' goals, as equally measurable as goals associated with, for example, build quality or production cost. The challenge now is to develop means of framing quantitative goals that relate to the user experience in a sense that is far broader than usability. The complementary approach is to involve others in the product development process as a way to understanding the users and their needs. Empathic techniques for helping others to understand the needs of the users are a means of gaining 'buy-in' from those involved in product development decisions.

Designing pleasurable products presents challenges that go beyond those associated with assuring a product's usability. A number of chapters presented case studies of designs which were created with the intention of eliciting particular emotional responses (e.g. Höök, Persson and Sjölinder; Chapter 20; Battarbee and Säde, Chapter 18). Others reported studies exploring the relationship between particular design decisions and the responses of users to a product (e.g. Bonapace, Chapter 15; Lee, Harada and Stappers, Chapter 16). As the field develops it may be possible to establish some generic guidelines with respect to the effect of particular design decisions on users' overall response to a product. In the meantime a number of qualitative and quantitative approaches may be used in order to establish such links on a case by case basis.

Pleasure-based human factors — the systematic study of the factors which will determine what makes a product pleasurable to own and use — is a discipline in its infancy. The chapters in this book report early work in the discipline, raising a number of important theoretical, methodological and implementation issues. The challenge over the next few years is to take these issues further, to gain an insight and an understanding of how to consistently create products which are a genuine joy to own and use.

Index